Graphene for Defense and Security

Graphene for Defense and Security

Andre U. Sokolnikov

CRC Press is an imprint of the
Taylor & Francis Group, an **informa** business

CRC Press
Taylor & Francis Group
6000 Broken Sound Parkway NW, Suite 300
Boca Raton, FL 33487-2742

First issued in paperback 2019

ISBN-13: 978-1-4987-2762-4 (hbk)
ISBN-13: 978-0-367-87334-9 (pbk)

Visit the Taylor & Francis Web site at
http://www.taylorandfrancis.com

and the CRC Press Web site at
http://www.crcpress.com

Contents

Foreword

Graphene is a crystal-like form of carbon in which one or several layers of carbon atoms are formed in hexagons. The material is very strong and light with respect to the thickness of layers, and is an excellent conductor of heat and electricity. Graphene layers are almost transparent. Andre Geim and Konstantin Novoselov received the Nobel Prize in 2010, boosting the development and publicity of graphene "for ground breaking experiments regarding the two-dimensional material graphene." It is the thinnest material in the world with a crystalline structure that can be stretched up to 20% of its length, sustaining electric current densities six orders of magnitude larger than that of copper. The charge density has the highest intrinsic mobility and the highest thermal conductivity and impermeability. Patenting in graphene has grown substantially over the period from 2005 to 2014 from almost zero to more than 9000.[1,2] China, Korea, and the United States are the three top countries with graphene patents. Overall, China has a prominent role in patenting graphene. Samsung possesses the largest graphene patent portfolio and Sungkyunkwan University (SKKU), Seoul, South Korea, holds the sixth largest patent portfolio, collaborating in research that grows rapidly (several research papers are published every day). The growth in the graphene literature shows no sign of decreasing. Despite the apparent dominance of Korean patent organizations, the leading country in the field of graphene patenting continues to be China. The continuation of the identified trends shows just how many countries and major multinational corporations are investing heavily to commercialize graphene and fulfill its theoretical potential. There has been some indication recently that Samsung alongside SKKU is becoming the first to grow a large-scale, impurity-free sheet of graphene capable of sustaining its electric properties.

This book does not have the goal of encompassing all aspects of the new material, as a reference volume collection may have, nor has it been intended to present the most fundamental approach to graphene. Rather, it is intended to guide a technical specialist, a scientist, an engineer, a business consultant, or a student through critical aspects of the current understanding of graphene's present and prospective developments, especially from a military and special application point of view. Incidentally, they are the most essential perspectives, because large research and implementation investments have restricted proliferation of commercial applications.

REFERENCES

1. Wallace, P.R., The band theory of graphite, *Phys. Rev.,* **71**, 9, 622–634 (1947).
2. Meyer, J.C., Geim, A.K., Katsnelson, M.I.., Novoselov, K.S., The structure of suspended graphene sheets, *Nature,* **446**, 60–63 (2007).

Foreword

Graphene is a crystallike form of carbon in which one to several layers of carbon atoms are arranged in hexagons. The material is very strong and light with respect to the thickness of layers, and is an excellent conductor of heat [illegible] layers are almost transparent. Andre Geim and Konstantin Novoselov received the Nobel Prize in 2010 boosting the development and publicity of graphene. Their ground-breaking experiments regarding the two-dimensional material graphene [illegible]

About the Author

Dr. Andre U. Sokolnikov is currently the president of Visual Solutions & Applications. He has worked in the field of electronics for more than 40 years, starting his professional carrier after completion of his MS degree in the Air Force, majoring in optoelectronic intercom systems. He then joined a counter organized crime and terrorism agency focusing on projects of microwave surveillance systems—the research and implementation that spanned eight years before the beginning of the PhD program. Since 2004, Dr. Sokolnikov has been an independent consultant for defense and special applications, such as THz identification and materials characterization, which resulted in the publication of the book *THz for Defense and Security Purposes* by World Scientific in 2013. He is the author of several dozen scientific articles and book chapters. His projects have covered the fields of mathematical modeling and designing of optoelectronic devices for the Chrysler, Inc, DARPA, and some other agencies and companies. He is an active participant and committee member of the SPIE "Defense and Security" symposia.

1 Introduction

Graphene is a new derivative that belongs to a big family of graphite-like materials. One major feature is that it is two dimensional. Usually, graphene is a single layer formed by lattices of carbon atoms that may be defined as a two-dimensional lattice of benzene rings without hydrogen atoms. Notwithstanding being one-layer thick, graphene is more robust than many conventional materials including steel. Another phenomenon that graphene has revealed is the possession of properties of electrons similar to those of photons and neutrinos. The new properties of electrons result from the symmetry of the atomic positions and the cone-shaped (rather than parabolic) regions in the electron energy distribution areas. The solid state physic laws are applicable to the graphene phenomenon. The Schrodinger theory explains the electron behavior in graphene as the Dirac expression do, but the whole phenomenon goes beyond the theoretical predictions of what condensed matter can be.

1.1 2D CRYSTALS

A graphene crystal may be envisioned as a hexagonal benzene ring with the following parameters: the distance between the opposite carbon atoms is 2a = 284 pm (picometers), where a is the carbon-carbon spacing, a = 142 pm (the hydrogen atom's radius, a_0 = 0.0529 nm). Benzene molecule C_6H_6 has one electron per atom with a hydrogen atom at the benzene ring location numbering from 1 to 6. In graphene, there is no hydrogen atom and the above electron is "smeared" over the crystal. For the usual 10 μm graphene sheet, there are 35 211 benzene rings arranged in well-ordered honey-comb lattice with the ring diameter of 284 pm. The facile three-fold planar bonding is governed by the Schrodinger equation for the carbon atom with 2s and 2p quantum states. The hexagonal structure provides the stability of the crystal very much the same way as it does for the microscopic structures such as the real honeycombs. The A and B atoms are translated by the vectors 1 ⟶ 3, 1 ⟶ 5 (with 60° angle between them) multiple times in either direction giving as a result a honeycomb lattice (see Fig. 1.1):

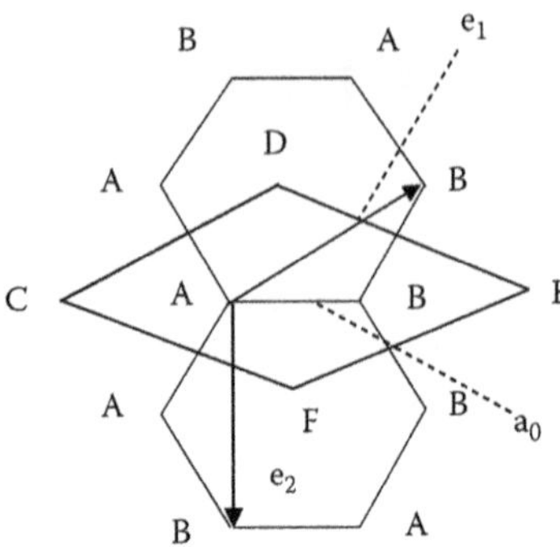

FIGURE 1.1 Vector translations of the basis atoms A and B resulting in a honeycomb lattice

The vectors of the above translations form triangles which may be perceived as overlapping triangles of A and B (Fig. 1.1)[1]. The real one-atom-lattice layer cannot remain stable without a support of some sort. Because of the sheet's instability there may be a question whether it is a crystal per se. However, the sheet is very tensile strong. Although, the chemical bonds between two $2p_z$ electrons try to retain the sheet planar form, the force is too weak so the word "crystal" may be put in quotation marks, also because long wavelengths flexural phonons with their large fluctuational displacements are much larger than the lattice constant. The real question is whether we have a perfect static crystal or a dynamic crystal that allows its deformation but remains fundamentally a crystal. On the other hand, any practical crystal existing at some arbitrary temperature and in an arbitrary environment cannot remain perfectly crystalline and from this point of view graphene is still a crystal. In real crystals, large in-plane motions of atoms do not influence the crystal structure on the whole but may cause different effects, including diffraction and smearing. On the whole, a 2D system such as graphene, remains largely within the limits of conventional 3D crystalline system. Early research efforts of Boem et al. in 1962, Van Bommel et al. in 1975, Forbeause in 1998 and Oshima et al. in 2000, to name a few, indicated theoretical probability of graphene. However from the point of view of receiving practical graphene samples, it has been an impossible task until the discoveries of the Nobel laureates, A. Geim and K. Novoselov in 2004 that made graphene sample available to use and study. As of right now, specimens on a scale of 10 nm to 10 μm are available and have the electron mobility that far surpasses the one of the current semiconductor devices based on calculations applicable to an infinite 2D array. The graphene conductivity, therefore, is excellent but more like a semimetal with conical but not parabolical electron energy bands close to the Fermi energy level. The energy has a linear dependence on crystal momentum, $k = p/\hbar$. Thus, energy $E =$ *"pc"* $= c^*\hbar k$. The band theory gives the electron speed (which is as if the electrons move like photons) at $c^* \sim 10^6\ m/s$ with vanishing effective mass. This is another important feature of carrier transport in graphene with almost no backscattering.

Mechanically, monolayer graphene is strong but easily bends by a small transverse force if not supported. Graphene, notwithstanding its large Young's modulus (4 ~ 1 TPa) is the most flexible material in the transverse direction. Engineering formulas are still applicable for vibration and bending of the material. "The softness" greatly depends on the dimensions: a square sample with the side of 10 nm has the spring constant of $K = 12.6\ N/m$ while a square sample with the side length of 10 μm (10^3 times longer) has $K = 12.6 \cdot 10^{-6}\ N/m$ (10^6 times smaller). Thus, the question of graphene "softness" or "flexibility" will depend on the dimensions of the material applications. In addition, on the micrometer-size scale, graphene adapts to an adjacent (e.g. supporting) surface under the influence of van der Waals forces. The necessity of a support layer for graphene, however, undermines the 2D dimensionality of the material, although no material can be authentically two-dimensional: the probability distribution $P(x, y, z)$ must expand in the z-direction by at least one Bohr radius. In the existing two-dimensional systems, e.g. electrons on the surface of liquid helium have quantum-well heterostructures to restrict a third dimension. Other two-dimensional systems are MoS_2, TaS_2, $NbSe_2$ and $BN(BN)_n(C_2)_m$, where n, m are integers. As a 2D system, graphene has the feature pertinent to such structures. One is that the attraction between the graphene atoms may exceed the attraction forces between the substrate and graphene. Sometimes, ripples may be formed on the graphene surface absorbants. Although irregularities do exist, on the whole the resistance to tension makes graphene one of the strongest materials known. If graphene is seen as an extension of benzene rings with 2s and 2p electrons supporting the crystal, the structure should resist bending and tend to remain planar, although rolled forms of graphene do exist, such as carbon nanotubes. 3D crystal forms are more stable: the sum of bonding energy of bonds in 2D may be loose here, to keep the crystal spatially firm. For carbon, the atoms in 3D have twice as many neighbors vs 2D formations (6-8 neighbors vs 3-4).

The micrometer scale of most of graphene samples results in noticeable influence of temperature and moisture on the material. More specifically, the carrier mobility depends on the presence of moisture in the sample. Heating to ~ 400° C can greatly improve the mobility. The "rippling" behavior may have something to do with the presence of adsorbents in the material.

1.2 GRAPHENE ELECTRON BANDS

The electron bands of graphene were described by Wallace and McClure[2]. The hexagonal honeycomb lattice has two sublattices A and B to form the honeycomb. Two atoms per unit cell represent two distinct groups of bands and allowed states. A two-component wavefunction describes a carrier in the above bands. The hexagonal Brillouin zone has non-equivalent corners K and K` (Fig. 1.2):

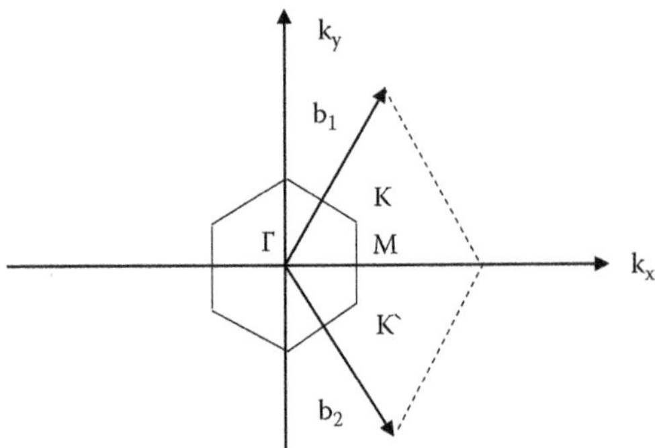

FIGURE 1.2 Honeycomb lattice's Brillouin zone. M, the zone boundary is $2\pi/3a$; $\mathbf{b}_1$ and $\mathbf{b}_2$ are reciprocal lattice vectors. The distance from Γ (origin) to K is $4\pi/(3\sqrt{(3a)})$. The valence and conduction bands meet at K` and $k_F = |K|$, and the Fermi wave-length, $\lambda_F = 2\pi/k_F = 3\sqrt{3a}/2 = 369$ μm.

In pure material, the Fermi energy level is close to K and K` points but in n- or p-type semiconductor materials the Fermi level is situated in the upper cones of electron or hole distributions that result from chemical doping or an electric field application (Fig. 1.3).

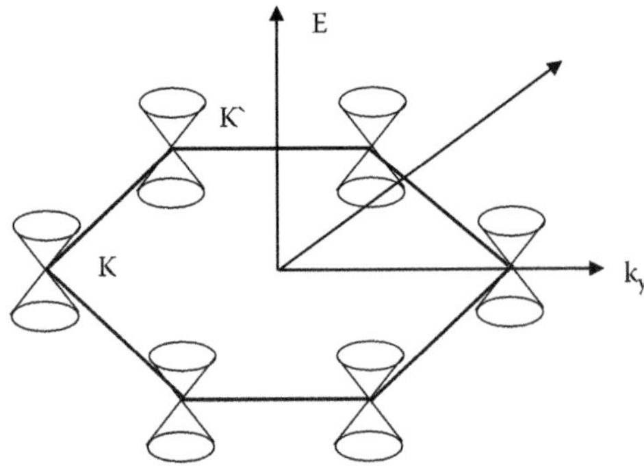

FIGURE 1.3 The conical electron bands of the honeycomb lattice. $\vec{E}$ shows the applied electric field at the gate of the tested sample.

In pure graphene, the electrons are located in the lower parts of the cones and the upper parts of the cones are empty. The two valleys at K and K` explain the graphene's special electronic characteristics, in particular, graphene being a zero-band gap material. In practice, the typical graphene's resistivity is lower than $h/e^2 \approx 25K$[1]. In typical semiconductor materials, Si and GaAs, the effective mass is $m^* = \hbar^2/d^2E/dk^2$ with their electron bands exhibiting parabolic minima. In contrast, with no such minima, graphene carriers move in a more linear fashion, $c^* = \upsilon_F = 3ta(2\hbar)^2$, where t is the nearest neighbor hopping energy (~ 2eV), and a is the distance between neighboring atoms (~ 142 pm)[3]. Undoped graphene has no free carriers and, therefore, has zero conductivity because of its conical band structure with the Fermi energy at the Dirac point (where the density of states, comes to zero). The conductivity, however, increases away from a minimum at the neutrality[4]. The sharp peak of resistivity is shown in Fig. 1.4.

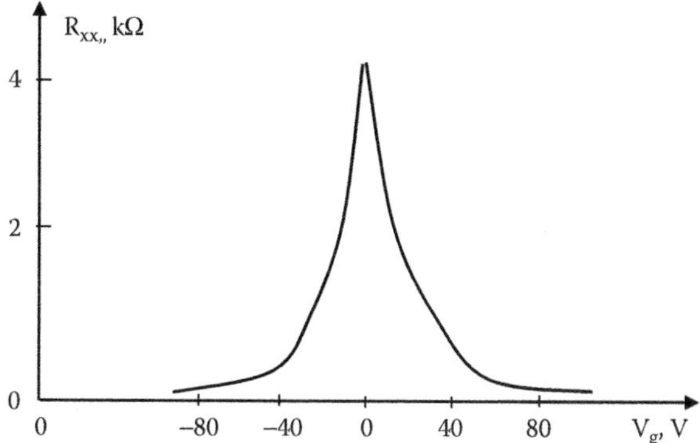

FIGURE 1.4 The dependence of resistance of a single-layer graphene on the applied gate voltage corresponding to neutrality point at 1.7 K.

The mobility in graphene varies with high values reaching 200,000 cm^2/Vs[5] while the maximum resistivity is close to $h/4e^2 \sim 6.45\ k\Omega$[6]. The early indications on the nature of the conductivity variation gave artifacts or "puddles" as an explanation. Also, increase of mobility was found by low gradual heating of the graphene samples to anneal scattering effects[5]. Another experiment to remove "puddles" was conducted as shown in Fig. 1.5[7]. A maximum value of 33 kOm is at 10 K. The temperature gives a steady insulating tendency and peaks at zero concentration.

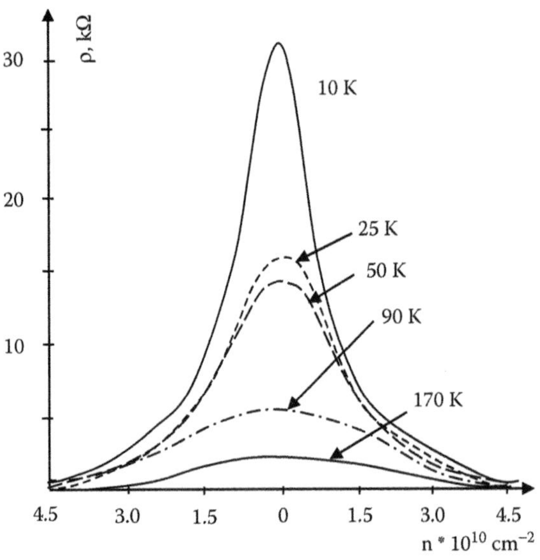

FIGURE 1.5 A comparison of resistivity dependence upon carrier concentration and ambient temperature.

An explanation for the "cone-conductivity" phenomenon came from the research by DiVincenzo and Mele[8] and later from Novoselov et al.[4] and Semenoff[9]. These authors showed that the double-sublattice origin of the K, K` cone states needs a concept of "pseudo-spin" electron wavefunction near the Dirac points. Semenoff's version suggests two "right-and left-handed degeneracy points" of the Brillouin zone where conduction and valence bands come together. Each hexagonal ring contains three A and three B atoms (see Fig. 1.1). The Dirac-form of the Schrodinger equation is applicable to the conical bands. The Hamiltonian used for the calculations takes into account only the interactions of the closest opposite superlattices ($t(a_i^* b_j + b_i^* a_j)$) and the next closest neighbors on the same superlattice ($t^{'}(a_i^* a_j + b_i^* b_j)$). Here, we use $a_i^* a_j$ operators that count for the presence or absence of electrons on sites i or j. The Hamiltonian (the Hamiltonian, the total energy of the system, is different for different situations or number of particles since it includes the sum of kinetic energies of the particles and the potential energy function) for the single-layer electron is given:

$$\breve{H} = \hbar c^* \begin{pmatrix} 0 & k_{x-ik_y} \\ k_{x+ik_y} & 0 \end{pmatrix} = hc^* \delta \cdot \vec{k}; \tag{1.1}$$

In (1.1) $\vec{k}$ is the quasiparticle momentum and δ is the two-dimensional Pauli matrix. The energy bands intersect in the cones at the sine corners of the Brillouin zone. Each C-atom has one $2p_z$ valence electron. π^* - anti-bonding with positive energy and π-bonding bands with negative energy form the above energy bands. The Pauli matrix δ uses the pseudo-spin but not the actual spin. The wavefunction for the Dirac points K` is given below[10]:

$$\Psi \pm, \vec{k}(k) = \frac{1}{\sqrt{2}} \begin{pmatrix} e^{i\Theta_k/2} \\ \pm e^{-i\Theta_k/2} \end{pmatrix}; \tag{1.2}$$

The π^* -states are above and π - states are below the Fermi energy. The *pseudo-spin* gives the projection of the spin δ on the direction of motion $\vec{k}$. The projection is called helicity or chirality. The helicity is positive for electrons and negative for holes. It is given as:

$$h = \frac{1}{2} \delta * p \left| \rho \right|; \tag{1.3}$$

The helicity is applicable close to the Fermi energy. In this case, we neglect the second-nearest-neighbor's interaction. The anomalous quasi-Hall effect happening in graphene confirms the described helicity. Different effects concerning the helicity have been reported including cancellation of backscattering of electron in a given valley and a reversal of the pseudo-spins. The latter effect is controversial since the reversal from k_x to $-k_x$ because of chirality is forbidden due to the fact that A and B sublattices contributions are orthogonal. On the other hand, the "spin"-part of the equation (1.2) shows that a particle takes a trajectory with the angle of $\Theta = 2\pi$, the wavefunction phase advances by $\frac{\Theta}{2} = \pi$ acquiring a minus sign and cancelling backscattering. The electron states have 4-fold degeneracy, in particular, the electron spin is degenerate and the degeneracy mentioned earlier in Semenoff's version of "pseudo-spin".

1.2.1 Landau Level Effects

If we apply a perpendicular magnetic field to a sample that has a 2D electron system, we shall see energy levels that are $E = (n+\frac{1}{2})(eB\hbar/m)$, where n = 0, 1, 2,…, and m is the electron mass. These energy levels are called the Landau levels that are a feature of the anomalous Hall effect (Fig. 1.6).

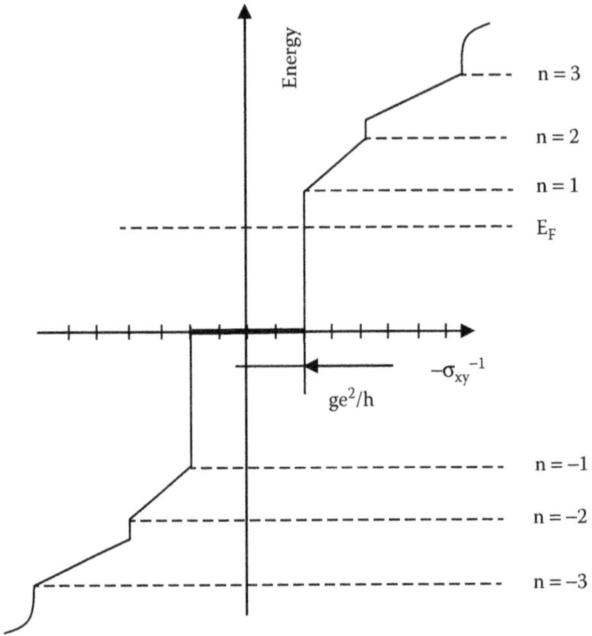

FIGURE 1.6 Landau levels characterizing the anomalous quantum Hall effect

The minimum energy is $\frac{1}{2}\hbar\omega_c$. $\omega_c = eB/m$ is the cyclotron energy and the total number of the orbital states is $N = AB/\varphi_0$, where $\varphi_0 = \frac{\hbar}{e}$ is the magnetic flux quantum per one electron. For the Hall effect to occur, the total number of mobile electrons should be comparable to N and all electrons should be in fully quantized states in a 2D system. In Fig. 1.6 the Landau levels form arrays with a gap at zero energy. The anomalous levels have energies given by[11]:

$$E_n = \pm\upsilon_F[2e\hbar B(n+\frac{1}{2}\pm\frac{1}{2})]^{1/2}; \tag{1.4}$$

when n = 0, 1, 2…

Two energy levels exist at zero energy with degeneracy which is half of the Landau level. The progressive separation of the Landau levels is shown in Fig. 1.7 where the nonlinearity of the dependence can be seen confirming the description Dirac fermions as massless.

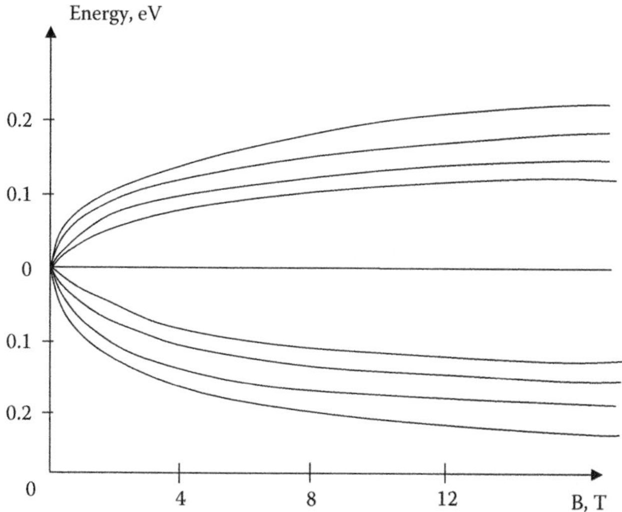

FIGURE 1.7 The Landau levels anomalous dependence[12].

Thus, the presence of the zero-energy level and the square root dependence of the energy (see Fig. 1.4) testify to the graphene anomalous spectrum.

1.3 ANDRE K. GEIM AND KONSTANTIN S. NOVOSELOV'S BREAKTHROUGH AND GRAPHENE DEVELOPMENT'S PERSPECTIVE

After several decades of hard preliminary work a breakthrough in graphene phenomenon's mastery was achieved. A simple technique of producing an isolated layer of graphene was elaborated by Andre K. Geim and Konstantin S. Novoselov who shared the Nobel Prize in Physics in 2010. Their experiments discovered the graphene's outstanding electronic properties, which they showed, were connected to the previous work on relativistic particles. Graphene with its unique mechanical properties has many current applications and promises to suggest more in the future. The Nobel laureates have also published a prolific number of articles on the 2D graphene phenomenon paving the way for numerous graphene-related publications.

One such possibility is to make transistors from a single-layer which may save money on the material. The possible disadvantage is the absence of the high-resistivity state necessary for switching applications. The carrier mobility is very high which promises the creation of a ballistic transport device working at room temperature. The other characteristics of value are the valley degeneracy and pseudo-spins that prevent back scattering. Nanoribbons on the basis of graphene (with space quantization across the ribbons) also have outstanding electronic properties. However, pattering of graphene is a big problem in itself.

REFERENCES

1. Wallace, P.R., "The band theory of graphite", *Phys. Rev.* **71**, 9, 622-634 (1947)
2. Meyer, J.C., Geim, A.K., Katsnelson, M.I.., Novoselov, K.S., "The structure of suspended graphene sheets", *Nature,* **446**, 60-63 (2007)
3. Snyman, I., Beenakker, C.W.J., "Ballistic transmission through a graphene bilayer", *Physical Review* B (*Condensed Matter and Materials Physics*), **75**, 4 (2007)
4. Novoselov, K.S., Geim A.K., Morozov, S.V., Jiang, D., Zhang, Y., Dubonos, S.V., Grigorieva, I.V., Firsov, A.A., "Electric Field Effect in Automatically Thin Carbon Films", *Science*, **22** (2004)
5. Bolotin, K.I., Sykes, K.J., Jiang, Z., Klima, M., Fuderberg, G., Hone, J., Kim, P., and Stormer, H.L., "Ultrahigh electron mobility in suspended graphene", *Solid State Communications*, **146** (2008)
6. Novoselov, K.S., Geim, A.K., Morozov, S.V., Jiang, D., Katsnelson, M.I., Grigorieva, I.V., Dubonos, S.V., Firsov, A.A., "Two-dimensional gas of massless Dirac fermions in graphene", *Nature*, **438** (2005)
7. Ponomorenko, L.A., Geim A.K., Zhukov, A.A., Jalil, R., Morozov, S.V., "Tunable metal-insulator transition in double-layer graphene heterostructures", *Nature Physics*, **7** (2011)
8. DiVincenzo, Mele, E.J., "Self-consistent effective mass theory for intra-layer screening in graphite intercalation compounds", *Physical Review* **B29** (4) (1984)
9. Semenoff, G.W., "Condensed-matter simulation of a three-dimensional anomaly", *Physical Review Letters*, **53** (1984)
10. Kane, C.L., Mele, E.J., "Size shape and electronic structure of carbon nanotubes", *Physical Review Letters* **78** (1997)
11. Novoselov, K.S., "Graphene. Materials in the Flatland" (Nobel lecture) (2011)
12. Nazin, G., Zhang, Y., Zhang, L., Sutter, E. and Sutter P., "Visualization of charge transport through Landau levels in graphene", Nature Physics, 6 (2010)

2 Physics of Important Developments That Predestined Graphene

The work of A Geim and K. Novoselov has made possible practical applications of graphene, a new class of two-dimensional (2D) that are large enough to be used in devices. The practicality comes from the fact that one-atom thick layers are able to conduct with virtually massless carriers with almost no scattering. 2D systems, however, do not exist as such in the real world and the notion of two dimensions apply only to certain physical effects such as spins, and the cases when a motion in the third dimension is restricted by a single quantum state.

One of the simplest 2D systems is the one that restricts spins to two dimensions with the spins being not self-supporting. The model consists of discrete variables that describe atomic spins' magnetic dipole moments. The spins can be in + 1 or – 1 state. The spins can interact with their neighbors. The Ising model presents phase transitions as elements of the model. The energy described by the model is given:

$$E = -J_{ij} \sum \sigma_i \sigma_j; \tag{2.1}$$

where $J_{ij} = J$ for all pairs i,j in Λ lattice sites, and E is the energy of the nearest-neighbor interaction of spins $\pm\frac{1}{2}$. The model is a set of lattice sites Λ (+ a set of adjacent sites) that forms a d-dimensional lattice. Each lattice site ($k \in \Lambda$) has a discrete variable σ_k ($\sigma_k \in \{+1,-1\}$ that represents the sites's spin.

$$M = (1/N) \sum \sigma_i; \tag{2.2}$$

where N = the number of paths of length L on a square lattice in "d" dimensions ($N(L) = (2d)^L$) and the transition temperature:

$$T_c = 2J[\log(1+\sqrt{2})]; \tag{2.3}$$

with phase transition taking place only in two dimensions.

Another example of 2D behavior may be the process of diffusion and random walks when the process is restricted in the third dimension. An example was provided by George Polya in 1921. A more important example of electrons in two dimensions are electrons on liquid helium. The first mention of this phenomenon was published by Grimes and Adams in 1979: crystallization of extremely high mobility electrons on the surface of liquid helium was observed at the temperature ~ 0.5 K. The experimental set-up was a capacitor 5 cm in diameter. One plate of the capacitor was positioned approximately 2 mm above the helium level. A tiny metal point emitted electrons to charge the surface of the helium. The equilibrium electron density between the electrodes and the helium surface is:

$$N_S = \varepsilon\varepsilon_0 V_T / ed; \tag{2.4}$$

where V_T = the upper plate voltage, ε = the permittivity of liquid helium, d = 1 mm; ε = 1.06.

The density of the emitted electrons is modulated by an ac voltage. A circular electron is used for the purpose. The impedance of the system is measured by a lock-in detector (Fig. 2.1).

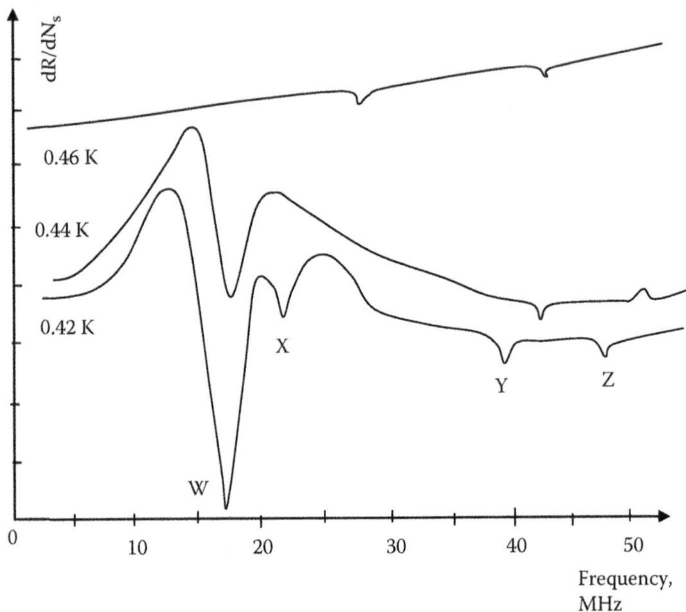

FIGURE 2.1 Dependence of impedance on frequency of the liquid helium system. The surface crystallization takes place at approx.. 0.45 K[1]. The electron density N_s = 4.6 x 10^8 e/cm^2.

The emitted electrons at the helium surface experience attraction by the surface image charge and, at the same time, repulsion from the electrons that are part of the helium atoms. The electrons have a quantum-bound state above the helium surface at ~ 4.1 nm and the binding energy of 0.7 meV. The electrons, however, are free to move horizontally. At 1K, the superfluid helium surface has characteristic capillary waves, called "ripplons". At the temperature of 1K, the ripplon's energy is smaller than 0.004 k_BT, frequency, f of appr. 80 MHz and the wave length of appr. 0.13 μm. The mobility of electrons on liquid helium is extremely high and the data[2] are given in Fig. 2.2.

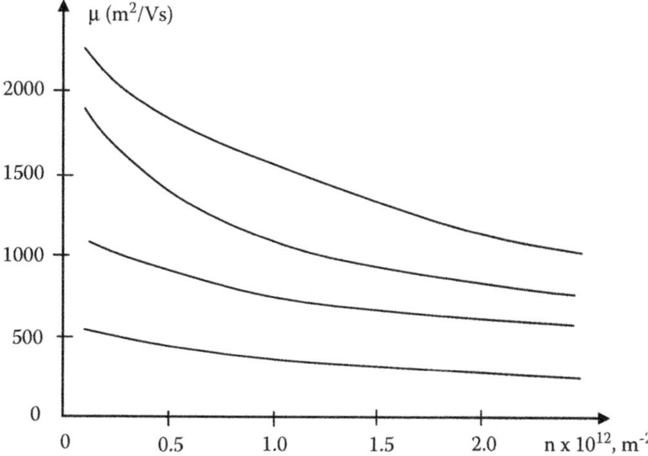

FIGURE 2.2 Mobility vs concentration of electrons on helium surface[2]

According to Durkop et al (2004)[3], the graphene mobility can reach 2090 m^2/Vs at T = 0.4K. This extremely high mobility allows the creation of a quantum computer that can function at T = 0.4 K. The cost refrigeration may be reduced by low power consumption of the computer, making it attractive first of all for special applications. A more practical and already existent 2D system example is GaAlAs/GaAs/GaAlAs – transistor (Fig. 2.3). The Fermi energy that corresponds to the Fermi level may be expressed as:

$$E_F = (\pi\hbar^2/m^*)N_S; \tag{2.5}$$

where N_S is the charge surface density, # of charges/cm^2.

In order to have a 2D system, all electrons must be in the first band. The probability (coefficient) of a possible excitation into the second band is:

$$P = \exp(-\Delta E/k_B T); \tag{2.6}$$

where $k_B T$= 0.147 meV for 1.7 K, $\Delta E \approx 4$ meV and $P = 7 \times 10^{-13}$ which is close to zero, thus leaving all the electrons in the first band (2D system).

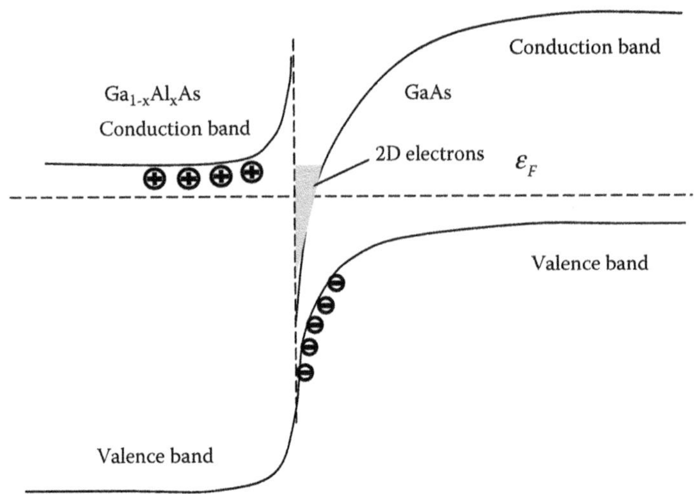

FIGURE 2.3 Band diagram of GaAlAs/GaAs heterojunction

2.1 LOW AND HIGH – FIELD EFFECTS IN 2D

Physical effects in 2D may be different from those in 3D. In particular, heterostructures of modern semiconductor devices brought to life specific features at low and high magnetic fields. At low magnetic fields the carrier type and density may be measured by the conventional Hall effect. For the channel of the length L, width ω and thickness t, with the magnetic induction B in Z direction, the Lorentz force in Fig. 2.4:

$$F = q\upsilon B; \tag{2.7}$$

where υ– carrier velocity, q – the electron charge, and B – magnetic field. The electron field caused by the Hall effect:

$$F_H = \upsilon B = jB/nq; \tag{2.8}$$

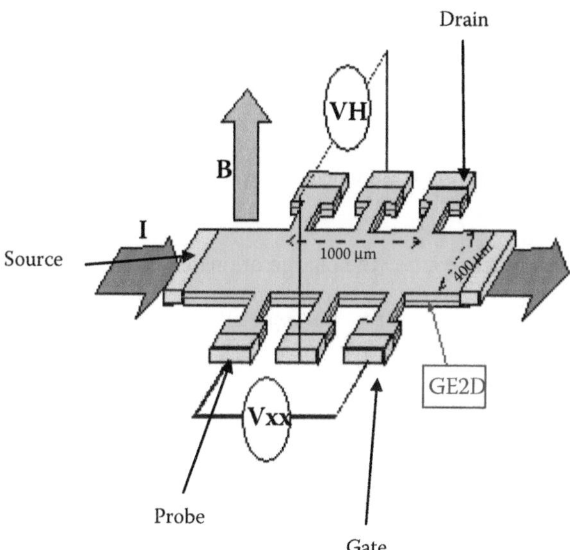

FIGURE 2.4 a) Measurement of the Hall resistance *(Continued)*

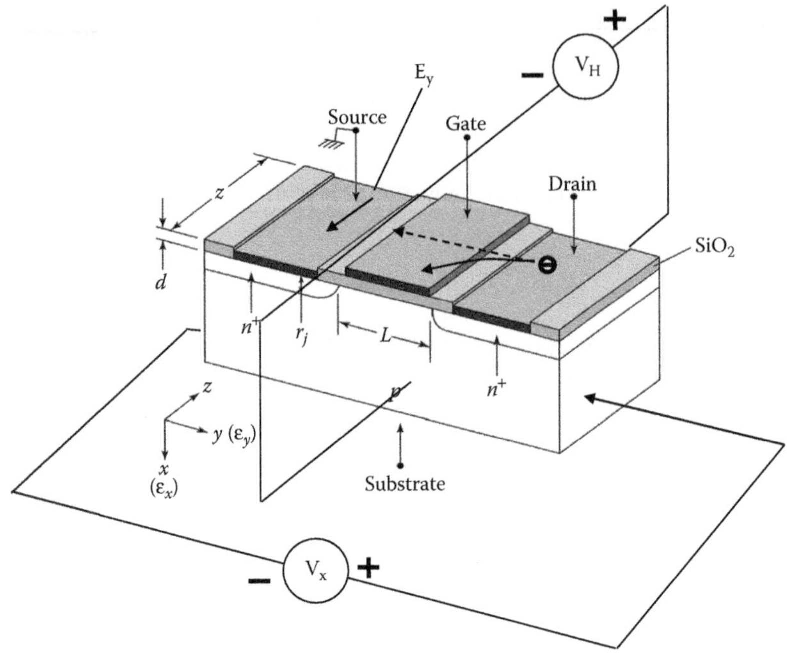

FIGURE 2.4 (Continued) b) Measurement of the Hall resistance

where n – the carrier density, q = the charge of one electron, j = current density

$$j = nq\upsilon; \tag{2.9}$$

The Hall coefficient is defined as:

$$E_H = R_H jB; \tag{2.10}$$

Comparing (2.10) and (2.8):

$$E_H = \frac{jB}{nq} = R_H jB \Rightarrow R_H = \frac{1}{nq}; \tag{2.11}$$

For the two-dimensional case, the measured resistance for the Hall effect:

$$R_{xy} = \omega E_H / I; \tag{2.12}$$

where I = carrier current.

$$I = jwt; \tag{2.13}$$

Inserting (2.11) in (2.12), R_{xy} becomes:

$$R_{xy} = \omega E_H/I = B/nqt; \tag{2.14}$$

For the 2D case

$$n_{2d} = \omega t; \tag{2.15}$$

where t = thickness.

And R_{xy} becomes:

$$R_{xy} = B/qn_{2d}; \tag{2.16}$$

R_H for the quantum Hall effect:

$$R_H = (h/e^2)(1/i); \tag{2.17}$$

where i = integer.

If the Hall effect probes are placed at the interval L, the corresponding longitudinal resistance ($R_{xx} = R_{pp}$):

$$R_{pp} = \rho L/t\omega; \tag{2.18}$$

For a high magnetic field, the ratio of the vertical and horizontal resistances:

$$\frac{R_{xy}}{R_{xx}} = (\omega/L)\omega_c\tau; \tag{2.19}$$

where $\omega_c = \dfrac{eB}{m}, eB/m;$

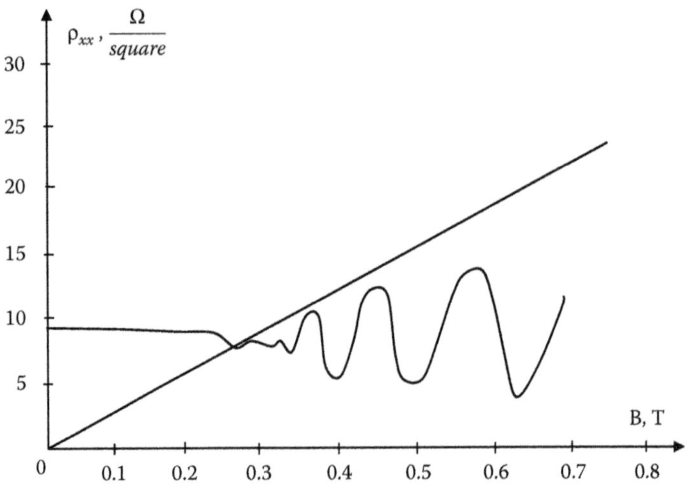

FIGURE 2.5 Dependence of tangential resistance on the magnetic field[4]

$\rho_{xy} = \frac{V_H}{I}$ is the Hall resistance, V_H is the voltage measured across the probes. The oscillation of $R_{xx}(\frac{R_{xy}}{R_{xx}} = \rho_{xx})$ in Fig. 2.5 begin for $B \geq 0.3T$. V_H is the voltage measured across the probes. The oscillations of R_{xx} in Fig. 2.5 begin for $B \geq 0.3T$. It is called the Shubnikov-de Haas effect (the R_{xy}increases linearly). The changes of the density of states as the Landau levels achieve the corresponding energy. The longitudinal resistance also oscillates as can be seen from the following expression that contains E_F:

$$\sigma_{xx} = e^2 D g(E_F); \tag{2.20}$$

where D is diffusion constant and $g(E_F)$ is electronic state density.

2.1.1 The Quantized Hall Effect[5]

Von Klitzing et al., published in 1980 a description of the quantum Hall effect. The set-up included the mentioned above MOS structure (Fig. 2.4). Fig. 2.6 shows the peaks corresponding to zero-voltages areas n = 0, 1, 2.... The electrons (that are carriers, in this case) are controlled by the gate voltage in the inversion layers on p-type Si. The longitudinal voltages are zero when E_F is between the Landau levels. The discovered effect was that the values of $R_H = (h/e^2)n^{-1}$ were identical for devices with different L to ω ratios. Thus, the effect takes place when all the mobile carriers are quantized. The quantization (distribution of the carriers by quantized states) is done by the confinement on z-direction and control by magnetic field of x and y distribution.

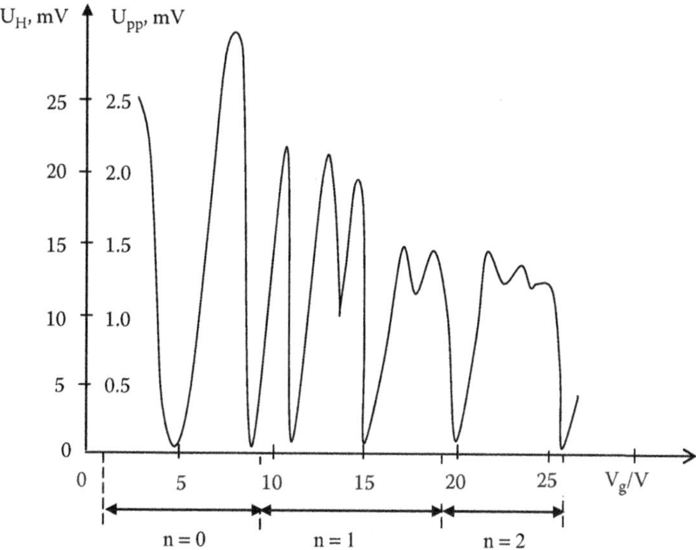

FIGURE 2.6 The quantum Hall effect voltages[5]

The plateaus are related to the peaks of the Raman level spectrum. The magnetic field changes the wavefunction of the electron current of the Hall effect:

$$\psi \sim \exp(ik_x x + ik_y y); \tag{2.21}$$

The wavefunction changes to:

$$\psi \sim L^{-1/2}\exp(ik_x x)\varphi(y); \tag{2.22}$$

when L is the distance between the probes in Fig. 2.4; $\varphi(y)$ is a wavefunction for a harmonic oscillator. The harmonic oscillator is centered at $y_0 = l^2 k_y$, where $e = (\hbar/eB)^{1/2}$, and B is magnetic field, T.

The fundamental magnetic length:

$$l = (\hbar/eB)^{1/2}; \tag{2.23}$$

From the Hermite polynomials component relation, the Christoffel-Darboux expression for Hermite polynomials:

$$\sum_{k=0}^{n} \frac{H_k(x)H_k(y)}{k!\,2^k} = \frac{1}{n!\,2^{n+1}} \frac{H_n(y)H_{n+1}(x) - H_n(x)H_{n+1}(y)}{x-y}; \tag{2.24}$$

For the magnetic flux:

$$\Phi_n k(y) = H_n[(y-y_0)/l]\exp[-(y-y_0)^2/4l^2]; \tag{2.25}$$

where n = 0, 1, 2….

From the solution of the time independent equation exponent $k = ik_x$ the delocalized wavefunction in y_0:

$$\varphi_{n,k}(x,y) = L^{-1/2}\exp(ik_x x)H_n[(y-y_0)/l]\exp[-(y-y_0)^2/4l^2]; \tag{2.26}$$

The distance between the quantum states is $2\pi l^2/L$ where L is the length (see Fig. 2.4) and l is magnetic length. Thus, each state has an area $2\pi l^2$ (Fig. 2.7).

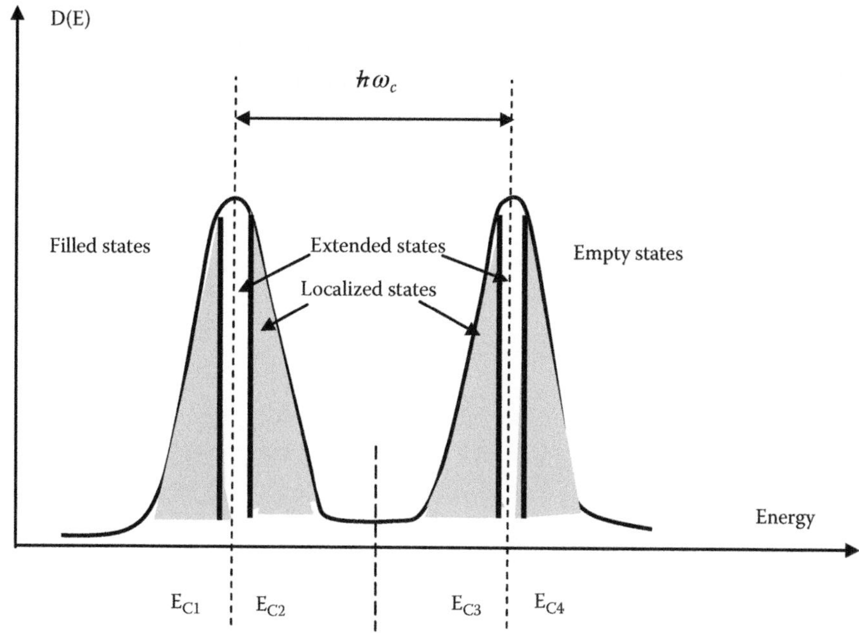

FIGURE 2.7 The densities of the localized and delocalized states[5]

Fig. 2.7 illustrates the integer quantum Hall effect plateaus. The Fermi level is positioned between the extended and localized states[6].

Each Landau level specifies a number of quantum states which is the number of the flux quanta crossing the area of surface A = LW. The Landau levels number from 0 to n – 1. It appears that as long as at least one Landau level peak has a delocalized orbit, the transverse Hall conductance exists. Also, the electrons in impurity bands do not contribute to the Hall current. However, the electrons at the Landau level do contribute to such a current[7]. In Fig. 2.7, the Fermi level is in the range of localized states. In this case, the longitudinal conductivity σ_{xx} is zero (from $\sigma_{xx} = \rho_{xx}/(\rho_{xx}^2 + \rho_{xy}^2)$) and the corresponding longitudinal resistance $R_{xx} = 0$ (the plateau regions). The Fermi level may shift with respect to the localized states (Fig. 2.7). With a magnetic field increase, the longitudinal resistance does not change within the limits of long stretches (Fig. 2.8).

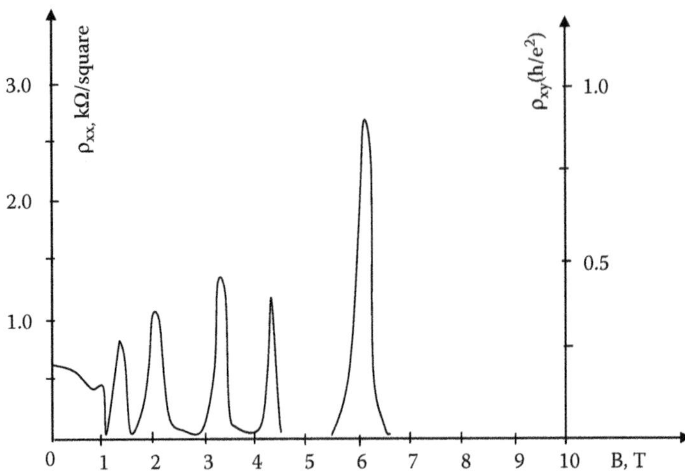

FIGURE 2.8 Region of zero longitudinal resistance of the quantum Hall effect[4]

In a silicon field-effect transistor (or in a heterostructure based on GaAs) the quantum Hall effect is imperfect, the fact that eliminates the possibility of Klaus von Klitzing's effect, the quantization of the Hall conductance to integral multiples of e^2/h. The imperfection is necessary, though, for the quantum Hall effect to occur. Let us consider the effect (Fig. 2.9).

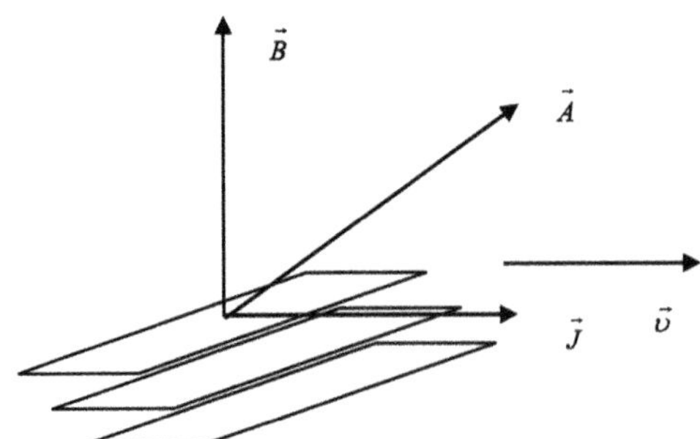

FIGURE 2.9 2D electron gas is placed in a magnetic field. The Lorentz force generates a current and a transverse electric field.

The Lorentz force provides a current (Fig. 2.9), $j = \upsilon\rho$, an electric field $E = \upsilon B/c$ and σ_{xy}, a Hall conductance:

$$\sigma_{xy} = \frac{J}{E} = \frac{\rho c}{B}; \tag{2.27}$$

where c = speed of light.

In real heterostructures ρ is fixed by doping and the value of the gate voltage is not precisely quantized. The above-mentioned imperfection in the quantum Hall experiment manifests itself in the Hall plateau formation. The Hall conductance in the plateau is equal to a multiplication of e^2/h (the parallel resistance is zero in a plateau). The explanation is that the new state adiabatically transformed into a filled Landau with the exception of excitation with partial charges[8]. It is argued that a perfect system is invariant along the x-direction and $\sigma_{xy} = P_c/B$; ρ is not quantized charge density. The high probability of the electron cloud in the z-direction makes the electron fully-quantized in a $f(z)$ bound state in a potential well narrower than the de Broglie wavelength for the electron, $\lambda = \frac{h}{p}$. The de Broglie wavelength is the wavelength, λ, associated with a particle and connected with its momentum, p.

Adding to the theory by von Klitzing, it has been argued[9] that a magnetic field creates at least several delocalized states. Summarizing the above discussion, we can see that electrons on the surface of liquid helium crystallize forming a lattice that exists immediately next to the liquid helium phase. In semiconductor layers, with an applied high magnetic field, electrons confined between the layers exhibit the qualities of a 2D state. Graphene that as a 2D structure has the same quantum Hall effect thus remains (in this aspect) a typical 2D crystal. Although graphene is generally a good conductor, at temperatures close to zero, it exhibits an exponentially decreasing conductance with increasing sample size[10]. Prospective field-effect transistors should be capable of current densities of 10^{12} A/m^2.

2.2 THERMAL MOTION IN A 2D STRUCTURE

A number of theoretical predictions determine the positional order of 2D crystals. For T > 0, the mean square deviation of an atom from its equilibrium position logarithmically increases with the size (L) of the system:

$$|\vec{u}_i|^2 \sim T\ln(l/a); \tag{2.28}$$

where a = lattice constant.

The increase of L brings the possibility of large systems. However, so far there has been experimental confirmation of the absolute divergence $|\vec{u}_i|^2 \sim T\ln(l/a)$. The relative local fluctuation $|\vec{u}_i - \vec{u}_j|^2$ of the local atoms i and j is finite[11]. The considered above liquid helium system does not exist in large extensions of L. The elaboration of the 3D case was provided by Landau and Lifshitz[12]. A large crystal may be described by a vector:

$$\vec{R} = n\vec{a} + m\vec{b} + l\vec{c}; \tag{2.29}$$

where n, m, l are integers.

At a finite temperature (2.29) is rather idealistic, not taking into account crystal inner structure displacement. However, the Fourier decomposition of the displacement may be calculated as:

$$\vec{u} = \sum_k \vec{u}ke^{ik \bullet r}; \tag{2.30}$$

The probability of fluctuation is evaluated by its free energy change (cost), ΔF :

$$P \propto \exp(-\Delta F / k_B T); \tag{2.31}$$

ΔF, the potential energy cost:

$$\Delta F = V/2 \sum_{kil} \vec{u}_{ik}\vec{u}_{lk}\varphi_{i,l}(k_x, k_y, k_z); \tag{2.32}$$

where $\varphi_{i,l}(k_x, k_y, k_z)$ are the components of the displacement; V is the entire body volume connected with the displacement $\vec{u}$. The displacement is analogous to the mechanical displacement of a spring with K, a spring constant in the classical mechanics expression $W = \frac{1}{2}Kx^2$ (spring force work). The expectation value of the displacement vectors:

$$\left\langle u_{jk}u_{lk}^{*} \right\rangle = T/V \sum_{il} A_{il}(\vec{n})/k^2; \tag{2.33}$$

where T is temperature and A_{il} depends on $\vec{n}$, the direction of the vector $\vec{k}(\vec{n} = \vec{k}/k)$, the value of $\left\langle u_{jk}u_{lk}^{*} \right\rangle$ is found from the wave vector k. The mean-square displacement vector may be calculated:

$$\left\langle \vec{u}^2 \right\rangle = T\int [d^3k/(2\pi)^3]A_{ll}(\vec{n})/k^2 = T\int [dkd\sigma/(2\pi)^3]A_{ll}(\vec{n}); \tag{2.34}$$

where σ = component of area;

If k is not large, the integral converges at $k = 0$ and the integral is linear. It means that the mean square fluctuation displacement is a finite quantity and does not depend on the size of the system.

Next, the authors (Landau and Lifshitz) approach the 2D problem:

$$\left\langle u_{ik}u_{lk}^{*} \right\rangle = T\int [dk_x dk_y/(2\pi)^2]A_{il}(\vec{n})/k^2; \tag{2.35}$$

The integral equals a constant times *ln(k)* and $dk_x dk_y = 2\pi k dk$. *k* is proportional to 1/L and 1/a where *L* is the size of the system and a = lattice constant. Thus, the integral diverges as *ln(k)* (logarithmically):

$$\left\langle \vec{u}^2 \right\rangle = const. \times T \ln(L/a); \tag{2.36}$$

Summarizing the above argument, the absolute thermal motion of a point in a 2D system with size *L* is proportional to *T ln(L/a)*, where *T* = temperature[12]. This logarithmic divergence is rather slow and even for the large size *L*, the displacement remains comparatively small. In such a case, the system has properties of a solid state crystal at low temperatures. The actual temperature is relevant to the at least Debye temperature which is comparatively high for graphene as it is for graphite. More specifically, at low temperature, the displacement $\vec{u}(\vec{r})$ of atoms is considered slowly changing at the distance close to the lattice constant distance *a*. Then, the variation in density and the local value of the displacement vector are considered[12]. Let $\rho_0(\vec{r})$ be the density distribution at $T = 0$. Then, the density may be expressed:

$$\rho(\vec{r}) = \rho_0[\vec{r} - \vec{u}(\vec{r})]; \tag{2.37}$$

The correlation function describes the relation between fluctuations at different locations:

$$\left\langle \rho(\vec{r}_1 \rho(\vec{r}_2 \right\rangle = \left\langle \rho_0[\vec{r}_1 - \vec{u}(\vec{r}_1)]\rho_0[\vec{r}_2 - \vec{u}(\vec{r}_2)] \right\rangle; \tag{2.38}$$

The expansion in a Fourier series:

$$\rho_0(\vec{r}) = \rho_{AV} + \sum_{b \neq 0} \rho_b e^{ib \bullet r}; \tag{2.39}$$

where ***b*** are vectors in the 2D reciprocal lattice. Inserting (2.42) into (2.41) yields:

$$|\rho_b|^2 \exp[i\vec{b} \cdot (\vec{r}_1 - \vec{r}_2)]\left\langle \exp[-i\vec{b} \cdot (\vec{u}_1 - \vec{u}_2)] \right\rangle; \tag{2.40}$$

where $\vec{u}_1 = \vec{u}(\vec{r}_1)$;

The right-hand side of (2.40) can be expressed by the expectation value of the reciprocal lattice parameters:

$$\left\langle \exp[-i\vec{b} \cdot (\vec{u}_1 - \vec{u}_2)] \right\rangle = \exp(-1/2 b_i b_1 \chi_{il}); \tag{2.41}$$

where $$\chi_{il}(\vec{r}) = T \int [dk_x dk_y / (2\pi)^2](1 - \cos \vec{k} \cdot \vec{r}) A_{il}(\vec{n}) / k^2; \tag{2.42}$$

Eq. (2.42) converges for small values of k since the upper part of the fraction of the integrand. The maximum value of k may be written as:

$$k_{\max} = T/(\hbar c); \tag{2.43}$$

where c = velocity of sound, $\hbar$= Plank's constant.

Then,

$$\chi_{il}(\vec{r}) = T/\pi A_{ilAV} \ln(k_{\max} r); \tag{2.44}$$

The result for the correlation function with a fictitious temperature T`:

$$\langle \rho(\vec{r}_1)\rho(\vec{r}_2)\rangle - \rho_{AV}^2 = \cos(\vec{b}\cdot\vec{r})/r^{T/T'}; \tag{2.45}$$

where $\vec{b}$ is a reciprocal lattice vector corresponding to the maximum value of $r^{T/T'}$.

For practical purposes, in the two-dimensional lattice, we can consider the motion at close locations to be correlated at low temperatures (although as ***r*** goes to infinity, so the correlation goes to zero). The temperature *T*` was defined[12]:

$$T' = 4\pi mc^2/(b^2 k_b); \tag{2.46}$$

where m is an 2D crystal atom's mass.

In the 3D lattice, the correlation approaches a constant but in a liquid, the correlation diminishes exponentially. The nearest-neighbor correlation is highly correlated. The question of thermal stability: at finite *T'* the correlations of the nearest are finite for any L determined the melting temperature of the 2D lattice[13]. It is stated the existence of a Lindemann criterion of 2D melting. The criterion, $\Upsilon_M^C \equiv \langle [\vec{u}(\vec{R}+\vec{a}_0) - \vec{u}(\vec{R})]^2/a^2\rangle$ was found independent of the nature of a 2D classical crystal, for 2D dipole and 2D Lennard-Jones crystals $\gamma_M^C = 0.12$ and that for a 2D electron crystal $\gamma_M^C = 0.10$. Using a differential criterion:

$$\gamma_M = \left\langle \left|\vec{u}(\vec{r}+\vec{a}) - \vec{u}(\vec{r})\right|^2 \right\rangle / a^2; \tag{2.47}$$

In Fig. 2.10 the simulated curves of the Lindemann criterion are presented. At T ~ 4900 K graphene melts. The lowest curve is received for three neighbors, the middle curve – for 12 neighbors and the upper curve was built for 9 nearest neighbors.

The next question is whether high temperature deformations may be large enough to destroy a 2D crystal (e.g. to melt it). A thermal displacement exceeding a lattice constant, if it is local, does not imply melting since melting presumes a long-range order destruction. The experiment with graphene annealing at T = 2300 K (with the melting point at about T = 3900 K) shows local order defects but not melting[14].

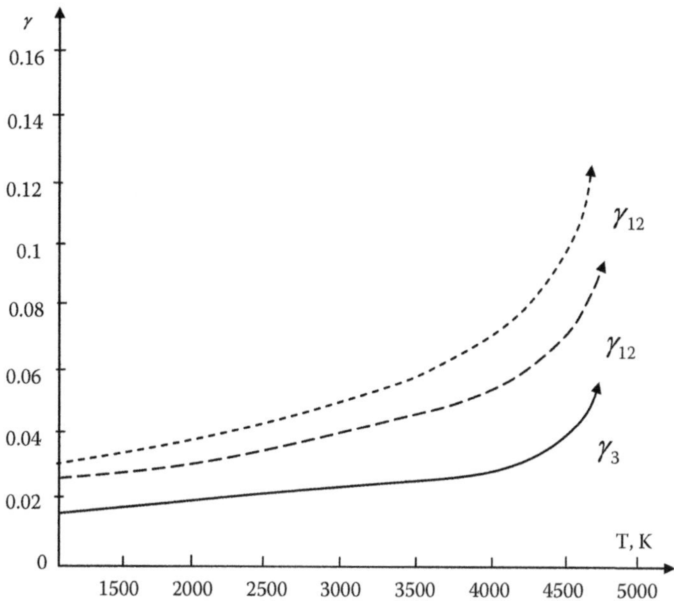

FIGURE 2.10 Dependence of the Lindemann criterion vs temperature[15]

The melting temperature dependence on lattice defects as predicted by Zakharchenko, et al[15] caused by clustering defects and octagon for motion leads to carbon chains. It is possible that melting occurs as the atomic lattice bends. Thus, unit cells that are located some distance from the source of thermal fluctuations of a large amplitude are no longer positioned according to the vector $\vec{R} = n\vec{a} + m\vec{b} + l\vec{c}$. Although, the cells at a distance from the source may be thermally dislocated, the local cell position remains stable. A one- square- meter graphene sample, for example, at room temperature will have a root mean square (rms) amplitude approximately 2 mm and the oscillation period of about 270 hours. Such a sample is not a crystal by definition since the unit cells in this sample move (oscillate) over distances exceeding their lattice constants. Three-dimensional semiconductor samples (such as *Si*, for example) will not have such oscillations and remain perfectly crystalline. The melting effect, therefore, is a pure matter of a two-dimensional system.

2.3 CRYSTALLINE LIMITATION PREDICTIONS

The further question of crystalline instability comes when symmetry stability in 2D crystals is considered. From the mechanical point of view the Hohensberg-Landau-Mermin-Peierls-Wagner (HLMPW) theorem of statistical mechanics shed some light on our 2D crystal investigation[16]. The theorem states that at any finite temperature continuous symmetry with arbitrary dimensions spontaneous magnetization is absent. In 3D spontaneous crystal growth is not limited by statistical mechanics, for 2D dimensional crystal (such as graphene, for example) limits exist. Carbon nanotubes have such a limit[17]. In practice, crystal dimensions are restricted by manufactured devices practical applications and seldom exceed one centimeter. Also, important question is whether the crystal is grown on a substrate or exists unsupported[18]. Some carbon-like materials, such as carbon nanotubes grown by CVD on a silicon substrate can reach 4 cm. The length was not a limit[19]. The described nanotube was stable on a substrate but could be unstable against thermal distortion. The structural changes can be noticeable if the thermal distortion approaches the sample size[20]. It was reported that flexural modes have low frequencies and large amplitudes with a large sample size[20]. At the moment, it is unclear what the thermal limit on the size of a 2D structure is before thermal displacement amplitude exceeds the interatomic distance. Similar to the 2D systems described above, there are so-called "membranes". For such a membrane the lateral displacement[21]:

$$\xi = a\exp(4\pi\kappa/3k_BT); \tag{2.48}$$

where the rigidity $\kappa = Yt^3$; a = the lattice constant, T = temperature, K.

Another view of a possible long-range order possible breakdown was suggested by Wei Gao, et al[22] who believe that for a large sample size L a breakdown will occur because of long-range rotational fluctuations that are formed by the entropy increase and low-energy rotations at finite temperatures. The critical length is given then:

$$L = a\exp(Ga^2/k_BT); \tag{2.49}$$

where a = the lattice constant and G = sheer stress.

Even for the lattice constant a = 0.1 nm and the sheer modulus for bulk graphite $\mu = G/a = 440$ GPA, L exceeds 10^{30} at room temperature which is quite suitable astronomical distance. Whether a graphene sample can remain flat is not obvious either. Castro Neto[22] maintains that a free graphene sample at T = 300 K (room temperature) due to the thermal fluctuations does not remain perfectly flat. Deviations from flatness may be ascribed to shear or tensile strain applied to a sample or come from random chemical impurities. For example, molecules OH may affect C-C bonds in graphene. Some of the impurities and defects may be eliminated by annealing.

Dislocations affect the melting temperature[23]. In 2D sample on liquid helium a model for the melting process was built: melting is based on the above phenomenon (presence of the defects)[15]. Graphene has a triangular lattice with instabilities causing atoms to move into the third dimensions. Fig. 2.11 shows a five-fold 2D lattice with disclinations (disclination is one of line defects that is characterized by violation of rotational symmetry) that are believed to cause melting of the solid on liquid helium surface.

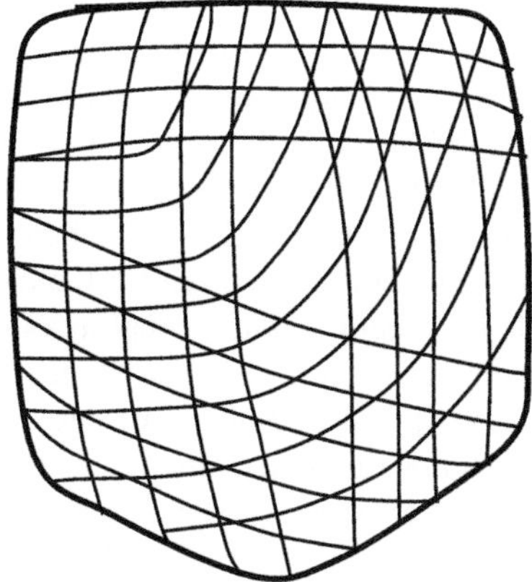

FIGURE 2.11 Equilibrated triangulated lattices of a five-fold disclination[24]. Disclinations are embedded in a 2D triangular lattice.

In polymers and some biological structures have configurations similar to the one on Fig. 2.11. The bonds between cells (e.g., triangular) are not elastic. The idea is based on a square-well potential that connects the neighboring atoms.

The stability of a 2D crystalline layer has been investigated by a number of researchers. One model represents hard spheres connected by strings with a fixed length. Further, a 3D crumpling is stimulated. An increased entropy causes deviation from an initial planar surface[25]. The authors (Kantor, et al) used a mechanical analogy of crumpled sheets of paper and aluminum foil when the resulting deviation from planarity was measured. In this case the membrane has a bending modulus $\kappa = Yt^3$. As it has been reported, for membranes with the bigger dimensions (larger than two), or if there are long-range forces, we may have a crumpling segment that separates a rigid low T phase from a crumpled high T phase. For their calculations the authors used a perturbation theory. Their experiments showed that the deviation from a flat surface becomes pronounced and comparable to the lateral displacement L. For such a condition, the coherence length is[26]:

$$\xi = a\exp(2\pi\kappa / k_B T); \tag{2.50}$$

At room temperature the coherence length is very high but at approximately the melting temperature (~ 4900^0 F), the coherence length is on the order of 50 μm. The Eq. (2.50) suggests that crumpling takes place only beyond ξ_p that equals a constant in an exponentially rising function. An introduction of a dislocation further promotes unbinding by reducing of the dislocation energy. The formation of a stable phase that has a six-fold orientation order is otherwise called a "hexatic" phase. The crystal orientation which is acquired helps to resist the thermal softening with an existing finite crumpling temperature.

Several authors consider crumpling transitions, such as of a polymerized membrane with a linear dimension L with respect to changing temperature. R_G is defined as a crumpling transition[25]. At low temperature, $R_G \approx \xi L$, where ξ is an order parameter that measures the shrinkage of the membrane due to small scale fluctuations. At high temperature $R_G \propto L^{0.8}$. This dependence was described in 1985[25]. The elastic energy due to undulations is used to define the bending energy for two elements in a triangular lattice. The bonds are given for the pairs of atoms connected by a square well potential. The claim is that a coupling between modes is enough to eliminate transverse fluctuations[25]. It is also possible that a transition from a corrugated state to a flat state takes place as temperature decreases.

The undulations described above at finite temperature put restrictions on synthesis of 2D crystals. The existence of 2D crystals and their growth depend substantially on temperature and growth of such crystals depends on high temperature. This dependence results in grown crystals of very limited size and stability. One of the possible restrictions comes from low bending rigidity of 2D crystals that may easily bend and crumple e.g. to form 3D structures. These are exceptions to this rule: the growth of long carbon nanotubes. One of such was reported by Zheng et al (2004)[27]. So it seems possible to grow a sheet of graphene as well, such as the growth of long graphene sheets on Cu or Ni foil substrate. A problem associated with this is separation of 2D and 3D grown structures. It may be done by "peeling" from the parent 3D crystal. The growth and graphene isolation may take place at a high and room temperature respectively.

2.4 VIBRATIONS OF THIN PLATES

The growth of graphene implies strain on the material. If graphene is to remain unbuckled on a substrate, the compressive stain should not exceed ~ 1 % (for SiC substrate). A van der Waals force plays a large role in keeping graphene from buckling on the substrate. The following discussion elaborates different aspects of elasticity of thin layers, such as those of graphene.

Vibrations of thin plates of elastic materials (more precisely those materials that satisfy the elasticity conditions) are well studied and applicable to a large extent to graphene. An elastic solid is described by a Young's modulus Y, a shear modulus G and a bending modulus $\kappa = Yt^3$, where t = thickness of material. The material coordinates in 3D are given by sets of coordinates $x \equiv (x_1, x_2, x_3 ...)$. To determine a position of a material bit in space a set of points may be assigned which constitute a vector $\vec{r} = (r_1, r_2, r_3 ...)$. The elastic behavior is caused by an applied force or energy associated with it, $E(\vec{r})$ for a particular embedding $\vec{r}(x)$. The minimal (or zero) energy for a particular position may be denoted as $\vec{r}_0(x)$. The interaction between separate points depend on the distance between the points. The spatial derivatives are:

$$\delta r_3 / \delta x_2 \equiv \delta_2 r_3 \text{ and } \delta^2 r_1 / \delta x_1 \delta x_2 \equiv \delta_1 \delta_2 r_1; \tag{2.51}$$

The distance between the points next to each other is calculated as a sum of distances:

$$ds^2 = \sum_i dx_i^2; \tag{2.52}$$

In homogeneous and isotropic materials, the Hook's law in 3D satisfies the following condition: $\delta = 2\mu\varepsilon + \lambda tr(\varepsilon) I$, where ε is the strain tensor, and I is the identity matrix.

The two parameters together constitute a parameterization of the elastic moduli for homogeneous isotropic media, popular in mathematical literature, and are thus related to the other elastic moduli, for instance, the bulk modulus can be expressed as $K = \lambda + (2/3)\mu$. Although the sheer modulus, μ must be positive, the Lame's first parameter, x, can be negative, in principle; however, for most materials, it is also positive. Gabrial Leon Jean Baptiste Lame (1795 – 1870) was a French mathematician who contributed to the theory of partial differential equations. He made use of curvilinear coordinates along with the elasticity theory. The Lame parameters (Lame constants) are λ, the Lame's first parameter and μ, the second parameter, also called the dynamic viscosity or sheer modulus. A metric tensor is a type of function that has input of tangent vectors v and w that produce a scalar (real number) $g(v, w)$ that generalize the properties of the dot product of vectors in Euclidian space.

For our case, the metric tensor is defined as:

$$g_{ij} = d\vec{r}/dx_i \cdot d\vec{r}/dx_j = \delta_{ij} + \gamma_{ij}; \tag{2.53}$$

The dimensionless strain tensor is defined as $\vec{\gamma} = (\gamma_{11}, \gamma_{12}, \gamma_{13} ... \gamma_{ij})$.

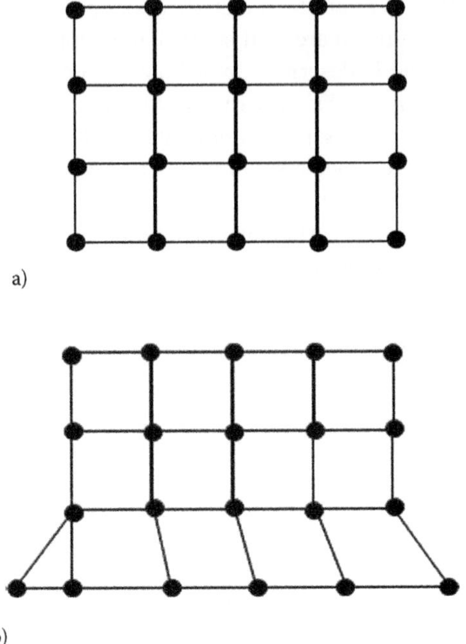

FIGURE 2.12 Elastic with crystalline structure changes: a) equilibrium with crystalline structure intact, b) equilibrium with the lower rows length increased.

The horizontal increase also causes an increase in the vertical direction. The upper links also become stretched but by a less amount. The derivation from the initial balanced position in Fig. 2.12 is described by the following equation based on a Gauss's theorem:

$$\mathrm{K}(y) = (1/\sqrt{g_{xx}})d^2(\sqrt{g_{xx}}))dy^2; \tag{2.54}$$

where K(y) – curvature resulted from an implied force. Any deviation γ_{ij} from the zero position results in quadratic strain, γ^2. In a symmetric tensor, $(Tr\gamma)^2$ and $Tr(\gamma^2)$ are equal. Then, the strain energy is:

$$E[\bar{r}] = \int_x \frac{1}{2}\lambda(Tr\gamma)^2 + \mu Tr\gamma^2; \tag{2.55}$$

where λ and μ - properties of the material (the Lame coefficients).

With the material uniformity strain and g's in (2.54) are not zero, the work $\delta E/\delta\gamma_{ij}$ is necessary to maintain the strained position per unit area. The derivative is

$$\frac{\delta E}{\delta \gamma_{ij}} = \sigma_{ij}; \tag{2.56}$$

From (2.55), $\sigma_{ij} \propto \gamma$. In Fig. 2.12, if the strain is applied in one direction only, then the stress is called the Young's modulus, Y:

$$Y = \sigma_{ij}/\gamma_{11} = \mu(3\lambda + 2\mu)/(\lambda + \mu) \tag{2.57}$$

If the material (e.g. crystal) is shifted in two orthogonal directions with the coefficients $\gamma_{22} = -v\gamma_{11}$. The Poison ration v is then:

$$v = \lambda/2(\lambda + \mu); \tag{2.58a}$$

Please note:

In general, if the material is stretched or compressed along the axial direction, the Poisson's ration is defined as:

$$v = -\frac{d\varepsilon_{trans}}{d\varepsilon_{axial}} = -\frac{d\varepsilon_y}{d\varepsilon_x} = -\frac{d\varepsilon_z}{d\varepsilon_x}; \tag{2.58b}$$

where ε_{trans} is transverse strain (negative for axial tension/stretching); ε_{axial} is axial strain (positive for axial tension but negative for axial compression).

In a 2-dimensional case of thickness, t, we have:

$$c_{ij} \equiv \vec{n}\cdot\delta^2\vec{r}/(\delta x_i \delta x_j); \tag{2.59}$$

We have only two coordinates, x_1 and x_2. The displacement $\vec{r}\cdot\vec{n}$ is normal to the plane of the 2-dimensional material and the curvature sensor c_{ij}:

$$c_{ij} \equiv \vec{n}\cdot\delta^2\vec{r}/(\delta x_i \delta_j); \tag{2.60}$$

The curvature sensor is a (image-plane) measurement of a local wavefront curvature derived from two specific out-of-plane images. The symmetrical tensor determines "the depth" of the curve: how much deviation from a straight line we have. For a two-dimensional surface, the stain tensor γ and the curvature tensor c work for the surface tensors[12]:

$$\begin{aligned} E(\vec{r}) = S(\gamma) + B(\vec{c}) = \int dx_1 dx_2 (\lambda/2(Tr\gamma)^2 + \mu Tr\gamma^2) \\ + \int dx_1 dx_2 [\kappa/2\{Tr\vec{c})^2 + \kappa_G/2[(Tr\vec{c})^2 - Tr(\vec{c})^2]\} \end{aligned} \tag{2.61}$$

where $S(\gamma)$ has surface components and $B(c)$ has curvature components. Also, $S(\gamma)$ is stretching energy and $B(c)$ – bending energy. Continuing with specific forces, the membrane stress is $\delta S/\delta\gamma_{ij} = \sum_{ij}$, when the units are force per unit surface or per unit length in j-direction, Q_i is one of the stresses applied to an element of the surface[12]. Using the Lame's coefficients, in 2D the Young's modulus is:

$$Y_{2D} = 4\mu(\lambda + \mu)/(\lambda + 2\mu),\ N/m \tag{2.62}$$

where $Y_{2D} + hY$, $Y =$ the Young's modulus, $h =$ the layer thickness. The Poison's ratio for the membrane is:

$$\nu = \lambda/(\lambda + 2\mu); \tag{2.63}$$

For graphene Y_{2D} was measured to be 342 N/m (+/- 50 N/m). $Y_{2D} = 1.02\ TPa$ if $h = 0.34$ nm[28]. Another measurement of the graphene's characteristics gave $Y_{2D} = 3.4\ TPa$ and $h \sim 0.1$ nm[29]. The physical graphene layer thickness, however, $h = t = 0.34$ nm. From the crystalline geometry, the radius of carbon in tetrahedral bonds is 0.77 angstrom and the graphene layer thickness, therefore, 0.154 nm. "Elastic thickness" is another graphene's characteristic; $h_e = (\kappa/Y)^{1/2} = (\kappa/hY)^{1/2}$, which is smaller than the layer thickness[28]. κ is a stretching coefficient. κ may be several eV for graphene. From the nominal one-layer graphene thickness of 0.34 nm and $\kappa = Yh^3 = 1eV$, we find that $Y_{2D} = 320\ N/m$. From Lame's coefficients:

$$\lambda = 2h\mu/(1 + 2\mu/\lambda) = hY_{2D}\nu/(1 - \nu^2); \tag{2.64}$$

$$\text{and } \mu = hY_{2D}/2(\nu + 1); \tag{2.65}$$

Then, the stretching coefficient is:

$$\kappa = h^3\mu[1 + (\mu/\lambda)^2]/3(1 + 2\mu/\lambda) = h^3Y_{2D}/12(1 - \nu^2); \tag{2.66}$$

The elastic thickness is:

$$h_e = (\kappa/Y_{2D})^{1/2} = h/[12(1 - \nu^2)]^{1.2}; \tag{2.67}$$

where $\nu =$ Poisson's ratio, which is approximately $1/3 < \nu < 1/2$ and $0.3\,h < h_e < 0.33\,h$. From $h = 0.34$ nm for graphene and $\nu = 0.165$, $h_e \sim 0.1$ nm. Such values for h_e have been reported recently[28].

2.5 SOURCES OF 2D LAYER STRAINS

The surface of a 2D layer may not be perfectly flat. In this case, we deal with curvatures. Classical mathematical approach is applied to characterize the curvature field[30]. The curvature tensor, $\delta_i c_{jk} = \delta_k c_{ji}$ is the product of the two principal curvatures (eigenvalues of the curvatures). For the Gaussian curvature C_G, the Gaus-Bonnet theorem[30] specifies that the stretching coefficient K_G is associated with C_G. The curvature energy produces a strain field γ impacts the membrane configuration.

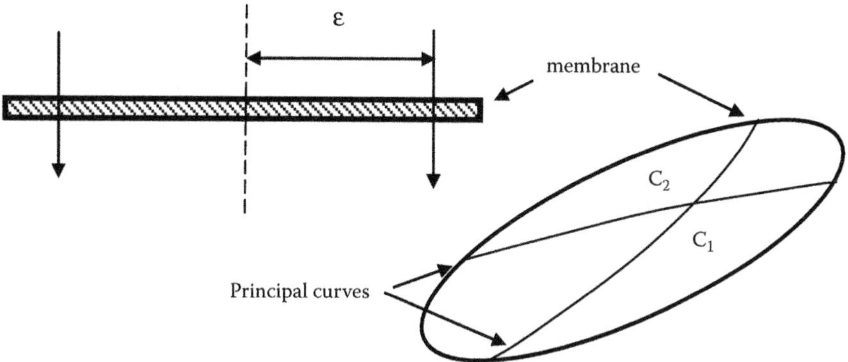

FIGURE 2.13 Membrane perimeter reduction. The initial position of the membrane is shown in: principal curves C_1 and C_2 are given; the membrane is still unbent.

For the principal curves C_1 and C_2, we can write[30]:

$$\vec{r} \cdot \vec{n} = 1/2c_1x_1^2 + 1/2c_2x_2^2; \tag{2.68}$$

where x_l, x_2 – coordinates for the membrane surface in x – plane and $\vec{r} \cdot \vec{n}$ space coordinates vector dot product.

The supposition is that the principal curves retain their lengths. The membrane initial radius ε is diminished by $(\varepsilon C)^2$. The strain associated with the curvatures grows quadratically with the size of the membrane. The Gauss's theorem of surfaces states[30]:

$$c_1c_2 = 2\delta_1\delta_2\gamma_{12} - \delta_1\delta_1\gamma_{22} - \delta_2\delta_2\gamma_{11}; \tag{2.69}$$

The importance of membrane mechanical characteristics comes from the fact that thin membranes (with thickness which is close to the thickness of one atomic layer) may be considered as non-stretchable sheets. The ratio of bending with respect to stretching is proportional to K/λ and K/μ[31]. Thin membranes require more energy to stretch them than to bend. Some specific forms (such as wedge) may change the situation when a new configuration can work against radial bending. The dependence of stretching on bending explains the anomalous rigidity of thermally fluctuating membranes. This energy necessary for stretching prevents isotropic bending. The amount of stretching depends on the ratio of thickness to the membrane fluctuation[31]. The law is that at least one direction remains unchanged[32]. This law imposes restrictions on an isometric membrane. E.g. a disk-shaped membrane specifies a sphere of a diameter that equals that of the disk. The Gaussian curves specify a point at the vertex of the membrane when it is transformed into a cone under the stretching forces[32].

Another form of deformation is vibrations when the applied force changes its amplitude and directions. The size of the sample is critical in such case. The effective spring constant $K = Yt^3/L^2$, where L is the square length and t is the thickness. For all practical purposes the graphene samples should be on the scale of micrometers. For a 2D sample, the following relation holds[12]:

$$\rho\delta^2\zeta/\delta t^2 + Yh^2/[12(1-v^2)][\delta^2\zeta/\delta x^2 + \delta^2\zeta/\delta y^2] = 0; \quad (2.71)$$

where h = the sample thickness, v = Poisson ratio, Y = Young modules, ρ = density, $\zeta(\vec{r},t)$ = vertical displacement.

Finding the vertical displacement as a periodic function $\zeta(\vec{r},t) = \zeta_0 \exp[i(\vec{k}\cdot\vec{r} - \omega t)]$, Eq. (2.71) may be written:

$$-\rho\omega^2 + Yh^2k^4/[12(1-v^2)] = 0; \quad (2.72)$$

In (2.72) the frequency ω is found from (2.72) as:

$$\omega = k^2[Yh^2/12\rho(1-v^2)]^{1/2}; \quad (2.73)$$

The frequency ω has a quadratic dependence, rather than being linear as it is the case in 3 dimensions. The velocity of the wave in 2D sample can be found as:

$$\vec{v} = \vec{k}[Yh^2/3\rho(1-v^2)]^{1/2}; \quad (2.74)$$

which is proportional to the wave vector and changes with time. Further, the mechanical characteristics of tree-standing membranes (e.g. graphene samples) are similar to the characteristics of thin filaments (ribbons and rods). The equation of motion for a wave propagating in the z-direction along a ribbon or a rod (thickness t, width ω and x – vertical displacement, S = ωt) is:

$$\rho S\delta^2 X/\delta t^2 = YI_y\delta^4\chi/\delta z^4; \quad (2.75)$$

Making the same assumption as we did in [2.73]:

$$X(z,t) = X_0 \exp[i(k_z - \omega t); \quad (2.76)$$

Solving for the frequency, ω:

$$\omega = \kappa^2 [YI_y / \rho S]^{1/2}; \tag{2.77}$$

where $I_y = \omega t^3/12$.

In case of a sample (e.g. rod, ribbon) supported or clamped at edges:

$$X(z,t) = X_0(z)\cos(\omega t + \alpha); \tag{2.78}$$

where $\cos(\omega t + \alpha)$ is the periodicity of a wave moving in z-direction.

Differentiating (2.78), we have:

$$d^4 X_0 / dz^4 = \kappa^4 X_0; \tag{2.79}$$

and then,

$$\kappa^4 = \omega^2 \rho S / YI_y; \tag{2.80}$$

The solution for (2.79)[12] is:

$$X_0 = A\cos\kappa z + B\sin\kappa z + C\cosh\kappa z + D\sinh\kappa z; \tag{2.81}$$

For the boundary $X_0 = dX_0/dz = 0$ which implies no stretching at z = 0, z = L, the solution is in the form of $\xi_0 = A\sin(m\pi x/a)\sin(n\pi y/b)^{12}$.

Then, the vertical displacement is:

$$X_0 = A[(\sin\kappa L - \sinh\kappa L)(\cos\kappa z - \cosh\kappa z) + \\ + (\cos\kappa L - \cosh\kappa L)(\sin\kappa z - \sinh\kappa z)]; \tag{2.82}$$

The corresponding frequency is then:

$$\omega_{\min} = 22.4/L^2 [YI_y / \rho^S]^{1/2}; \tag{2.83}$$

where $t_y = \omega t^3/12$. ρS implies $m/L = \rho\omega t$, i.e. mass per unit length (not density since we deal with 2D structures). In case, only one sample's edge is fixed, the vertical displacement:

$$X_0 = A[(\cos\kappa L + \cosh\kappa L)(\cos\kappa - \cosh\kappa z) + \\ +(\sin\kappa L - \sinh\kappa L)(\sin\kappa z - \sinh\kappa z)]; \tag{2.84}$$

The minimal frequency in this case is:

$$\omega_{\min} = 3.52/L^2 [YI_y / \rho S]^{1/2}; \tag{2.85}$$

Thus, the frequency beam of a free sample (such as a membrane) is higher than that of a clamped sample. A free position sample, a beam of length L oscillates with the frequency of a one-end-clamped beam of the length $L/2$. And the minimal frequency:

$$\omega_{\min}(xylophone) = 14.08/L^2[YI_y/\rho S]^{1/2}; \tag{2.86}$$

Thus, the highest frequency corresponds to the free oscillating beam with the lowest corresponding to the clamped version. Another deformation possibility is buckling (see Fig. 2.14).

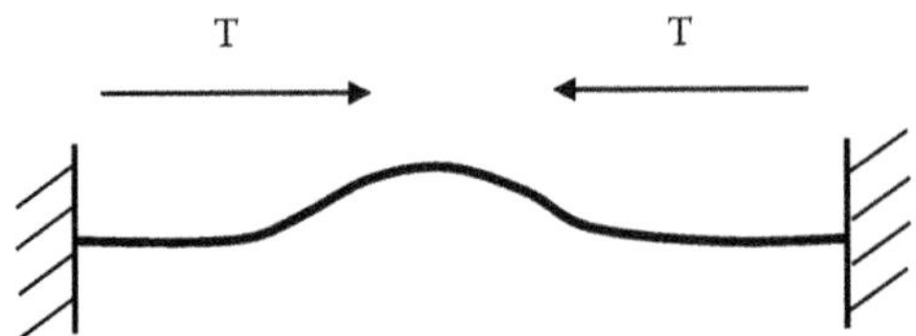

FIGURE 2.14 Buckling of a sample with compressive forces. The sample's edges are stationary fixed.

If the sample has a cross section area $S = \omega t$, where ω = width and t = thickness. The moment for this area $I = \omega t^3/12$ with the applied compressive force T. Then, the critical compression force is[33]:

$$T_{crit.} = 4\pi^2 YI_y/L^2; \tag{2.87}$$

Considering a square sample with dimensions $L \; x \; L \; x \; t$ that buckles under a force $F = mg$ that may buckle under its own weight. The dimension L comes to:

$$L^3 = 4\pi^2 Yt^2/12\rho g; \tag{2.88}$$

For graphene, $L = 0.3$ mm. The resulting strain may be expressed:

$$u_{zz} = F'(1+\nu)(1-2\nu)/Y(1-\nu)^{12}; \tag{2.89}$$

where ν = Poisson's ratio, $\nu = 0.17$ for graphene[34]. If $F' = mg/Lt$, then $u_{zz} = 5.4 \cdot 10^{-12}$.

Graphene buckles easily as indicates the above reasoning. Some substrates may prevent graphene from buckling. The van der Waals attraction keeps a graphene layer on the substrate (e.g. SiC). In general, graphene can withstand compressive strain of approximately 1% when grown on substrates with lattice constants close (also of about 1% difference) to that of the material. If all sides of a sample plate are supported (e.g. a rectangle with sides a and b and the thickness t) then the sample's corresponding frequency is:

$$\omega = [Yt^2/12\rho(1-\nu^2)]^{1/2}\pi^2[m^2/a^2 + n^2/b^2]; \tag{2.90}$$

The frequency of oscillation for a sample greatly depends on the sample's size, for example, a graphene plate fixed on all sides with the area of 1 m^2 will oscillate with a period of longer than 10 days (270 hours) based on (2.90) calculations. More practical example (for a device purpose) yields $\omega = 41 \cdot 10^6$ rad/s for a square of 1 mm^2 area. In order to see the thermal dependence of the sample frequency, we consider again a free-standing sample, the oscillation frequency which is (see [2.86]:

$$\omega_{\min} = 14.08 / L^2 [YI_y / \rho S]^{1/2}; \tag{2.91}$$

Using a spring constant (2.69) ($\kappa = Yt^3/L^2$), we can write a simplified equation: $\omega_{\min} = (\kappa^*/m)^{1/2}$, where $\kappa^* \propto L^4 YI_y$ and $m \propto \rho S$. Assuming $\frac{1}{2}\kappa^* \cdot x_{rms}^2 = \frac{1}{2}k_B T$, then

$$\kappa^* = \frac{k_B T}{x_{rms}^2} \text{ and } \frac{x_{rms}}{L} = \sqrt{\frac{L}{L_{cr}}} \text{ then } \frac{x_{rms}^2}{L^2} = \frac{L}{L_{cr}}, \text{ then } L_{cr} = \frac{L}{x_{rms}^2}, \text{ thus } L_{cr} = \frac{k_B T}{\kappa^*}; \tag{2.92}$$

At room temperature, carbon tube critical length may be on the order of tenths of mm. For a tube fixed at one end, the amplitude of the movement from the balance[35]:

$$\sigma = 0.849 L^3 k_B T / [YWG(W^2 + G^2)]; \tag{2.93}$$

where W – the tube's diameter and G is the wall's thickness.

At room temperature, for typical carbon tube parameters of 0.34 nm (G, the wall thickness) and the length = 0.53 mm (L) and Y = 1 TPA, the geometrical shift $\sigma = 4.6 * 10^{-6}$ m. For the square-sample case, (L = square's side):

$$x_{rms} / L = (T / T_0)^{1/2}; \tag{2.94}$$

For graphene $T_o = 4.7 * 10^6$ K, $\sigma = 2.5$ mm.

The thermal motion at room temperature exceeds the lattice spacing which means that the crystal as such does not exist but is not destroyed, nevertheless. For practical reasons, the vibrations and movements described above may be considered crystal oscillations.

REFERENCES

1. Peeters, F., "Electron crystallites floating on superfluid helium", *American Physical Society,* **2**, 1979
2. Kono, K., "Electrons take their places on a liquid helium grid", *American Physical Society",* **4** (2011)
3. Durkop, T., Kim, B.M., and Fuhrer, M.S., "Properties and applications of high-mobility semiconducting nanotubes", *J.Phys. Condens. Matter,* **16** (2004)
4. Tsui, D.C., Stormer, H.L., Gossard, A.C., "Two-dimensional magnetotransport in the extreme quantum limit", *Phys. Rev., Lett.,***48** (1982)
5. von Klitzing, K., Dorda, G., Pepper, M., "New method for high-accuracy determination of the fine fine-structure constant based on quantized Hall resistance", *Phys. Rev., Lett.* **45** (1980)

6. Meier, F., von Bergmann, K., Feiriani, P., Wiebe, J., Bode, M., Hashomoto, K., Heinze, S., Wieendanger, R., "Spin-dependent electronic and magnetic properties of Co nanostructures on Pt (111) studied by spin-resolved scanning tunneling spectroscopy", *Phys. Rev.* **B11** (2006)
7. Ando, T., "Voltage distribution and phase-breaking scattering in the quantum Hall regime", *Proc. Of the 11th International Conference on the Electronic Properties of Two-Dimensional Systems",* Vol. 361-362, pp. 270 – 273 (1996) (Elsevier "*Surface Science*")
8. Laughlin R.B. *Nobel Prize lecture, Physics* (Dept. of Physics, Stanford Univ.) 1998
9. Levine, H., Libby, S.B. and Pruisken, A.M.M., "Electron delocalization by a magnetic field in two dimensions", *Phys. Rev. Lett.* **51**, 1915 – 1918 (1983)
10. Abrahams, E., Anderson, P.W., Licciardello, D.C., Ramakrishnan, T.V., "Scaling theory of localization: Absence of Quantum Deffusion in two Dimensions", *Phys. Rev. Lett.* **42,** 673 – 676 (1979)
11. Jancovici, B., "Infinite Susceptibility without long-range: The Two-Dimensional Harmonic Solid", *Phys. Rev. Lett.* **19** (1967)
12. Landau L.D., Lifshitz E.M., "*Theory of elasticity*", Interworth-Heinemann, Oxford, MA (1986)
13. Bedanov, V.M., Gadiyak, G.V., Lozevik, Yu.E., "On a modified Lindermann-like criterion for 2D melting", *Phys. Lett.*, Vol. 109, Issue 6, pp. 289-291 (1985)
14. Lin, Y.C., Lu, C.C., Yeh, C.H., Suenaza, K., and Chin, P.W., "Graphene annealing: how clean can it be?", *Nano Letters,* Vol. 12, #1, pp. 414-419 (2012)
15. Zakharchenko, K.V., Fasolino, A., Los, J.H., and Katsnelson, M.I., "Melting of graphene: from two to one dimension", *J. Phys.: Condens. Matt.* **23** (2011)
16. Ioffe, D., Shlosman, S. B., Velenik, Y., "2D models of statistical physics with continuous symmetry: the case of singular interactions", *Comm. Math. Phys.* **226** (2002)
17. Klein, A., Landau, L. J., Shucker, D. S., "On the absence of spontaneous breakdown of continuous symmetry for equilibrium states in two dimensions", *J. Statist. Phys.* **26** (1981)
18. Zhang, J., Feng, M., Tachikawa, H., "Layer-by-layer fabrication and direct electrochemistry of glucose oxidase on single wall carbon nanotubes", *Elsevier Biosensors and Bioelectronics,* Vol. 22, Issue 12, pp. 3036-3041 (2007)
19. Zheng, L. X., O'Conell, M. J., Doorn, S. K., Liao, X. Z., Zhao, Y. H., Akhadov, E. A., Hoffbauer, M. A., Roop, B. J., Jia, Q. X., Dye, R. C., Peterson, D. E., Huang, S. M., Liu, J., and Zhu, Y. T., "Ultralong single-wall carbon nanotubes", *Nature Materials,* **3** 613-676 (2004)
20. Schelling, P. and Keblinski, P., "Thermal expansion of carbon structures", *Phys. Rev.* **B68** (2003)
21. Kleinert, H., "Thermal softening of curvature elasticity in membranes", *Phys. Lett.* A, Vol. 114, Issue 5, pp. 263-268 (1986)
22. Neto, A. C. "The electronic properties of graphene", *Reviews of Modern Physics,* Vol. 81, Issue 1 (2009)
23. Kleinert, H., "Disclinations and first other transitions in 2D-melting", *Phys. Lett.* A, Vol. 95, pp. 381-384 (1983)
24. Seung, H. S., Nelson, D. R., "Defects in flexible membranes with crystalline order" *Phys. Rev.* A, **38**, pp. 1005-1018 (1988)
25. Kantor, Y., and Jaric M. V., "Triangular lattice foldings- a transfer matrix study", *Europhys. Lett.,* **11**, pp. 157-161 (1990)
26. Leibler, S., Lipowsky, R., Peliti, L., "Curvature and fluctuations of amphiphilic membranes", *Physics of Amphiphilic Layers,* Springer Proc. *Physics*, Vol. 21, pp. 74-79 (1987)

27. Zheng, L. X., O'Connell, M. J., Doorn, S. K., Liao, X. Z., Zhao, Y. H., Akhadov, E. A., Hoffbauer, M. A., Roop, B. J., Jia, Q. X., and Dye, R. C., "Ultralong single-well carbon nanotubes", *Nature Materials,* **3**, pp. 673-676 (2004)
28. Lee, C., Wei, X., Li Q., Carpick, R., Kysar J.W., and Hone, J., "Elastic and frictional properties of graphene", *Phys. Status Solidi* **3** 246 (2009)
29. Sakhaee-Pour, A., "Elastic properties of single-layered graphene sheet", *Solid State Communications,* Vol. 149, Issue 1-2, (2009)
30. Millman, R. S., Parker, G. D., "*Geometry: A Metric approach with models*", Springer-Verlag, N.Y., (1981)
31. Bowick, M. and Travesset, A., "The Statistical mechanics of membranes", *Phys. Rep.* **344**, 255 (2001)
32. Radzihovsky, L. and Toner, J., "Elasticity, shape fluctuation and phase transitions of anisotropic membranes", *Phys. Rev. Lett.* **57** (1998)
33. Li, Chunyu, "Modeling of elastic buckling of carbon nanotubes by molecular structural mechanics approach", *Mechanics of Materials,* **11** (2004)
34. Kis, A., Zettl, A., "Nanomechanics of carbon nanotubes", *Phil. Trans. R. Soc.* A, **366**, pp. 1591-1611 (2008)
35. Wang, L., Hu, H., Guo, W., "Thermal vibration of carbon nanotubes predicted by beam models and molecular dynamics", The Royal Society Publishing (2010)

3 Graphene
Physical Properties

Graphene is one of the crystalline forms of carbon related materials. In particular, graphene is similar to the crystalline structure of diamond and graphite. In fact, graphene is essentially one-atom thick layer of graphite. The term itself ("graphene") was coined by Hanns-Peter Boehm[1] who described single-layer carbon foils (1962). The word "graphene" is a variation of the word "graphite". Graphene is related structurally to carbon allotropes (including graphite) such as carbon nanotubes, fullerenes and even charcoal. Important feature of graphene is its one-dimensionality which sets it apart from three-dimensional forms, such as e.g. graphite.

ALLOTROPES OF CARBON: DIAMOND AND GRAPHITE

The electronic structure of carbon gives rise to its ability to bond in many different configurations and form structures with distinctly different characteristics. This is clearly manifested in diamond and graphite, which are the two most commonly observed forms of carbon. Diamond forms when the four valence electrons in a carbon atom are sp^3 hybridized (i.e., all bonds shared equally among four neighboring atoms), which results in a three-dimensional 3-diamond cubic structure. Diamond is the hardest material known to humankind due to its 3D network of carbon-carbon (C – C) bonds. It is also special by the fact that it is one of the very few materials in nature that is both electrically insulating and thermally conductive. On the other hand, graphite is the sp^2 hybridized form of carbon and contains only three bonds per carbon atom. The fourth valence electron is in a delocalized state, and is consequently free to float or drift among the atoms, since it is not bound to one particular atom in the structure. This creates a planar hexagonal structure (called graphene) and gives rise to the layered structure of graphite that is composed of stacked two-dimensional (2D) graphene sheets. Graphite contains strong covalent bonds between the carbon atoms within individual graphene sheets, which gives rise to its outstanding in-plane mechanical properties. However, the van der Waal's forces between adjacent graphene sheets in the layered structure are relatively weak and, therefore, graphite is much softer than diamond. Similar to diamond, graphite (in-plane) is a good conductor of heat, however, the free electrons present in graphite also endow it with high in-plane electrical conductivity, unlike diamond. The crystal structure of diamond and graphite is depicted schematically in Fig. 3.1.

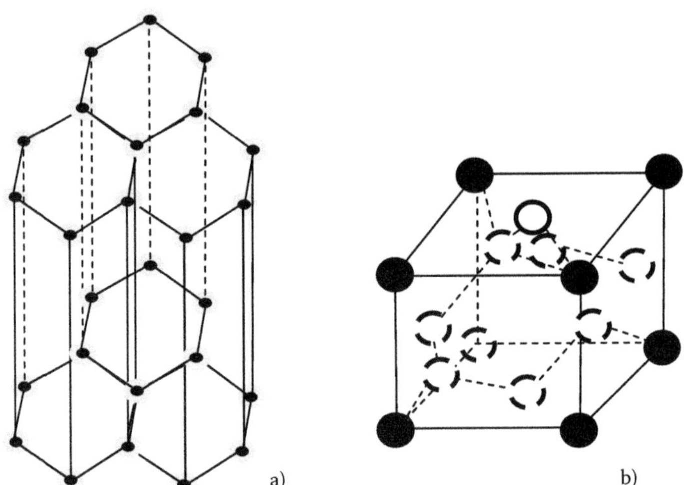

FIGURE 3.1 Schematic of the atomic structure for a) graphite, showing the sp^2 hybridized hexagonal lattice, and b) sp^3 hybridized diamond, which consists of the 3-dimensional diamond cubic lattice.

THE PROPERTIES OF GRAPHENE

As a result of its unique two-dimensional crystal structure and ultra-strong sp^2 carbon bonding system, graphene offers a promising blend of electrical, thermal, optical and mechanical properties that paves the way to a variety of possible applications. The elastic modulus of an individual graphene sheet was predicted[1] to be ~ 1 TPa (or 1,000 GPa). The data has been confirmed[1] by atomic force microscopy (AFM) – based indentation experiments performed on suspended graphene. The exceptionally high modulus of graphene, along with its low density (~ 1 – 2 g/cm^3), means that the specific modulus (i.e., modulus normalized by density) of graphene far exceeds that of all other structural materials, including aluminum, titanium, or steel. In addition to its very large elastic modulus, graphene also displays a fracture strength of ~ 125 GPa[1], which is higher than most structural materials demonstrate. Graphene has a very interesting electronic band structure. It may be called a semimetal with zero electronic band gap; the local density of states at the Fermi level is also zero and conduction is only possible by the thermal excitation of electrons[1]. However, an energy gap may be engineered in graphene's band structure in different ways. The possible methods are based on the breaking of graphene's lattice symmetry, such as defect generation[1], water adsorption[1], applied bias[1], and interaction with gases (e.g., nitrogen dioxide or ammonia). Among other outstanding properties of graphene that have been reported are exceptionally high values of its in-plane thermal conductivity (~ 5,000 $Wm^{-1}K^{-1}$), charge carrier mobility (200,000 $cm^2V^{-1}s^{-1}$), and specific surface area (2,630 m^2g^{-1}) (i.e. the total surface area per unit of mass), plus remarkable features such as the quantum Hall effect, spin resolved quantum interference, ballistic electron transport, and bipolar super-current and others. It should, however, be noted that the exceptional thermal and electronic properties of graphene (that are listed in literature) are usually true only for the free-standing graphene.

In practice, the achieved 30-inch, one-atom-thick graphene sample certainly would be so unstable that it would have to be placed on some surface. This is the real question as to whether graphene is a crystal. If we imagine the honeycomb sheet as unsupported, we realize it is very susceptible to being bent out of its flat planar condition. The chemical bonds ("π-bonds" between two $2p_z$ electrons) tend to return it to a flat planar condition, but the resulting force is too weak against flexing motion. It is like a sheet of paper, in being flexible but, unlike a sheet of paper, it retains a weak restoring force toward a perfectly planar condition. In the crystal form, inherent in two dimensions (2D) (embedded in three-dimensional space), we have long-wavelength flexural phonons that allow large root-mean-square (rms) fluctuational displacements much larger than a lattice constant. The sheet size determines how unstable it is. The statement, whether the graphene sheet is crystalline can be debated. By formal definition, long-range order does not take place, but in practice the local distortions may be so small, so that it is still acceptable to consider the sample a crystal, even if it is slightly distorted. For graphene in practice, the out-of-plane deflections pose the main problem: as to whether the system is crystalline. In addition, there are more subtle points, mostly of academic interest. One is that 2D array, even if arbitrarily kept absolutely planar, cannot have long-range (infinite) order except at $T = 0$. In this case, the planarity would have to be imposed without transverse pinning. The closest system of this type may be electron crystals on the surface of liquid helium. An infinitely large 2D array would exhibit, at any finite temperature, large absolute in-plane motions (but without tangibly affecting local inter-atom distances). The above motion may cause smearing of diffraction patterns of electron or x-rays on a large sample.

Since the phonon wavelengths (now in 2D) involved are large, local regions may move without disrupting of the local order. In this case, the melting point connected with local order is not influenced by changes in cohesive energy of the system. While there had been earlier suggestions that the single planes of graphite might be extracted for individual study (contrary to a theoretical literature that suggested that crystals in two dimensions should not be stable), Novoselov et al. (2004, 2005)[2] were the first to demonstrate that such samples were viable, and indeed represented a new clan of 2D materials with useful properties and potential applications. (Hints toward isolating single layers had earlier been given by Boehm et al., Van Bommel et al. (1975), Forbeaux et al. (1998) among others)[3]. On small size scales, (approximately from 10 nm to 10 μm), the graphene array of carbon atoms is "crystalline" and has sufficient local order to provide electronic behavior as predicted by calculations based on an infinite 2D array. Micrometer-size samples of graphene show some of the best electron mobility values ever measured. In microscopy, on scales 10 nm to 1 μm, it sometimes appears that the atoms are not entirely planar, and undulate slightly out of plane. It has been suggested that such "waves" are intrinsic due to the usual response of the thin membrane to inevitable deformation from its mounting, or as a result of adsorbed molecules, since in graphene every carbon atom is exposed. Monolayer graphene is strong (the space per layer in graphite is 0.34 nm that is widely quoted as the nominal thickness of the graphene layer). An equivalent elastic thickness of graphene, closer to the actual thickness is about 0.1 nm and $t \sim 0.34$ nm. All but the shortest samples are extremely "soft" and may be bent with a small transverse force. This can be understood from the definition of classical "spring constant K" (the spring constant K is a macroscopic dimension-related engineering quantity in SI units of N/m). It is related to the "bending rigidity" or "rigidity" $k = Yt^3$, a microscopic property usually measured in eV which is about 1 eV in graphene. The Young's modulus Y, an engineering quantity, is defined as pressure. It is about 10^{12} N/m^2 = 1 TPa for graphene. The rigidity k has units of energy, as force times distance. We can see that the rigidity k of graphene, by virtue of the minimal atomic value of thickness t, is the lowest of any possible material. With extension of a chemical bond, the spring constant K relates to the bond energy E as $K = d^2E/dx^2$. For deflection x of a cantilever of width ω, thickness t and length L on (with Young's modulus Y) under a transverse force F: $F = -Kx$. Since $K \sim Y\omega t^3/L^3$, with t near a single atom size, we can see that graphene, (in spite of a large value of Young's modulus, $Y \sim 1$ TPa), is the softest material against transverse deflection. Graphene length L, width ω and thickness t, quantitatively bend and vibrate as predicted by classical mechanics formulas. For example, the spring constant K defined for deflection and applied force at the center of a rectangle clamped on two sides depends strongly on the linear dimensions as $K = 32Y\omega t^3/L^3$. A square of graphene, of size $L = \omega = 10$ nm, from the above expression, yields $K = 12.6$ N/m, while a square of size 10 μm gives $K = 12.6*10^{-6}$ N/m. If the sample is short, approaching atomic dimensions, the spring constant is large and the object appears to be rigid. For example, the spring constant of a graphene square (ten benzene molecules along a side) as it is bent is ~ 156 N/m, using the above expression. At the same time, the spring constant of a carbon monoxide (CO) molecule (in extension), deduced from its measured vibration at 64.3 THz, is known to be 1860 N/m. The next quantity in the graphene characteristics is Yt, a 2D rigidity that has a value of about 330 N/m. If the graphene sample is longer than a few micrometers, with the spring constant K of a square sample reduced to $1/L^2$, the sample is exceedingly soft.

Accordingly, graphene, on a micrometer-size scale, adapts to any surface under the influence of attractive van der Waals forces. In an electron micrograph, graphene on a substrate appears adherent, more like a flat piece of cloth or "membrane" than a piece of a cardboard, and quite unlike a 10-inch diameter silicon wafer[4]. Graphite and diamond are grown in the depths of the earth at high temperature, but graphene 2D "crystals" cannot, at present, be grown from a melt, similar to silicon. Graphite crystals can only be obtained by extraction from an existing crystal of graphite, or by being grown epitaxially on a suitable surface such as SiC or catalytically on Cu or Ni from a carbon-bearing gas such as methane. Notably, graphene is an excellent electronic conductor, much like a semimetal, but with conical rather than parabolic electron energy bands near the Fermi energy and with a characteristic *linear* dependence of energy on crystal momentum, $k = p/h$; i.e. E = "pc" = c^*hk. These electrons move like photons, at speed $c^* \approx 10^6$ m/s and with vanishing effective mass.

An explanation results from the band theory for a particular crystal lattice. This aspect also presents a new paradigm in the field of condensed matter physics. Not only is graphene structure comes the closest to a two-dimensional (2D) self-supporting material, but it also has charge carriers moving in a different way, as if their mass were zero. The physics of the phenomenon also implies that "back-scattering" is "forbidden" and gives much larger carrier mobility.

In the real world of atoms, no material can be mathematically two dimensional: the probability distribution P(x, y, z) must extend in the z-direction by at least one Bohr radius[1]. The well-known examples of 2D subsystems of particles are electrons on the surface of liquid helium and the "2 – DEG" two-dimensional electron gases engineered into prospective semiconductor devices. The latter useful electron systems are supported by quantum well heterostructures. The remarkable difference, in graphene, is that there is no external supporting system: the layer of carbon atoms is the mechanical support, as well as the medium exhibiting light-like propagation of electrons. The above situation was not entirely clear before the discoveries of Geim and Novoselov: indeed the existence of free-standing graphene layers with new and superior electronic properties was a surprise, worthy of a Nobel Prize in Physics. Other one-layer materials include $BN(BN)_n(C_2)_m$, with n, m, integers; MoS_2, TaS_2, $NbSe_2$ and the superconductor $Bi_2Sr_2CaCu_2O_x$, although the last is seven atoms thick (Novoselov et al., 2005)[5]. So Geim and Novoselov, in fact, confirmed the practical reality of a new class of 2D locally crystalline materials.

The binding energy of a crystal, an extended periodic array of atoms, for temperatures below melting temperatures, T_M, is a subject of solid state physics[6]. The methods of this discipline do not always predict binding of an infinite 2D crystal. Indeed, thin layers of many substances are found to break up into small pieces (or "islands") as their thickness is reduced, especially if the attraction of an atom to substrate exceeds the attraction atom-to-atom. This island breakup definitely does not occur with graphene: on the contrary, graphene is found to be among the strongest known materials with force applied. Recently, tenth-millimeter scale sheets of one-atom-thick graphene have been studied as elastic beams and sheets. At a lattice constant of 0.246 nm, a 20 μm graphene sheet (80,000 unit cells) looks flat, if suspended across a trench, but may bend in response to van der Waals forces from the mounting. In some cases, 10 nm-scale "waves" or "ripples" of ~ 1 nm amplitude have been influenced judging by transmission electron microscope measurements, with a likely origin in a combination of molecular surface adsorbates and mounting strain. Notwithstanding some unclear physical mechanism experiments confirm the fact; these "crystals" are large enough to be useful under many circumstances. Molecules have vibrations: in an extended crystal these are called phonons. The vibrational motions of molecules are 3D in nature and any real 2D crystal should have a useful notation (for a "real 2D crystal" is "2D-3") meaning that motion into the third dimension is available. A "pure" 2D system is one, (like electrons on the surface of liquid helium), where no motion into the third dimension is allowed. The z-motion, is represented by a single quantum state. We take the electron system inside graphene to be "pure 2D" as confined by the graphene lattice, even though that lattice may slightly undulate or flex into the third dimension. The lattice has vibrational motion in the z-direction (called flexural). In an extended real 2D sample the flexural motion extends to low frequency and large amplitude, at any finite temperature T. Even when restricted completely to planar motion, the methods of solid state physics have predicted that thermal vibrations at any finite temperature lead to excessive transverse motion and destroy long-range (as distinct from short-range) order[4].

Transmission electron microscopy (TEM) reveals a honeycomb lattice. The bond length of the crystal is approximately 0.142 nm. The graphene layers form a stack, the interplanar distance is about 0.335 nm. The lattice has characteristic "rippling" of the graphene sheet, with amplitude of about one nanometer. The ripples may be attributed to the instability of two-dimensional crystals or due to contamination[7,8]. Ripples of graphene on the SiO_2 substrates may also be explained by conformation of graphene to the SiO_2 substrate which, in this case, is rather an extrinsic than an intrinsic effect[9].

Like some single-wall carbon nanostructures, graphene exhibits (002) plane preferable stacking. On the other hand, the unlayered graphene tends to form (h k o) rings that were reported to be in presolar graphite onions[10]. Naturally – occurring unlayered solidified graphene exits in the universe. In particular, unlayered graphene was found in the center of carbon spheres that were extracted from the Australian meteorite. Man-made counterparts of the natural graphene include bulk assemblies of single graphene sheet carbon molecules. Since carbon has a higher melting temperature than that of any other solid substance, it is a valuable quality for graphene mechanical applications. That is where the term "presolar graphite onions" comes from. The hexagonal shape of pure carbon atoms easily accepts carbon atom if any vacancies exist. Such an effect is achievable during bombardment with pure carbon atoms, or in case of being exposed to, e.g., hydrocarbons[11].

As it was already mentioned, intrinsic graphene does not belong to either metals or semiconductors. It may be classified as a "zero band-gap" semiconductor (or semimetal as it was suggested above). Its electronic structure has a number of features that drew attention of the early researchers of the material. The early studies (as early as 1947)[12] revealed zero effective mass for electrons and holes, in particular, that the energy momentum relation (dispersion relation) is linear in the two-dimensional hexagonal Brillouin zone. The electrons and holes near the six points of the Brillouin zone behave like relativistic particles (Dirac equations for spin ½ particles)[13]. The Dirac points are located in the six corners of the Brillouin zone, where the electrons and holes are called Dirac fermions. The electron linear dispersion is given as:

$$E = \hbar v_F \sqrt{\left(k_x^2 + k_y^2\right)}; \tag{3.1}$$

where v_F, the Fermi velocity (~ 10^6 m/s); k is the wave vector which is measured from the Dirac points. In the Dirac points, the energy is zero[14]. For the electron and hole mobility is very high, reaching 15,000 $cm^2.V^{-1}.s^{-1}$ (ref. 13). Between the temperatures of 10 K and 100 K the mobility is almost constant which means that the scattering is caused mainly by defects. The room temperature mobility has the limit of 200,000 $cm^2.V^{-1}.s^{-1}$ for intrinsic graphene. The resistivity is $10^{-6}\Omega \cdot cm$ at a carrier density of $10^{12} cm^{-2}$. It is still bigger than the resistivity of silver (which has the lowest resistivity of a solid at room temperature)[15]. In practice, the graphene resistivity is lower, e.g. up to 40,000 $cm^2.V^{-1}.s^{-1}$ on SiO_2 substrates[16].

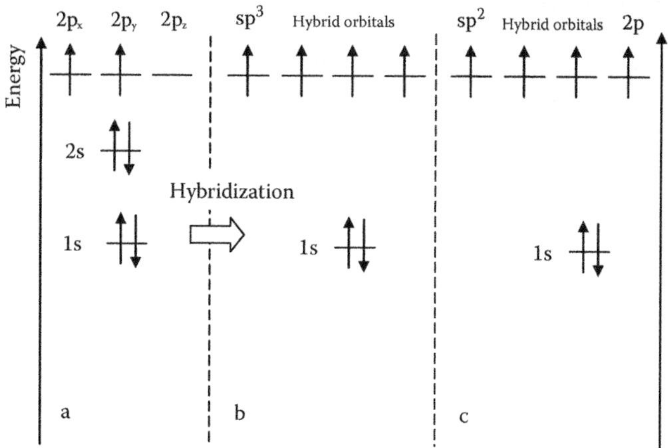

FIGURE 3.2 Atomic orbital diagram of a carbon atom

The electrons occupy the $1s^2, 2s^2, 2p_x^1$ & $2p_y^1$ atomic orbitals (Fig. 3.2). As a tetravalent element, only its four exterior electrons form covalent chemical bonds. The ground state in particular is given in a) forming bonds with other atoms. Carbon gives out one of the 2s electrons to the vacant $2p_z$ orbit. As a result, a hybrid orbital forms. For example, diamond 2s-energy level and the three $2p$ level form four sp^3 -orbitals that have one electron in any one of them (Fig. 3.2 b). The four sp^3-orbitals form imaginary tetrahedron. The sp^3-orbitals of different atoms overlap thus creating the 3D diamond structure. The strong binding energy of the C-C bonds gives the diamond structure its hardness. In hybridization, only two of the three $2p$-orbitals participate (Fig. 3.2 c). The remaining $2p$-orbitals are situated in the X-Y plane at 120° angles. The sp^2-orbitals are perpendicular to the above $2p$-orbitals. Thus, the formed covalent in-plane bonds define the hexagonal structure of the graphite. The in-plane σ-bonds within the graphene layers are even stronger than the C-C bonds in sp^3-hybridized diamond. At the same time, the interplane π-bonds (formed by the remaining 2p-orbitals) posses a substantially lower binding energy. It causes an easy shearing of graphite along the layer plane. The measurements performed by Haering R.R. still in 1958 give the following parameters: a single graphene layer has a lattice constant $a = \sqrt{3a_0}$; where $a_0 = 1.42\ \mathring{A}$. The distance between two adjacent layers is 3.35 A in AB-stacked graphite.

The presence of H (hydrogen), O (oxygen), or other C (carbon) atoms is advantageous to excite one electron from the 2s to the third 2p orbital in order to form covalent bonds with the other atoms. In such a scenario, the gain in energy exceeds 4 eV which is necessary for the electron excitation. Thus, in the excited state, we have four equivalent quantum-mechanical states, I2s>, I2p_x>, I2p_y>, and I2p_z>. It is important to note that a quantum-mechanical superposition of the state I2s> with **n** I2p_j> states is called sp^n hybridization that is substantial in covalent carbon bonds.

sp[1] HYBRIDIZATION[17]

The considered here sp[1] hybridization (often called just "sp hybridization" for simplicity) mixes I2s> state with one of the 2p orbitals. We'll consider the I2p$_x$> state as an example. A combination of the symmetric and anti-symmetric combinations produces:

$$\text{Isp}_+> = \frac{1}{\sqrt{2}}\left(\text{I2s}> + \text{I2p}_x>\right) \text{ and } \text{Isp}_-> = \frac{1}{\sqrt{2}}\left(\text{I2s}> - \text{I2p}_x>\right);$$

In this case, the other states, I2p$_y$> and I2p$_z$>

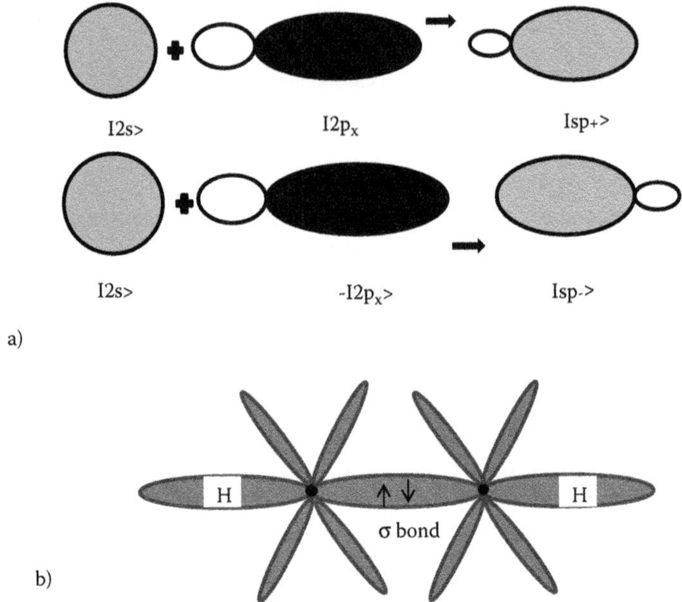

FIGURE 3.3 a) sp[1] hybridization. On the r.h.s. the hybridized orbitals formed from electronic densities of the I2s> and I2p$_x$>. b) Acetylene molecules ($H-C\equiv C-H$).

The electronic density of the hybridized orbitals form drop-like configuration as shown in Fig. 3.3. The drops are elongated in the +x (-x) direction for the Isp$_+$> (Isp$_-$>) states. The described above process of hybridization plays a role, for example, in the formation of the acetylene molecule $H-C\equiv C-H$ (as illustrated in Fig. 3.3 (b)). The overlapping sp[1] orbitals of the two carbon atoms form a strong covalent σ-bond. The 2p orbitals then participate in the formation of the two additional π-bonds that are weaker than the σ-bond.

In sp^1 hybridization, the s-orbital and one of the p – orbitals from carbon's second energy level are combined together to form two hybrid orbitals. These hybrid orbitals form a straight line. The angle between the orbitals is 180^0. The orbitals are opposite from one another with respect to the center of the carbon atom. Since this type of *sp* hybridization uses only one of the p – orbitals, these are still two p – orbitals left which the carbon can use. The p – orbitals have 90^0 – angle between them and with respect to the line formed by the hybrid orbitals. The described type of hybridization takes place when a carbon atom is bonded to two other atoms.

GRAPHENE'S HONEYCOMB LATTICE[17]

If we have a superposition of the 2s and two 2p orbitals (which may be chosen as the $|2p_x\rangle$ and $|2p_y\rangle$ states), we can obtain the planar sp^2 hybridization. Their orbitals are in the x-y plane and have mutual 120^0 angles (Fig. 3.4a).

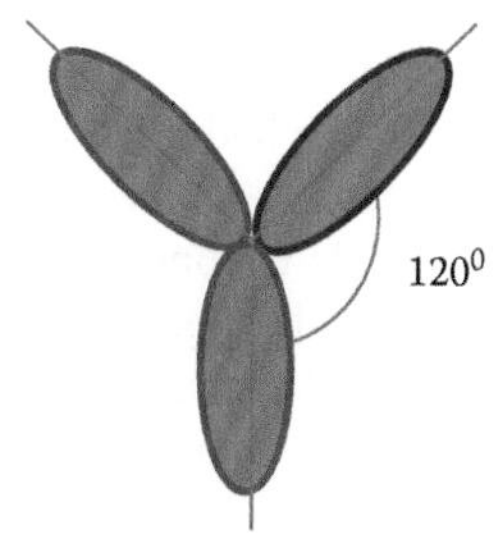

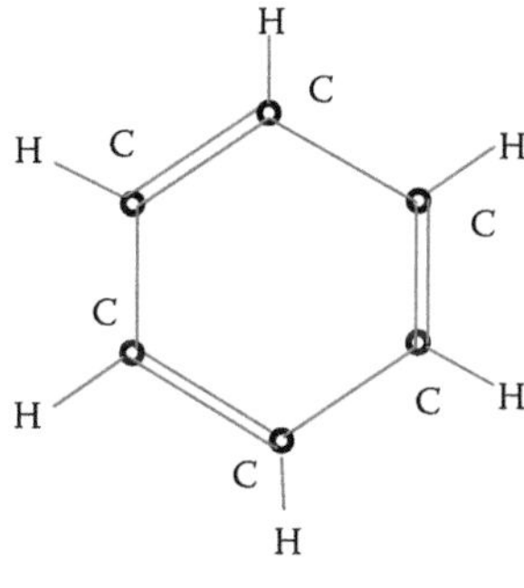

FIGURE 3.4 Hybridization of carbon *(Continued)*

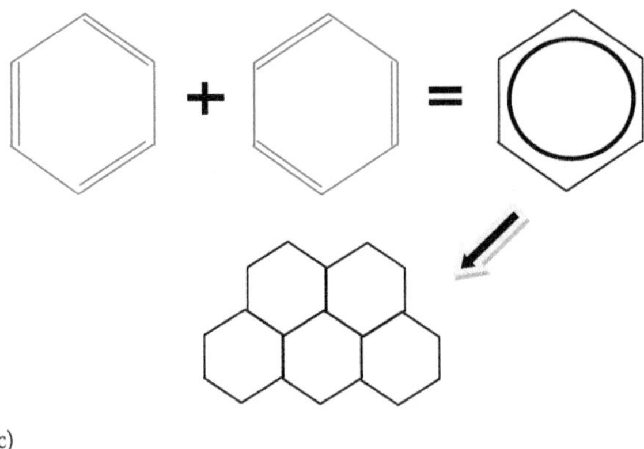

FIGURE 3.4 (Continued) Hybridization of carbon

The third (unhybridized) $2p_z$ orbital is perpendicular to the plane. One characteristic example for the above hybridization is the benzene molecule which is a hexagon with carbon atoms at the corners connected by σ bonds (Fig. 3.4 b). Each carbon atom has a covalent bond with one of the hydrogen atoms situated in a star-like arrangement. In addition to the above 6 σ-bonds, the remaining $2p_z$ orbitals form 3 π-bonds. Double bonds alternate with single σ bonds in the hexagon. Since the double bond is stronger than the single one, the hexagon is not uniform. Also, the lengths of the bonds are different: a double bond has a carbon-carbon distance of 0.135 nm and 0.147 nm for a single σ-bond. Thus, a graphene sheet may be simply considered as tilted benzene hexagons with the hydrogen atoms replaced by carbon atoms to form a carbon hexagon (Fig. 3.4 c) – graphene honeycomb lattice. The honeycomb lattice due to the sp^2 hybridization is not a Bravais lattice (the neighboring sites are not equivalent).

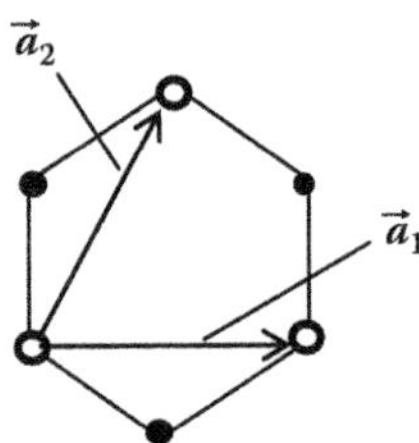

FIGURE 3.5 Graphene's honeycomb lattice. The vectors $\vec{a}_1$ and $\vec{a}_2$ are basis vectors of the triangular Bravais lattice.

Please note that the distance a = 0.142 nm is an average of the above-mentioned distances of a = 0.135 nm and 0.147 nm. Fig. 3.5 depicts the triangular Bravais lattice, taking it as a basis, we can visualize the honeycomb lattice with a two-atom basis. The average distance between carbon atoms is 0.142 nm which is the average of the single (C-C) and double (C=C) covalent σ bonds. The following vectors connect a site as the A sublattice with the nearest neighbors on the sublattice B:

$$\vec{\delta}_1 = \frac{a}{2}(\sqrt{3}\vec{e}_x + \vec{e}_y); \qquad \vec{\delta}_2 = \frac{a}{2}(-\sqrt{3}\vec{e}_x + \vec{e}_y); \qquad \vec{\delta}_3 = -q\vec{e}_y; \tag{3.2}$$

and the basis vectors of the triangular Bravais lattice:

$$\vec{a}_1 = \sqrt{3}a\vec{e}_x; \qquad \vec{a}_2 = \sqrt{3}\frac{a}{2}(\vec{e}_x + \sqrt{3}\vec{e}_y); \tag{3.3}$$

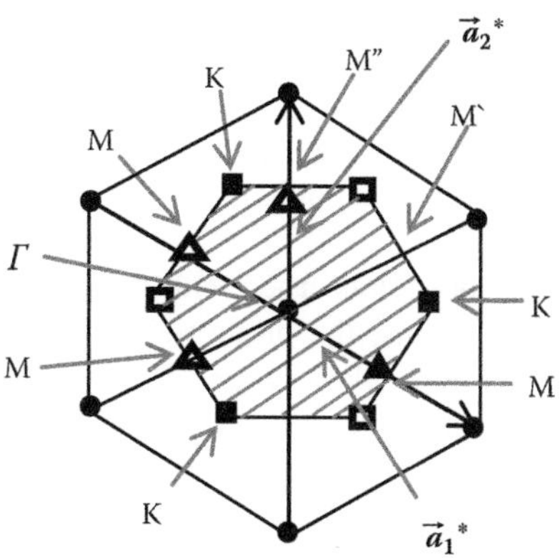

FIGURE 3.6 The reciprocal lattice for the graphene's honeycomb lattice.

The reciprocal lattice in Fig. 3.6 is defined with respect to the triangular Bravais lattice. The vectors of the lattice are:

$$\vec{a}_1^{\,*} = \frac{2\pi}{\sqrt{3}a}(\vec{e}_x - \frac{e_y}{\sqrt{3}}); \qquad \vec{a}_1^{\,*} = \frac{4\pi}{3a}e_y; \tag{3.4}$$

All sites of the reciprocal lattice represent equivalent wave vectors. If a quantum-mechanical electron wave packet or vibrational lattice excitation propagate in the lattice, wave vector difference equals a reciprocal lattice vector's phase with a 2π factor:

$$\vec{a}_i \cdot \vec{a}_j^{\,*} = 2\pi\gamma_{ij}; \tag{3.5}$$

where i, j = 1,2 (the angle between direct and reciprocal lattice vectors).

The first Brillouin zone (the shaded region and thick lines of the hexagon) represents a set of points in the reciprocal space which may not be connected to one another by a reciprocal lattice vector. In the propagation of any type of wave motion through a crystal lattice, the frequency is a periodic function of wave vector k. This function may be complicated by being multi-valued; that is, it may have more than one branch. Discontinuities may also occur. In order to simplify the treatment of wave motion in a crystal, a zone in k – space is defined which forms the fundamental periodic region, such that the frequency or energy for a k outside this region may be determined from one of those in it. This region is known as the Brillouin zone (as it was mentioned above). It is also called the first Brillouin zone. It is usually possible to restrict attention to k values inside the zone. Discontinuities occur only on the zone's boundaries. The Γ point is the location for long wavelength excitations. Further, there are six corners of the first Brillouin zone which consists of non-equivalent points K and K` represented by the vectors:

$$\pm K = \pm \frac{4\pi}{3\sqrt{3}a}\vec{e}_x; \tag{3.6}$$

The four other corners may be connected by reciprocal lattice vectors. These points at these corners play a role in defining the electronic properties of graphene. The points K and K` are centers of low-energy excitations. The form of the first Brillouin zone is an intrinsic property of the Bravais lattice regardless of the presence of other atoms in the unit cell. The rest of the points in the reciprocal lattice in Fig. 3.6 are shown for completeness.

The above discussion has been concerned with a single-layer graphene. If there are several layers of carbon atoms, the properties will be different from single-layer electronic properties. Only single-layer graphene (SLG) and bi-layer graphene (BLG) correspond to semiconductors with a zero gap with only one type of electrons and holes. If the number of layers is fewer than 10, the conduction and valence bands overlap which results in charge carriers[16]. If the number of graphene layers exceeds 10, we deal with a thin graphite film.

Graphene oxide (GO) and reduced graphene oxide (rGO) are alternative versions to graphene. In the case of GO, diffraction patterns reveal a nonlinear behavior. The nonlinearity may be attributed to so-called "short-wavelength" ripples or distortions comparable to inter-atomic distance which is ~ 10 % of the carbon-carbon distance[18]. The distortions are connected with the strain in the lattice. This may indicate fundamental structural and topographical differences between different layers. Using atomic force microscopy (AFM), direct visualization of graphene rippling is achievable. In particular, GO is substantially rougher when it is free-standing in comparison with the cases when GO is deposited on (an atomic scale) smooth surfaces. The observed distortions were on the order of 10 nm (approximately the size of the AFM tip).

3.1 GRAPHENE SEVERAL-LAYER THICK

As it was mentioned earlier, the properties of graphene depend substantially on the number of atomic layers of the material. Usually, the following classification is adopted: a single-layer, a bi-layer, and a few layer graphene (the last one implies the number of layers should not exceed 10). The second aspect that has substantial influence on the electronic band structure of graphene is how the graphene layers are stacked[6].

In AB-Bernal stacking the third layer is above the first layer. In Rhombohedral ABC stacking (Fig. 3.7) the third layer is shifted with respect to the first and second layers. The fourth layer is located, in this case, above the first layer. For AB-Bernal stacking (Fig. 3.7), alternate layers have the same projections on the X-Y plane. The distance between the layers is 3.35 $\mathring{A}$. In ABC-stacked Rhombohedral graphite, the third layer is shifted with respect to the first and second layers. The fourth layer and the first layer have the same projection on the base layer. The separation between the layers becomes 3.37 Å now[6]. The adjacent graphene layers are parallel. They may contain rotational stacking faults (since they can rotate relative to each other without a preferred orientation). If multilayer graphene is grown on the carbon-terminated $(000\bar{1})$ face of SiC, it obtains a high density of rotational disorder[6]. Yet, these multilayered structures are of interest because introduction of rotational stacking faults into the ABC may cause a separation of adjacent layers.

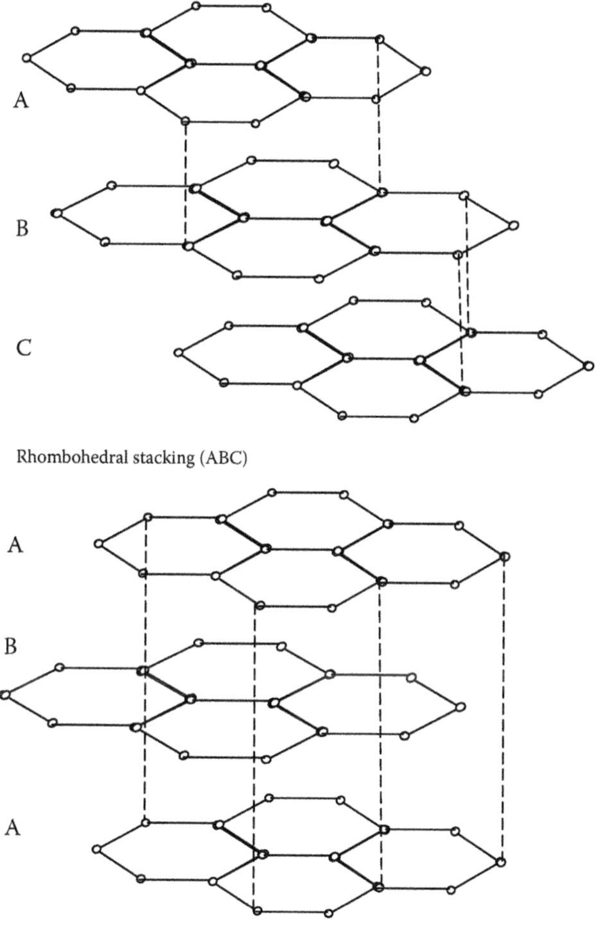

FIGURE 3.7 AB-Bernal stacking and ABC Rhombohedral stacking

Near the Dirac points, graphene has a minimum conductivity equal approximately to $4e^2/h$. Taking into account the zero carrier density at the above points, the origin of the conductivity is not clear at the moment. The mentioned before rippling and the presence of defects may explain the local density of carriers. Many measurements confirm the factor of $4e^2/h$ or even greater depending on the impurity concentration[19].

The chemical dopants also have influence on the carrier mobility in graphene. However, the influence is less than could be expected: even for chemical dopant concentration of 10^{12} cm^2 there is no observable change in the carrier mobility. The noticeable results have been reported by Chen, et al[19]. They found that potassium ions could reduce the mobility 20 times as long as the potassium was present in the graphene. The band gap of graphene can be changed within the limits of ~ 0 to 0.25 eV using a dual gate bi-layer graphene FET (field effect transistor) at room temperature. The optical radiation frequency to produce the above effect is about 5 μm (Zhang Y. et al).

The double-layer graphene (double-GL) structures have been studied and fabricated recently[20]. The structure consists of two GLs separated by a thin layer. Such a structure deserves a considerable attention due to its potential applications for optical transparent transistor circuits, THz lasers, THz detectors, frequency multipliers, and THz photomixers. In particular, the double GL structures exhibit tunneling or thermionic conductance, the effects that be utilized in resonant detectors and photomixers[21]. The experiments prove that the inter-GL resonant tunneling in the double-GL structures results in inter-GL negative differentiated conductivity - valuable property for the creation of a new generation of transistors with multi-valued current-voltage characteristics.

3.2 OPTICAL PROPERTIES

One-atom-thick layer is almost invisible to the naked eye since only $\pi\alpha \approx 2.3\%$ of white light is actually absorbed. Where α = fine structure constant (fine structure constant is the coupling constant characterizing the strength of the electromagnetic interaction). However, even such a low percentage is high enough in comparison to other materials this thin. This opacity comes from the unusually low-energy structure that a graphene monolayer possesses. The electron and hole conical bands connect with each other at the Dirac points. The electronic structure is different from more usual quadratic massive bands[22]. The measurement of the fine structure constant nevertheless remains rather difficult, often with insufficient precision. The optical conductance can be calculated using the traditional Fresnel equations for the thin-film limit. The basis for the calculation is provided by the Slonczewski-Weiss-McClure (SWMcC) graphene band model.

It has been reported that the optical response of graphene (graphene nanoribbons, in particular) may extend into THz range[23] by an applied magnetic field as well as graphene may change its properties for both linear and ultrafast regimes[24]. A practical application example may be a graphene-based Bragg grating (which is a one-dimensional photonic crystal) that has been announced as capable of excitation of surface electromagnetic waves in a periodic structure (such as a grating). As a light source a 633 nm He-Ne laser may be used[25]. In addition, not only the magnetic field but an electronic current can be used to influence the optical response. Another effect is saturable absorption in the THz and microwave ranges. It occurs when the input optical intensity exceeds the threshold value. The saturable absorption is a nonlinear effect. Graphene may be saturated under strong excitation due to its zero-band gap. This saturation effect relates to mode-locking of fiber lasers.

The full-band locking is achieved by a graphene-based saturable absorber. Under more intensive laser radiation, there may be a nonlinear phase shift (optical nonlinear Kerr effect). Under open and closed aperture z-scan measurement, graphene may be able to produce a big nonlinear Kerr coefficient of $10^{-7} cm^2 W^{-1}$, which is almost 10^9 times larger than that for bulk dielectrics (in nonlinear optics, a z-scan measurement is used to measure the nonlinear index n_2, Kerr nonlinearity and the nonlinear absorption coefficient $\Delta\alpha$ using "closed" and "open"). Such an effect opens new possibilities for graphene-based nonlinear Kerr photonics, e.g. solition in graphene[26]. All of the above-mentioned phenomena may contribute to different devices and their improvements: fiber laser, mode-locking (full-band mode locking is achieved by graphene-based saturable absorber), microwave saturable absorber, polarizers, modulators in the microwave range, and broad-band wireless access networks to name a few[27].

3.3 THERMAL PROPERTIES

Graphene's thermal properties are different from those of other carbon materials, in particular, nanotubes or diamond. The room temperature thermal conductivity is within the limits of $(4.84 \pm 0.44) \cdot 10^2$ *to* $(5.30 \pm 0.48) \cdot 10^3 \ W \cdot m^{-1} \cdot K^{-1}$. These measurements were performed by non-contact optics. The isotropic composition has an influence on the graphene thermal non-contact optics development. The isotropically pure 12G graphene has a higher conductivity than the natural graphene occurring 99:1 ratio, or 50:50 isotropic ratio. The thermal conductance of graphene is isotropic, provided it is ballistic. In general, the conductance is not ballistic, i.e. there is scattering by impurities, defects, or by the atoms or molecules composing the medium. In our case, the ballistic conductance (conduction or transport) may exist due to the medium of negligible electrical resistivity and scattering. Thermal conduction can be phonon or electron dominated. The nature of the thermal conduction depends on whether the applied voltage causes an increase in electronic contribution (caused by a Fermi energy shift greater than $k_B T$) over the phonon contribution at low temperatures.

In graphite, the vertical (the c-axis, out of plane) thermal conductivity is more than a hundred times smaller than basal (i.e. parallel to the lateral or horizontal axis) plane thermal (of over 1,000 $W \cdot m^{-1} \cdot K^{-1}$). Also, the ballistic thermal conductance of graphene is lower than that of other carbon structures[28]. As the temperature decreases, the resistivity's magnitude diverges from a power-law (i.e. a functional relationship between two quantities where one quantity varies as a power of another) behavior and becomes as high as several megoohms per square at about 20 mK as opposed to the commonly observed saturation of the conductivity. With an applied perpendicular field, the graphene layer remains insulating with direct transitions to the broken-valley-symmetry, $\nu = 0$ quantum Hall state, which means that the insulating behavior at zero magnetic field is a result of the broken-valley-symmetry. At zero magnetic field, the disorder landscape dominates[29]. Recently, the disorder influence is reported to be reduced by keeping the graphene sheet on automatically supplied flat boron nitride (BN)[30]. Anomalous patterns, in this case, were seen as to be caused by a close alignment of the graphene and boron lattices.

The nature of the conductivity at the charge-neutrality point, σ_{GNP} has been discussed in connection with graphene-based devices fabrication. As mentioned earlier, theory of ballistic transport in graphene yields a value of $4e^2/\pi h$. Experimentally, however, the value of σ_{GNP} is between 2 and $12e^2/\pi h$[18]. σ_{GNP} depends on the carrier density produced by static charges close to the graphene surface. However, the conductivity saturation in (particularly in suspended graphene) still occurs at low temperature and remained higher than $4e^2/\pi h$. As it was mentioned earlier, graphene exhibits the qualities of a two-dimensional material, exhibiting a high crystal quality (electrons can move distances of less than a micron without scattering), on the other hand, two dimensional crystals cannot exist in the free state. This may be avoided if graphene sheets are supported by a bulk substrate or embedded in a three-dimensional matrix. Individual graphene sheets may be freely suspended on a microfabricated scaffold in the air or vacuum. This fact, brings us to the concept of "suspended graphene". Further work was done by screening of potential fluctuations by placing an additional sheet of doped graphene nearby the tested sheet. In this case[6], instead of saturating at the values close to $e^2/\pi h$, σ_{GNP} decreased with a power law dependence temperature dependence T^α, where $\alpha = 2$ for the most insulating graphene sheets (samples) for the temperatures as low as 4 K. In addition, a strong magneto-resistance was observed at the temperatures higher than 10 K which was attributed to weak localization. The implication was that if the sample was absolutely clean, then the graphene may be an Anderson insulator (it implies the possibility of electron localization inside a semiconductor, provided that the degree of randomness of the impurities or defects is sufficiently large). Also, it was concluded that this temperature dependence is explained by increasing order in the sample crystal and, as a result, the sample becomes more insulating. Hence, $\sigma_{GNP} \propto T^\alpha$ is the conduction temperature dependence in the presence of electron and hole puddles[31]. Obviously, complete understanding of graphene conductivity is still lacking at the present moment. The temperature dependence of the graphene resistance can be used to measure the entropy level in graphene samples. Less contaminated samples show a higher resistivity which increases with the temperature drop. The saturation takes place when k_BT is smaller than the Fermi energy's fluctuation[32].

Graphene has three acoustic phonon modes: two in-plane modes (LA, TA) (that have a linear dispersion relation) and the out-of-plane mode (ZA) (that has a quadratic dispersion relation). At low temperatures, the thermal conductivity T^2 contribution of the linear modes is dominated by out-of-plane modes, $T^{1.5}$. In addition, at low temperatures, the Gruneisen parameters dominate the thermal conductivity: the thermal coefficient is negative under such conditions. Some of the Gruneisen parameters may be negative for the graphene (most optical modes with positive Gruneisen parameters are still not excited at low temperatures). Please note that Gruneisen parameter, γ, (named after Eduard Gruneisen) determines how the changing volume of a crystal lattice influences its vibrational properties and, consequently, how the changing temperature influences the size or dynamics of the lattice. Thus, the thermal expansion coefficient is negative at low temperatures[25]. The negative Gruneisen parameters correspond to the lowest transversal acoustic (ZA) modes. In-plane frequencies in x-y (in-plane) direction are greater than in z-direction since the atoms are more restricted to move in z-direction. This phenomenon is similar to the string stretching and vibrating where the latter takes place mostly along the string. This "membrane effect" was predicted in 1952 by I.M. Lifshitz[28].

3.4 MECHANICAL PROPERTIES

At the moment, the graphene is one of the strongest materials from the mechanical point of view. If it were possible to experimentally compare the breaking strength of steel and graphene films of the same thickness, the graphene would exhibit over a 100 times greater breaking strength than steel. For an industrial purpose in particular, or for many other practical applications, the separation a few-atom layer sheets from graphite is difficult. Technological innovations are, thus, necessary before graphene becomes an economically usable material for series- production devices. Graphene weighs only $\sim 0.77 \cdot 10^{-6} kg/m^2$ (which is approximately 0.001 % of the weight of 1 m^2 of paper). Since graphene sheet (or graphene paper as it is sometimes called) can undergo different reshaping from its original state it gives the promise of future manufacturing out of graphene. The easiness of shaping and firmness of thin graphene layers may facilitate making lighter vehicles, airplanes, etc. that use less fuel and generate less pollution. As carbon-based fibers are taking place of the aircraft materials so could graphene in time[33].

According to the Mermin-Wagner theorem, the amplitude of long-wave fluctuations in regions of graphene grows logarithmically with the scale of a 2-D structure and is unbounded in structure of infinite size. The implication is that long-range fluctuations can be created with little energy since these fluctuations increase the entropy of the structure. However, the local deformations and elastic strain are negligibly affected for relative displacement on a large scale. Thermal fluctuations can cause ripples in graphene. Also, it is believed that a sufficiently large (2-D) sheet of graphene will deform to become a fluctuating 3-D structure. The above properties cast doubt on the notion that graphene is a genuine 2-D structure[29].

3.5 QUANTUM HALL EFFECT IN GRAPHENE

The transverse conductivity in graphene depends on a magnetic field which is perpendicular to the current. The quantum Hall effect can usually be observed in *Si* or *GaAs* solids almost absolutely devoid of contamination or foreign additives and at very low temperatures (around 3 K). The quantization of the Hall conductivity σ_{xy} implies a

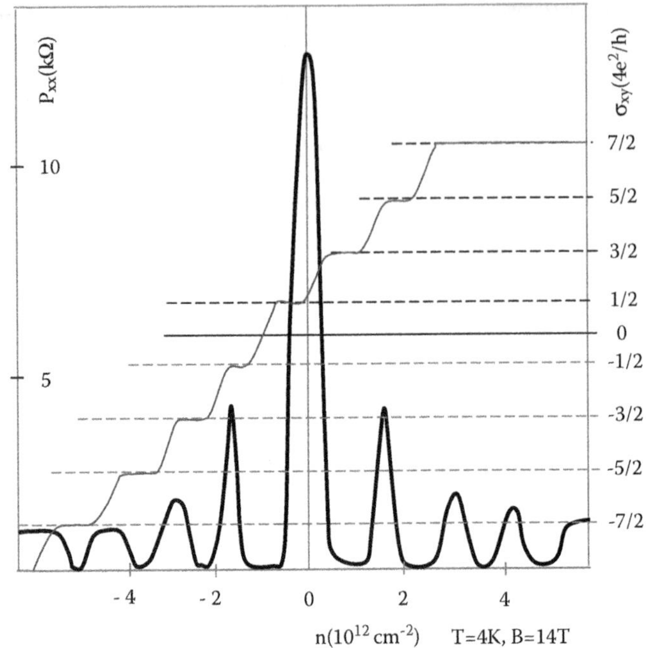

FIGURE 3.8 Quantum effect in graphene

multiplication of the basic quantity e^2/h, where e is the charge of the electron and h is Plank's constant. Graphene exhibits an anomalous quantum Hall effect which can be measured[34]. The conductivity is $\sigma_{xy} = \pm 4 \cdot (N + \frac{1}{2})e^2/h$, where N is the "Landau level" index (see Fig. 3.8). The double valley and double spin degeneracies give the factor of 4.

Thus, the usual sequence of steps is shifted by ½ and with an additional factor of 4. The described anomalous behavior is a result of Dirac electrons in graphene with no mass. In general, Dirac electrons relate to the "Dirac sea" theoretical model of the vacuum as an infinite sea with negative energy. The theory served to explain the anomalous negative-energy quantum states predicted by Dirac equation for relativistic electrons. When exposed to a magnetic field the half-filled Landau level (in quantum mechanics, the quantization of the cyclotron orbits of charged particles in magnetic fields results in the charged particles being able to occupy only orbits with discrete energy levels known as "Landau levels") corresponds to an energy level exactly at the Dirac points that are defined by the Atiyah-Singer theorem (in differential geometry the Atiyah-Singer theorem states that for an elliptic differential operator on a compact manifold (which is a type of topological space without a boundary)), the analytical index is equal to the topological index (which comes from the topological data). As a result, we have + ½ in the Hall conductivity. The quantum Hall effect is also valid for bilayer graphene, however, with only one anomality, i.e. ($\sigma_{xy} = \pm 4 \cdot N \cdot e^2/h$). In this case, the first plateau at N=0 is absent, meaning that bilayer graphene exhibits metallic qualities at the neutrality point[5].

The longitudinal resistance of graphene is maximum (minimum for normal metals) for integral Landau filling factors in measurements of the Shubnikov-DeHaas oscillations (unlike in bulk materials, in a 2-dimensional system the electrons can only move in one plane and may not travel perpendicular to this plane). Such measurements show a phase shift (equal to π) known as Berry's phase. In classical and quantum mechanics the Berry's phase is a phase acquired over the course of a cycle, when the system is subjected to cyclic adiabatic processes, which results from the geometrical properties of the parameter space of the Hamiltonian. The Berry's phase occurs when the amplitude and phase are changed simultaneously but very slowly (adiabatically) and eventually are brought back to the initial configuration. Close to the Diracs points, the Berry's phase is explained by the zero carrier mass. However, the carriers still have a non-zero cyclotron mass[35].

With sufficiently strong magnetic fields (more than 10 Teslas), additional plateaus of the Hall conductivity appear at $\sigma_{xy} = \nu e^2/h, with\, \nu = 0, \pm 1, \pm 4$. The Hall effect was observed also at $\nu = 3$ and partially at 1/3[33]. The quantum Hall effect is responsible for eliminating the degeneracy of the Landau levels with $\nu = 0, \pm 1, \pm 3, \pm 4$. The degeneracy is fourfold (two valley and two spin degrees of freedom).

3.6 ACTIVE GRAPHENE PLASMONICS

Graphene's physical properties make it a promising material to achieve high laser efficiency at room temperature. In particular, plasmonic dynamics exploit the optical properties of graphene to design a new type of a solid state laser in the THz range[36]. THz solid-state lasers are commonly used at the present for THz imaging and spectroscopy. Their efficiency, however, is low and they do not work at room temperature. Notwithstanding the designers' efforts in the past decade or so to produce a compact, tunable and coherent THz source, there are still no commercially available sources that cover the whole THz range. In a photonic device, such as a QCL (semiconductor quantum cascade laser), a decrease in operating frequency takes place by scaling down the energy of transition from the interband level (where electrons and holes transfer between two adjacent energy state bands in a semiconductor) to the inter-sub-band. The QCLs, however, produce substantial thermal noise at room temperature making them unusable as efficient THz sources. In the conventional field-effect transistor (FET), in order to increase the operating frequency, the distance between the source and the drain sides of the electron channel is scaled down but this reduces the output power of the devices.

In order to overcome the described difficulties, a graphene structure with unique transport of carrier, and unique optical properties has been proposed[1]. The conduction and valence bands are symmetrical around the edges of the Brillouin zone. The edges of the Brillouin zone correspond usually to the minimum energy gap in a semiconductor. In graphene, this bandgap is zero. The carries behave in this case as relativistic particles (they behave as if they have no mass), or relativistic fermions. Characteristically for fermions, each energy state can be occupied only by one electron or a hole as well as they (the fermions) can be transported at ultrafast speed without back-scattering. The graphene crystal structure, the honeycomb symmetry results in very high optical phonon energy at the zone edges K and K` for transverse mode[37]. The energy values are ~ 198 meV at the zone center Γ and ~ 163 meV at the zone edges K and K` (Fig. 3.9).

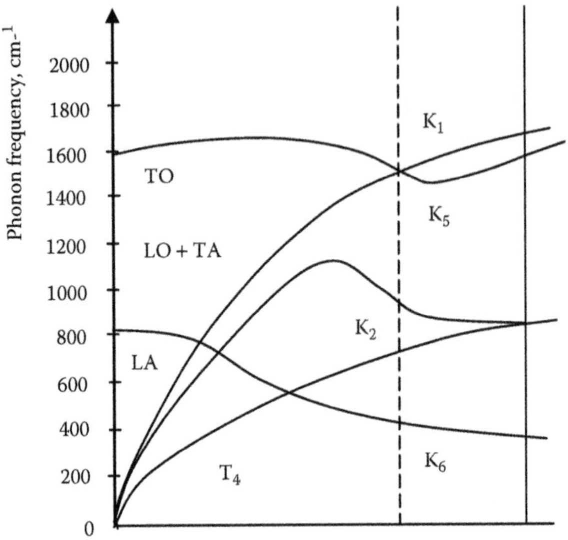

FIGURE 3.9 Phonon band diagrams of graphene[37]

The process of nonequilibrium energy relaxation in graphene takes place through carrier-phonon interaction through intra-valley, inter-valley, intra-band and inter-band transitions (Fig. 3.10).

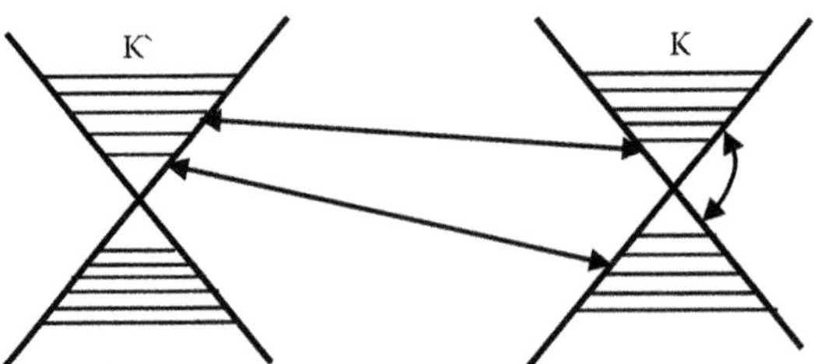

FIGURE 3.10 Interaction of optical phonons and carriers through inter-valley, intra-valley, intra-band and inter-band transitions[37].

Under optical pulse excitation, the intrinsic graphene has dominant carrier-carrier scattering that takes place in quasi equilibrium. Both, intra and inter-band optical phonons are taken into account but the inter-band Auger-like carrier-carrier scattering is neglected. The carrier distribution for the total energy and concentration of carriers is given in (3.7)[37]:

$$\frac{d\Sigma}{dt}=\frac{1}{\pi^2}\sum_{i=\Gamma,K}\int d\vec{k}[(1-f_{h\omega_i-v_\omega hk})(1-f_{v_\omega hk})/\tau^{(+)}_{iO,\text{inter}}-f_{v_\omega hk}f_{h\omega_i-v_\omega hk}/\tau^{(-)}_{iO,\text{inter}}];$$

$$\frac{dE}{dt}=\frac{1}{\pi^2}\sum_{i=\Gamma,K}\int d\vec{k}v\vec{k}_{\omega hk}[(1-f_{h\omega_i-v_\omega hk})(1-f_{v_\omega hk})/\tau^{(+)}_{iO,\text{inter}}-f_{v_\omega hk}f_{h\omega_i-v_\omega hk}/\tau^{(-)}_{iO,\text{inter}}]+$$

$$+\frac{1}{\pi^2}\sum_{i=\Gamma,K}\int d\vec{k}h\omega_i[f_{v_\omega hk}(1-f_{v_\omega hk+h\omega_i})/\tau^{(+)}_{iO,\text{intra}}-f_{v_\omega hk}(1-f_{v_\omega hk-h\omega_i})/\tau^{(-)}_{iO,\text{intra}}]; \quad (3.7)$$

where Σ and E are the carrier concentration and energy density, f_{index} = quasi-Fermi distribution, $\tau^{(+/-)}_{iO,inter}$ and, $\tau^{(+/-)}_{iO,intra}$ are the inverse coefficients for the scattering rates for inter and intra-band optical phonons (where i = Γ for optical phonons close to the zone boundary with ω_Γ = 198 meV and i = K for optical phonons close to the zone boundary with ω_Γ = 163 meV), "+" stands for absorption and "-" for emission that are calculated by the Fermi golden rule, and υ_ω is the Fermi velocity. The latter may be defined as following: even if we extract all possible energy from a Fermi gas by decreasing the temperature to absolute zero, the fermions can still move at a comparatively high speed. The fastest fermions move, in this case, at the velocity that corresponds to the Fermi energy, and this is the Fermi velocity. After the graphene is pumped with a photon energy of 0.8 eV, ε_F instantly falls down because of carrier coding and of the emission of optical phonons carrier at high-energy tails.

Further, ε_F becomes positive. When the pumping exceeds a certain threshold level, the population becomes inverted. Following the crossing of the threshold, the recombination process slows down (at ~ 10 ps). Although the population inversion is necessary condition it is not sufficient to achieve gain since we have the Drude absorption by carriers in graphene. It is the real part of the net dynamic conductivity Reσω (that consists of the intraband $\text{Re}\,\sigma_\omega^{\text{intra}}$ and the interband $\text{Re}\,\sigma_\omega^{\text{inter}}$ that determines the gain). In particular, the negative values of Reσω produce the gain. Reσω is given by (3.8):

$$\text{Re}\,\sigma\omega=\text{Re}\,\sigma_\omega^{\text{inter}}+\text{Re}\,\sigma_\omega^{\text{intra}}\approx\frac{e^2}{4\hbar}(1-2f_{h\omega})+ \\ +\frac{(\ln 2+\varepsilon_F/2k_BT)e^2}{\pi\hbar}\frac{\tau k_BT}{\hbar(1+\omega^2\tau^2)}; \quad (3.8)$$

Reσω is proportional to the absorption of photons at frequency ω, where e = the electron charge; $\hbar$ is the reduced Planck constant, k_B is the Boltzmann constant, and τ is the momentum relaxation time of carriers. The value $\text{Re}\,\sigma_\omega^{\text{intra}}$ is always positive and is a loss contribution[37].

The graphene conductivity presents substantial difficulties for its mathematical description which may be different at different frequencies in the THz range, in the upper THz range and up to far infrared, the expression derived from the Kubo formula[10]:

$$\sigma_{k\omega} = \frac{ie^2}{\hbar\pi^2} \sum_{a=1,2} \int \frac{d^2 p \upsilon_x^2 \{ f[\varepsilon_a(p_-)] - f[\varepsilon_a(p_+)] \}}{[\varepsilon_a(p_+) - \varepsilon_a(p_-)][\hbar\omega - \varepsilon_a(p_+) + \varepsilon_a(p_-)]} + \tag{3.9}$$

$$+ \frac{2ie^2\hbar\omega}{\hbar\pi^2} \int \frac{d^2 p \upsilon_{21}\upsilon_{12} \{ f[\varepsilon_1(p_-)] - f[\varepsilon_2(p_+)] \}}{[\varepsilon_2(p_+) - \varepsilon_1(p_-)][(\hbar\omega)^2 - [\varepsilon_2(p_+) - \varepsilon_1(p_-)]^2]};$$

where the conduction band correspond to index 1 and the valence band to index 2, $\upsilon_F \approx 10^6 ms^{-1}$, $\varepsilon_1(p) = |p|\upsilon_F$ and $\varepsilon_2(p) = -|p|\upsilon_F$; $f(\varepsilon)$ - the electron distribution function; $\upsilon_x = \upsilon_F \cos\theta_p$ and $\upsilon_{12} = i\upsilon_F \sin\theta_p$ are the matrix elements of the velocity operator. It is assumed that the condition for the Fermi function are satisfied, i.e. $f(\varepsilon) = \frac{1}{1 + e^{(\varepsilon - \varepsilon_F)/T}}$. The first part of Eq. (3.9) corresponds to the intra-band transitions and the second – to inter-band transitions.

The intra-band Drude conductivity is characteristic of the THz range. The frequency dependence of the conductivity is given in Fig. 3.11. The Fermi level at 100 meV is depicted by the dashed lines. The intra-band conductivity increases with an increase of doping. The solid lines represent this higher amount of doping at 200 meV. The intra-band Drude conductivity can be adjusted by doping and structural changes, the same way as it is done in usual semiconductor heterostructures.

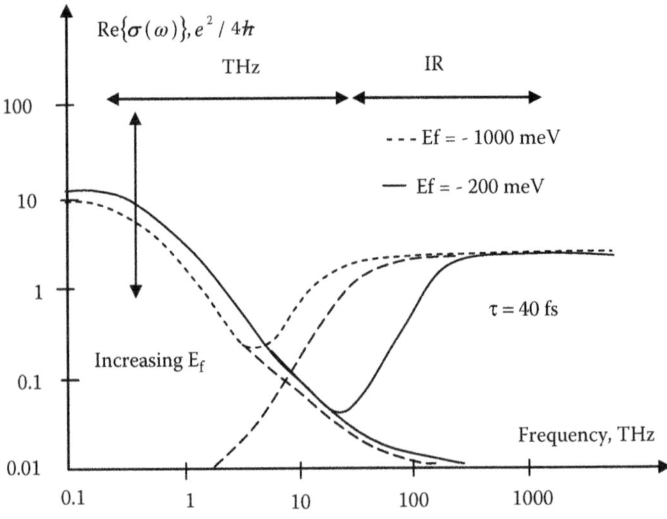

FIGURE 3.11 Frequency dependence of the real part of the optical conductivity in graphene[38]

Excitation of 2D plasmons in graphene causes extremely high plasmonic absorption and/or a giant plasmonic gain in the THz range. The gated plasmons are of particular significance for use in tunable graphene-based devices. The plasmons in gated devices have a super linear dispersion (Fig. 3.12).

In the semi-classical Boltzmann model describing the electron-hole plasma-wave dynamics in graphene, the plasmon phase velocity is proportional to the power of minus four of the gate bias and of the gate-to-graphene distance d (Fig. 3.13). The plasmons behavior in this case is quite different from those in conventional semiconductor quantum wells. The velocity always exceeds the Fermi velocity υ_F. The equal densities of electron and holes in graphene lead to two branches of carrier waves: the waves with neutral charge (sound-like waves) and the plasma waves. In the gates (heavily doped) both the majority and minority carriers exist. Subsequently, the minority carriers are damped and, consequently, only unipolar plasmons modes of the majority survive.

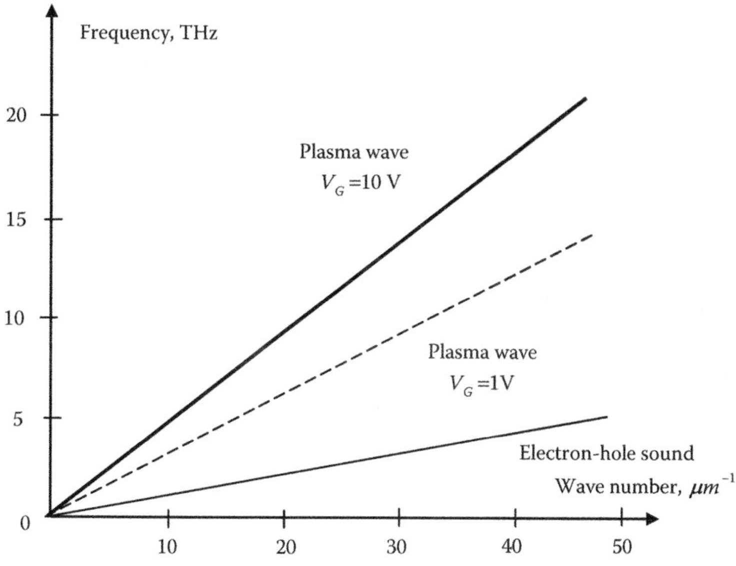

FIGURE 3.12 Dispersion of plasmons in graphene in gated devices[38]

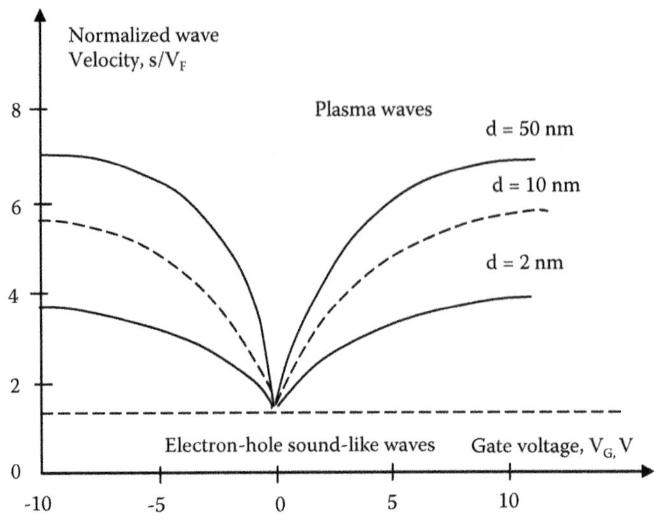

FIGURE 3.13 Plasma-wave velocity vs gate bias. The dashed line corresponds to the electron-hole sound-like waves measured in the vicinity of the neutrality point[38]

An important feature is the possibility of tuning the plasmon frequencies over the whole THz range. There are several factors that are taken into account:

1) The direction of the plasmon propagation;
2) The width of the micro-ribbons;
3) The modulation of the carrier density;
4) The splitting of the Landau levels with an applied magnetic field.

The presence of 2D plasmons in graphene substantially increases light-matter interaction improving the quantum efficiency[39]. If we pump graphene electrically or optically, we can achieve gain at THz frequencies and build lasers with higher power output than is available at the moment. In order to achieve this effect, there are two ways; one is propagation of the surface plasmon polaritons (SPPs) in graphene and the other is resonance of the SPPs[40]. In the former case, it is analytically revealed in Ref. 40 that when THz photons are incident with a TM mode to population-inverted graphene under pumping the THz photon could excite the SPPs so that they could propagate along with increasing the gain.

In such a situation, the absorption coefficient α is:

$$\alpha = \text{Im}(q_z) = 2\,\text{Im}(\rho \cdot \omega / c); \tag{3.10}$$

where z is the direction of the SPP propagation and q_z is the SPP wave vector component of z-direction. Fig. 3.14 depicts simulated α for a monolayer of graphene on a SiO_2/Si substrate at room temperature. In order to create a population inversion in graphene with negative dynamic conductivity, the quasi-Fermi energy values are chosen in the range of ε_F = 10 – 60 meV as discrete units. The carrier momentum relaxation time is assumed to be τ_m = 3.3 p.s. Fig. 3.15 shows the achieved gain of approximately 10^4 cm^{-1}. These values of the simulated gain are by three to four orders of magnitude higher than those in cases without excitation of the SPPs[39]. Since the absorption and gain coefficients are related to the dynamic conductivity, the gain spectra are similar in character on momentum relaxation graphs.

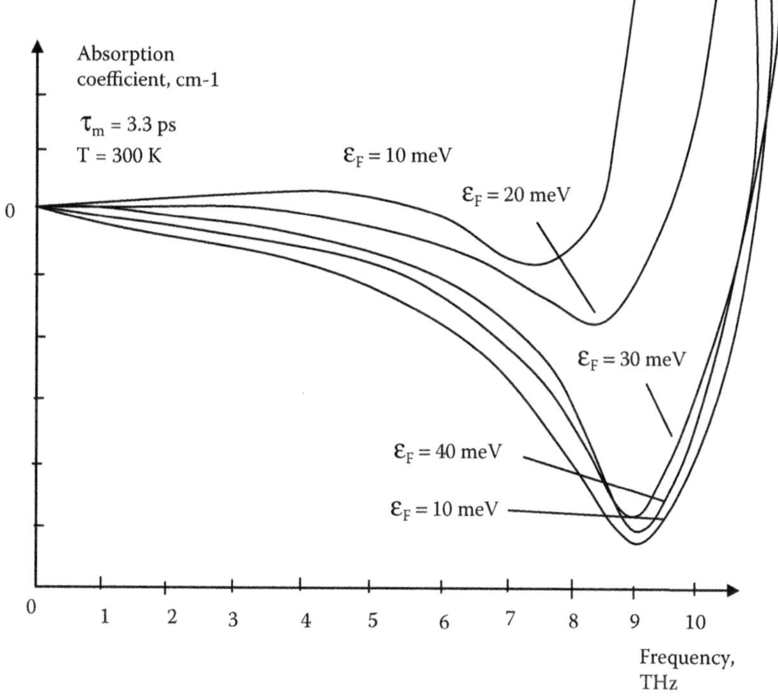

FIGURE 3.14 Frequency dependence of surface plasmon polaritons (SPP) absorption for monolayer-inverted graphene on SiO_2/Si for quasi-Fermi energies[41]

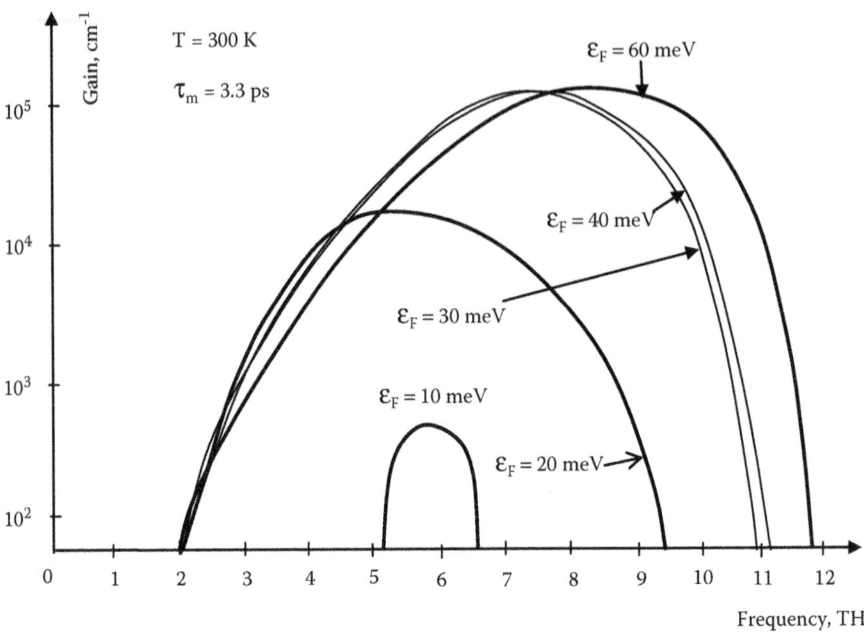

FIGURE 3.15 Frequency dependences of SPP gain for a monolayer of graphene[41]

Regarding the latter way that exploits the SPP resonance, graphene is structured with dimensions on the order of micrometers or submictometers which is smaller than the wavelength of THz radiation. These structures acting as metamaterials, have plasmonic responses in the THz range[39]. Plasmons play a large role in the possibility to achieve negative conductivity or gain in graphene. Plasmons (in the classical interpretation) may be described as oscillations of free electron density. They are also a quantization of this kind of oscillation.

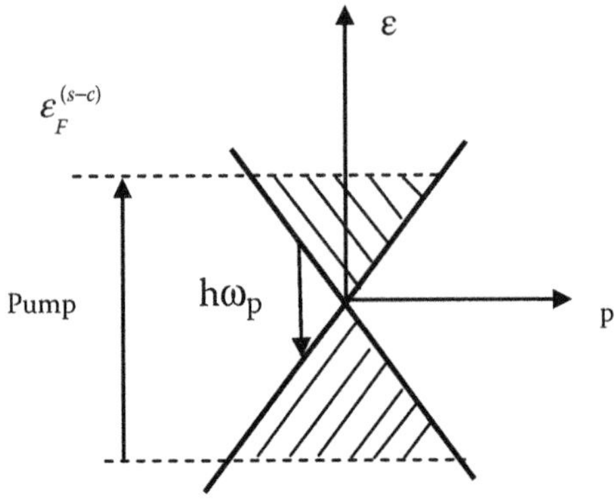

FIGURE 3.16 a) Optical pumping of graphene[42]. *(Continued)*

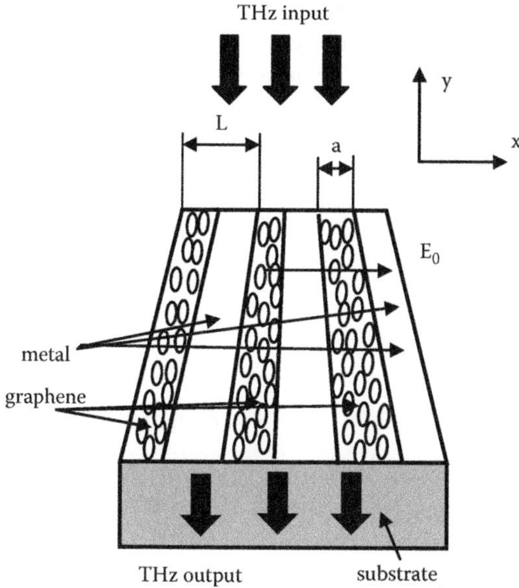

FIGURE 3.16 (Continued) b) A graphene-metal ribbon array. Silicon carbide may be used for the substrate[42]

The incident THz radiation (Fig. 3.16) is amplified in graphene. Optical or electrical pumping inverts the carrier population. (In Fig. 3.16 an array of graphene microcavities has the parameters of L = 4 μm and a = 2 μm[42]). The surface plasmons are excited by the incident THz photons with an effective field intensity vector component perpendicular to the direction of graphene ribbons with a – width of the ribbon and L – the period of graphene microcavities and stripes of metal. Surface plasmon polaritons (SPPs) produce significant gain at the frequencies that correspond to the SPP modes. An external THz wave incident on the planar array of the graphene microcavities is perpendicular to the substrate plane. The polarization of the electric field is directed across the metal strips. However, the stimulated emission of near IR (also possible range for the incident radiation) and of THz photons is limited by the quantum conductivity in the population-inverted graphene $(e^2/4\hbar)$[43]. The reason the limit imposed on absorbance by $\pi e^2/\hbar c \approx 2.3\%$ is because the photons available for stimulated emission are produced only by the inter-band transition process[44]. In order to overcome the limit, shorter wavelength surface plasmon polaritons (SPP) were introduced. The resonant plasmon absorption in graphene materials from the SPP excitation is utilized to achieve superradiant THz emission. The pumping of the graphene is done either by optical illumination or by electron and hole injection from the p-and n-doped areas through the opposite sides of the metal stripes. The amplification is achieved by using a number of microcavities[45]. If the quasi Fermi energy level (ε_F) increases, the energy gain matches and then exceeds the net loss of energy due to the electron scattering in graphene[41]. In Fig. 3.17, a dependence of absorption as a function of the quasi-Fermi energy and the THz frequency for an array of the graphene microcavities is shown. The negative values of absorbance means that we have a gain.

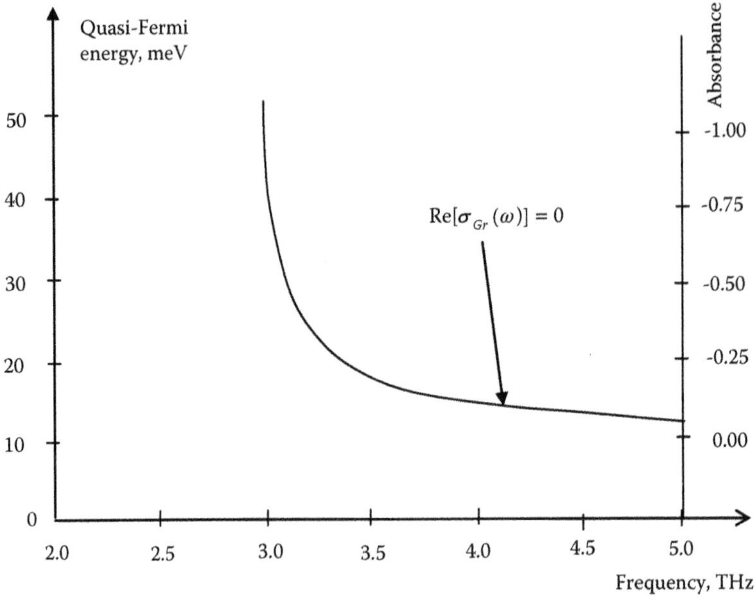

FIGURE 3.17 Contour map of the absorbance as a function of the quasi-Fermi energy and the frequency of the THz excitation radiation[42].

The negative value of the absorbance gives the amplification coefficient. The value $Re[\sigma_{Gr}(\omega)] = 0$ corresponds to the transparent graphene (when the loss equals gain). Also, at the transparency boundary, the THz wave amplification at the plasma resonance frequency becomes several orders of magnitude stronger. The behavior of the amplification coefficient close to the self-excitation regime is given in Fig. 3.18.

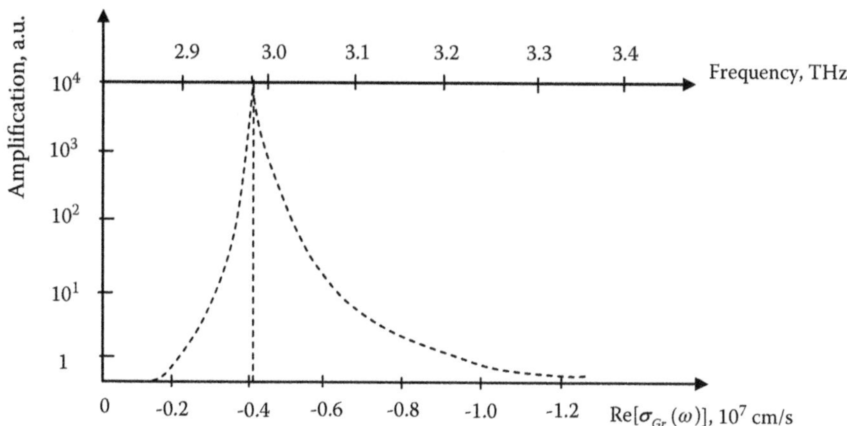

FIGURE 3.18 Variation of the power amplification coefficient[42]

The plasmon oscillations are highly coherent and lasing takes place as soon as the gain at least equals the loss. The metal stripes in Fig. 3.16 act as synchronizing elements so that the plasmons in adjacent microcavities oscillate in phase. As a result, a single plasmon mode covering the whole array area, contributes to superradiant THz radiation[46]. The reported gain exceeds 10^4 cm^{-1} in the THz range[42].

Compared to the current state-of-the-art THz sources, the graphene injection lasers[8] have relatively high quantum efficiency (~ 1), a higher output power (order of mW) and can operate at the room temperature, the advantages that are due to 1) the crystal structure of graphene that has no band gap and does not require any specific conditions for carrier depopulation which is not the case for the QCLs; 2) extremely long carrier relaxation times; 3) high optical phonon energy.

EXPERIMENTAL OBSERVATIONS[41]

An intrinsic monolayer graphene sample was used for measurements at room temperature on a $SiO_{2/}Si$ substrate. A schematic description of the experiment is given in Fig. 3.19. The procedure is based on a time-resolved near-field reflective electro-optic sampling. The optimal pumping employs a fs – IR laser pulse. Simultaneously, a THz pulse is generated for probing the THz characteristics of the sample. As the emitter of THz probe pulse a 140 μm-thick CdTe crystal was used. An exfoliated monolayer-graphene on SiO_2/Si substrate had an electro-optic sensor on its surface.

The optical pump and probe source was a femtosecond-pulsed laser with average power of about 4 mW with the duration of pulse of ~ 80 fs and repetition rate of 20 MHz. The laser beam used for optical pumping was mechanically chopped at ~ 1.2 kHz and linearly polarized. The chopping was necessary for the subsequent detection. The laser probing beam was cross-polarized with respect to the pumping beam's polarization. In order to receive the envelope THz probe pulse, the CdTe layer was used. The primary pulse (#1 in Fig. 3.19) was detected from the primary

THz beam at the top surface of CdTe layer. The THz pulse reflects from the interface at SiO_2/Si after having penetrated inside and transmitted through the graphene. The pulse is electro-optically detected as a THz echo signal (#2 in Fig. 3.19). Thus, the temporal response consists of the first THz pulse that propagates through CdTe without interacting with the graphene and of a THz photon echo signal that actually probes the graphene. The delay between the above two pulses is equal to the roundtrip propagation time of the probe pulse through the layer of CdTe. The bandwidth of the THz response is restricted by the Restrahlen band of the CdTe crystal and is estimated to be approximately 6 THz. The observed gain spectra show qualitative coincidence with analytical calculations having threshold property against the pumping intensity[41, 43]. The intensity of the observed THz echo pulse has a remarkable spatial dependence reflecting its polarization with a giant peak at the focused area where the THz probe pulse takes TM modes so that it can excite the SPPs in graphene. The observed gain enhancement effect is about 50 times as high as the other cases in which the SPPs could not be excited[41].

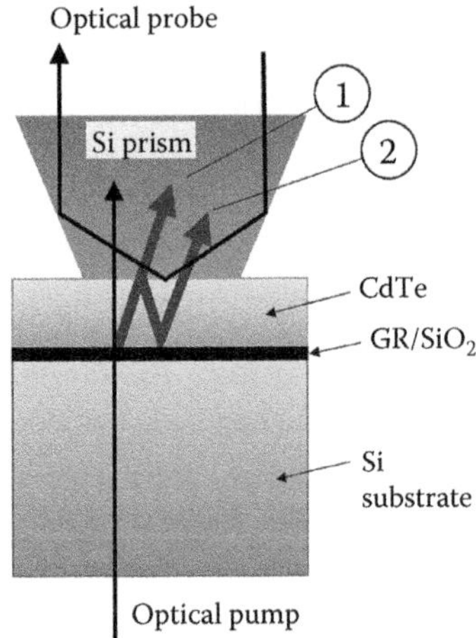

FIGURE 3.19 Cross-sectional image of the experimental set-up, THz probe, and optical probe[41]

3.7 QUANTUM SCARS IN GRAPHENE[47]

A remarkable physical effect in graphene is a concentration of wave functions around classical periodic orbits or so-called "quantum scars". In graphene whose dynamics, from the classical point of view, are chaotic, the semiclassical regime of a wave function can be regarded locally as a superposition of a number of plane waves. A nominal approach to the distribution of the wave functions states that the wave functions tend to concentrate on the paths analogous to unstable periodic orbits in classical systems. Bogomolny and Berry provided theoretical explanations based on the semiclassical Green's function in nonlinear physics.

In addition to the existing works on nonrelativistic quantum mechanical systems where the dependence of the particle energy is quadratic, a new research described that scarring can occur also in relativistic quantum system described by the Dirac equation (as in the previous case, the Schrodinger equation is applicable)[47]. The latter phenomenon is important for device physics and device applications. In graphene, in particular (because of its hexagonal lattice structure), the band structure exhibits a linear dependence of the energy on the wave vector. The motion around the Dirac points is relativistic. The electron behavior in graphene resembles that of fermions which have the Fermi velocity $\upsilon = 10^6$ cm/s[48]. The advantage lies in a potential capability of device made of graphene to operate at much higher speed than silicon devices can. Specifically, in quantum dots, at pointer states associated with scars, the conductance dependence exhibits narrower conductance resonant peaks. From the theoretical point of view, the relativistic quantum scars can be derived from the Dirac equation:

$$-i\hbar\upsilon_F\sigma - \nabla\Psi = E\Psi; \tag{3.11}$$

where σ is Pauli's matrices and $\Psi = [\Psi_A, \Psi_B]^T$ is a two component spinor describing the two types of nonequivalent atoms in graphene. On the whole, the scars in a relativistic quantum system have the same origin as the scars associated with ultrasonic fields, microwave and quantum dot billiards. Relativistic scars, however, are different from the conventional quantum scars.

In the presence of a magnetic field, the band structure of graphene changes which in its turn changes the scar pattern. In particular, it was found[47] that in a perpendicular weak magnetic field, the classical scar orbits remained almost unchanged, but in a strong field (with energies quantitized) the scars smear out. As a result, new orbits, both curved and straight, make appearance. The scar orbits could be more complicated, however, when spin effects take place.

Because of the relativistic nature of electron motion in graphene, a study of electronic transport is important for understanding quantum transport properties of the material from the fundamental and device design point of view[49]. In particular, the electron transport in quantum dots has a number of prominent features. The principal characteristics of quantum transport is conductance. Since the conductance is determined by transmission, it is sufficient to calculate the quantum transmission. With the change of electron energy, quantum dot transmission exhibits fluctuations. A correlation was found between the fluctuations and the appearance of scars inside the dots (such correlation may be found by investigating the local density of states (LDS)). L. Huang et al found that the LDS tends to concentrate in particular regions, around periodic orbits of quantum dots when such dots are treated classically. The local maxima or minima on the transmission curve indicate the above phenomenon, analogous to a one-dimensional finite square potential well. Similar to the finite square potential well, depending on the wave function phase, the resonances can either enhance or suppress transmission. Significantly, in conventional quantum dots in semiconductors, the conductance fluctuations are related to the presence of the scarring states.

For graphene quantum dots and nanoribbons, in particular, the standard Landauer formula applies. It relates the conductance $G(E_F)$ to the overall transmission $T_G(E_F)$[49]:

$$G(E_F) = \frac{2e^2}{h} T_G(E_F); \tag{3.12}$$

where $T_G(E_F) = \int T(E)\left(-\frac{\delta f}{\delta E}\right) dE$; T(E) – transmission of the device and $f(E) = 1/[1+e^{(E-E_F)/kT}]$ is the Fermi distribution function.

In order to calculate the transmission T(E) for graphene quantum dots, L. Huang et al used the low temperature conductance and non-equilibrium Green's function (NEGF) formalism, as well as the division of the system into three parts (see Fig. 3.20).

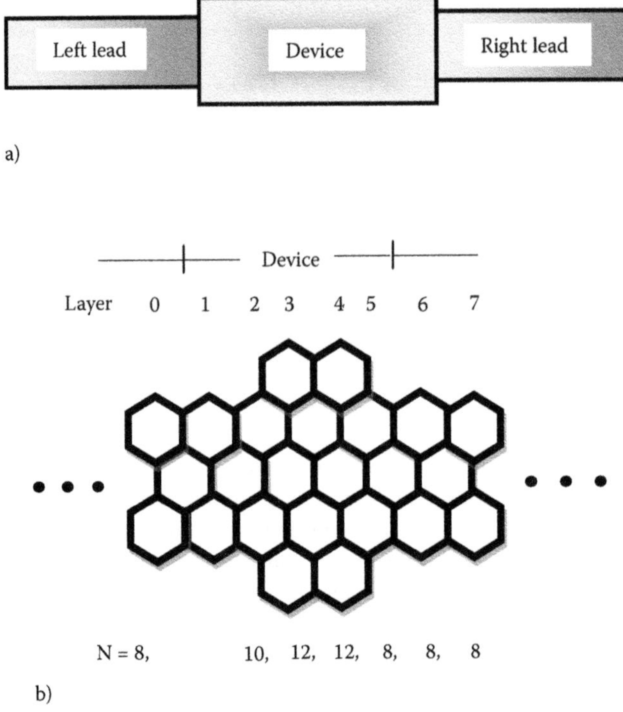

FIGURE 3.20 a) Schematic representation of a quantum dot[49] b) Tight-binding Hamiltonians for the Green's function calculation. The left lead goes from -∞ to layer 0; the right lead – from layer 6 to ∞. N= 8, etc is the number of atoms in each layer[49].

REFERENCES

1. Walt A. de Heer, "The invention of graphene electronics and the physics of epitaxial graphene on silicon carbide", *IOP Science*, 2012 T 140, (2012)
2. Geim, A.K. and Novoselov, K.S. "The rise of graphene", *Nature Materials*, 183-191 (2007)
3. Forbeaux, I., Themlin, J., Langlais, V., Yu, M. Belkhir, H. and Debever, J., The reconstructed 6H-SiC (0001): A semiconducting surface", *Surf. Rev. Lett,.* **5**, 193 (1998)
4. Mermin, N.D. and Wagner, H., "Absence of ferromagnetism or antiferromagnetism in one-or two-dimensional isotropic Heisenberg models", *Phys. Rev. Lett.*, **17**, 113 (1966)
5. Geim, A.K., and Novoselov, K.S., "The rise of graphene", *Nature Materials* **6** (3) 183-191
6. Das Sarma, S., Hwang, E.H., and Li, Q., "Disorder by order in graphene", *Physical Review B*, **85** 195451 (2012)
7. Boehm, H.P., Setton, R., Stumpp, E., "Nomenclature and terminology of graphite intercalation compounds", *Pure and Applied Chemistry*, **66** (9): 1893-1901, (1994)
8. Fasolino, A., Los, J.H., and Katsnelson, M.I., "Intrinsic ripples in graphene", Nature Materials (11) 858-861 (2007)
9. Ishigami, Masa, et.al. "Atomic structure of graphene on SiO_2", *NanoLett*, **7** (6), 1643-1648 (2007)
10. Fraundorf, P. and Wackenhut, M. "The core structure of presolar graphite onions", *"Astrophysical Journal Letters"* **578** (2): 153-156 (2002)
11. Kasuya, D., Yudasaka, M.Takahashi, K., Kokai, F., Lijima, S., "Selective Production of Single-Wall Carbon Nanohorn Aggregata and the formation mechanism", *J. of Physical Chem.* B 106 (19) 4947 (2002)
12. Wallace, P.R. "The Band Theory of Graphene", *Physical Review*, **71** (9): 622 (1947)
13. Charlier, J.C., Eklund, P.C., Zhu, J., and Ferrari, A.C. "Electron and Phonon Properties of Graphene: Their Relationship with Carbon Nanotubes" from Carbon Nanotubes, *Advanced Topics in Synthesis, Structure, Properties and Applications,* ed. A. Jorio, G. Dresselhaus, and M.S. Dresselhaus, Berlin/Heidelberg: Spirng-Vorlag
14. Avouris, P., Chen, Z., and Perebeinos, V., "Carbon-based electronics", *Nature Nanotechnology* **2** (10): 605-615
15. Morozov, S.V. et al "Giant Intrinsic Carrier Mobilities in Graphene and its Bilayer", *Physical Review Letters* **100** (1) (2008)
16. Morozov, S.V. et al "Strong Suppression of Weak Localization in Graphene", *Physical Review Letters* **97** (1) (2006)
17. Mounet, N. and Marzari, N., "First-principles determination of the structural, vibrational and thermodynamic properties of diamond, graphite and derivatives", *Physical Review* B **71** (2005)
18. Chen, J.H., Jang, C., Adam, S., Fuhrer, M.S., Williams, E.D., and Ishigami, M., "Charged-impurity scattering in graphene", *Nature Physics*, **4**, 377 (2008)
19. Chen, J.H. et al., "Charged impurity scattering in Graphene", *Nature Physics,* **4**(5) 377-381 (2008)
20. M. Ryzhii, V. Ryzhii, T. Otsuji, P.P. Maltsev, V.G. Leiman, N. Lyabova, and V. Mitin, "Double injection, resonant-tunneling recombination, and current-voltage characteristics in double-graphene-layer structures", *J. Appl. Phys.,* **115**, (2014)
21. V. Ryzhii, A. Satou, T. Otsuji, M. Ryzhii, V. Mitin and M.S. Shur, "Dynamic effects in double graphene-layer structures with inter-layer resonant-tunneling negative conductivity", *J. Phys. D: Appl. Phys.,* **46**, (2013)
22. Nair, R.R, et al "Fine Structure Constant Defines Visual Transparency of Graphene", *Science* **320** (5881): 1308 (2008)

23. Lin, J., Wright, A.R., Zhang, C. and Ma, Zh., "Strong terahertz conductance of graphene nanoribbons under a magnetic field", *Appl. Phys. Lett.* **93** (4) (2008)
24. Kurum, U. et al "Electrochemically tunable ultrafast optical response of graphene oxide", *Appl. Phys. Lett.* **98** (2) (2009)
25. Sreekanth, K.V., et al "Excitation of surface electromagnetic waves in a graphene-based Bragg grating", *Scientific report* **2** (2012)
26. Zhang, H., et al., "Large energy solition erbium-doped fiber laser with a graphene-polymer composite mode locker", *Appl. Phys. Lett.* **96** (11) (2010)
27. Zheng, Z., et al., "Microwave and optical saturable absorption in graphene", *Optics Express* **20** (21): 23201-23219 (2012)
28. Mingo N., Broido, D.A., "Carbon Nanotube Ballistic Thermal Conductance and its limits", *Physical Review Letters* **95** (9) (2005)
29. Frank, I. W., Tatenbaum, D.M., Van Der Zande, A.M., and McEuen, P.L., "Mechanical properties of suspended graphene sheets" *J. Vac. Sci. Technol.* B **25** (6): 2558-2561 (2007)
30. Dean, C.R., Young, A., Meric, I., Lee, C., Wang, L., Sorgenfrei, S., Watanabe, K., Tamiguchi, T., Kim, P., Shepard, K., and Hone, J., "Boron nitride substrates for high-quality graphene electronics", *Nature Nanotechnology,* **5**, 722 (2010)
31. Martin, J., Akerman, N., Ulbricht, G., Lohmann, T., Smet, J.H., von Klitzing, K., and Yacoby, A. "Observation of electron-hole puddles in graphene using a scanning single-electron transistor", *Nature Physics.* **4**, 144 (2008)
32. Lifshitz, I.M. "Journal of Experimental and Theoretical Physics" (in Russian) **22,** 475 (1952)
33. Sanderson, B., "Toughest Stuff Known to Man. Discovery Opens Door to Space Elevator", "Breakthrough in Developing Super Material Graphene", *Science Daily* **20** (2010)
34. Du, X., et al., "Fractional quantum Hall effect and insulating phase of Dirac electrons in graphene", *Nature*, **462** (72 70), 192-195 (2009)
35. Novoselov, K.S. et al., "Two-dimensional gas of massless Dirac fermions in graphene", *Nature* **438** (7065) 197-200 (2005)
36. V. Ryzhii, M. Ryzhii, V. Mitin, and T. Otsuji, "Toward the creation of terahertz graphene injection laser", *J. Appl. Phys.,* **110** (2011)
37. T. Otsuji, S. B. Tombet, A. Saton, M. Ryzhii, and V. Ryzhii, "Terahertz-Wave Generation using Graphene: Toward New Types of Terahertz Lasers", *IEEE Journal of Selected Topics in Quantum Electronics,* **19**, (2013)
38. T. Otsuji, V. Popov, and V. Ryzhii, "Active graphene plasmonics for terahertz device applications", *Journal of Physics D: Applied Physics,* **47**, (2014)
39. V. Ryzhii, A. Dubinov, T. Otsuji, V. Mitin, and M.S. Shur, "Terahertz Lasers Based on Optically Pumped Multiple Graphene Structures with Slot – line and Dielectric Waveguides", *J. Appl. Phys.,* **107**, (2010)
40. A.A. Dubinov, Y.V. Aleshkin, V. Mitin, T. Otsuji, and V. Ryzhii, "Terahertz Surface Plasmons in Optically Pumped Graphene Structures", *J.* Phys.: Condens. Matter, **23** (2011)
41. T. Watanabe, T. Fukushima, Yu. Yabe, S. Tombet, A. Satun, A.A. Dubinov, B. Aleshkin, V. Mitin, V. Ryzhii, and T. Otsuji, "The gain enhancement effect of surface plasmon polaritons on terahertz stimulated emission in optically pumped monolayer graphene", *New Journal of Physics,* **15**, (2013)
42. T. Otsuji, V. Popov, and V. Ryzhii, "Superradiant terahertz lasing with graphene materials, *SPIE Newsroom* 10.1117/2.1201305004892 (2013)
43. T. Otsuji, S.A. Boubanga Tombet, A. Saton, H. Fukidome, M. Suemitsu, E. Sano, V. Popov, M. Ryzhii, and V. Ryzhii, "Graphene materials and devices in terahertz science and technology", *Mater. Res. Soc.* Bull. **37**, (2012)

44. F. Bonaccorso, Z. Sun, T. Hansan, and A.C. Ferrari, “Graphene photonics and optoelectronics”, *Nat. Photonics,* **4**, (2010)
45. V.V. Popov, O.V. Polischuk, A.R. Davoyan, V. Ryzhii, T. Otsuji, and M. Shur, “Plasmonic terahertz lasing in an array of graphene nanocavities”, *Phys. Rev.* **B** 86, (2012)
46. M. Benedickt, A. Ermolaev, V. Malyshev, I. Sokolov, and E. Trifonov, “Superradiance: Multiatomic Coherent Emission”, *Inst. Of Phys.,* Bristol, 1996
47. L. Huang, Y.-Cheng, D. Ferry, S. Goodnick, and R. Akis, “Relativistic Quantum Scars”, *Phys. Rev. Letts,* (2009)
48. K.S. Novoselov, A.K. Geim, S.V. Morozov, D. Jiang, M.I. Katsnelson, I. V. Grigorieva, S. V. Dubonos, and A.A. Firsov, “Two-dimensional gas of massless Dirac fermions in graphene”, *Nature* (London), **438**, 197 (2005)
49. L. Huang, Y.-C. Lai, D.F. Kerry, R. Akis, and S.M. Goodnick, “Transmission and scarring in graphene quantum dots”, *J. Phys.: Condens. Matter*, **21**, (2009)

4 Quantum Mechanics of Graphene

The next step in learning about graphene may be a more detailed analysis of its atomic structure. Graphene consists of carbon atoms which is the basic unit for numerous organic and non-organic forms: diamonds, graphite, biological tissues to name a few. The carbon atom's characteristics may be described by the Schrodinger equation:

$$[-(\hbar^2/2m)\nabla^2 + U(r)]\Psi = E\Psi; \tag{4.1}$$

where the first term in parenthesis is the kinetic energy, *U(r)* – the potential energy; $E\Psi$ = the total energy.

The potential energy *U(r)* implies the energy confined within the atom itself (carbon atom in our case):

$$U(r) = -k_C Ze^2 / r; \tag{4.2}$$

where $k_C = 1/4\pi\varepsilon_o$, Z = 6 for carbon, r = radius from the nucleus to the electron, $\varepsilon_o = 8.85 * 10^{-12}$ F/m – the permittivity of free space.

4.1 CARBON ATOM AND ITS STRUCTURE

The atomic structure of carbon may be given as:

$$C: (1s)^2(2s)^2(2p)^2; \tag{4.3}$$

or schematically (Fig. 4.1):

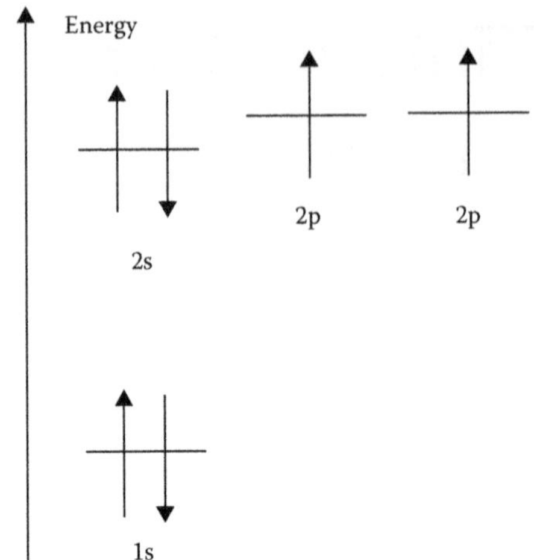

FIGURE 4.1 The carbon atom with atomic levels.

We can consider the carbon atom to be spherical since the potential energy has only one variable, *r*. In spherical coordinates, the Schrodinger equation becomes:

$$-\frac{\hbar}{2m}\frac{1}{r^2}\frac{\delta}{\delta r}\left(\frac{r^2\delta\Psi}{\delta r}\right)-\frac{\hbar^2}{2mr^2}\left[\frac{1}{\sin\theta}\frac{\delta}{\delta\theta}\left(\sin\theta\frac{\delta\Psi}{\delta\theta}\right)+\frac{1}{\sin^2\theta}\frac{\delta^2\Psi}{\delta\varphi^2}\right]+U(r)^{\Psi}=E^{\Psi}; \quad (4.4)$$

where $x = r\sin\theta\cos\varphi$, $y = r\sin\theta\sin\varphi$, $z = r\cos\theta$, θ = the polar angle, φ = the azimuthal angle.

The complexity of solving the Schrodinger equation in spherical coordinates may be somewhat simplified by the symmetry considerations:

$$\Psi = R(r)f(\theta)g(\varphi); \quad (4.5)$$

4.2 WAVEFUNCTION SOLUTIONS

In order to find a particular wave function we need to consider wavefunctions with principal quantum numbers $n = 1$ and $n = 2$. The principal quantum number n governs the energy of a state:

$$R_{n,l}(r) = (r/a_0)\exp(-r/na_0)L_{n,1}(r/a_0); \quad (4.6)$$

The solution (4.6) is associated with the radial equation. $L_{n,1}(r/a_0)$ is a Laguerre polynomial in $\rho = r/a_0$. The *n-l-1* nodes belong to the radial function. The orbital quantum number l is limited to the values of $l = 0$ to $n - 1$. a_0 (the Bohr radius $\hbar/k_c mc^2$) is equal to 0.0529 nm. The Bohr radius here does not stand exactly for the radius of an orbit of the Bohr planetary model. The energies of the electron states in such a model are:

$$E_n = -Z^2 E_0/n^2; \tag{4.7}$$

where
$$E_0 = k_c e^2/2a_0 = 13.6, \text{ eV}; \tag{4.8}$$

Each state has the energy:

$$E_n = -K_C Z e^2 / 2r_n; \tag{4.9}$$

where $r_n = n^2 a_0 / Z$;

In a sphere of the carbon atom, the lowest energy wavefunction ($\Psi_{n,l,m}$) is for $n = 1$:

$$\Psi_{100} = (Z^{3/2} / \sqrt{\pi})\exp(-Z_r / a_0); \tag{4.10}$$

The probability of finding the electron in a carbon sphere for $n = 1$ and the radius 1 is:

$$P(r) = 4\pi r^2 \Psi_{100}^2; \tag{4.11}$$

Eq. (4.11) has a maximum at $r = a_0/Z$;

This is the radius at which the probability of finding electrons is minimal. For the state $n = 2$ the wavefunction $\Psi_{200}(2s)$ is similar to Ψ_{100} being spherically symmetrical. All the further wavefunctions are anisotropic:

$$\Psi_{210} = R(r)F(\theta)g(\varphi) = C_2 \rho \cos\theta e^{-\rho/2}; \tag{4.12}$$

$$\Psi_{21\pm1} = R(r)f(\theta)g(\varphi) = C_{2\rho} \sin\theta e^{-\rho/2} \exp(\pm i\varphi); \tag{4.13}$$

For the real form of a wavefunction $\rho = Zr/a_0$ and $C_1 = Z^{3/2}/\sqrt{\pi}$. A feature of the wavefunctions of the form $\Psi_{21,\pm1}$ is that they have an orbital angular momentum. The projection of the orbital angular momentum of the electrons in the z-direction is (still for the spherical coordinates) is:

$$g(\varphi) = \exp(\pm im\varphi); \tag{4.14}$$

where m = the magnetic quantum number.

The orbital angular momentum L itself has a restricted range of allowed values:

$$l = 0, 1, 2, \ldots n - 1; \tag{4.15}$$

where l is the orbital angular momentum quantum number.

The orbital angular momentum L is also described by the magnetic quantum number m that depends on "n" and "l":

$$m = -l, -l + 1, \ldots, (l - 1), l; \tag{4.16}$$

Another characteristics of the orbital angular momentum L is its length $[l(l + 1)]^{1/2}$h, where h is the unit of the projection (4.14). The term in the parenthesis represents probability. An electron may have and orbital and spin angular momentum. The spin angular momentum vector S has the length $[S(S-1)]^{1/2}\hbar$, where S = ½ and the propagation of S:

$$m_s \pm (1/2)\hbar; \tag{4.17}$$

For a number of states and some other cases, there are two angular momenta, i.e. $\vec{J} = \vec{L} + \vec{S}$ and for its magnitude: $\vec{J} = \sqrt{(j(j+1)\hbar}$.

The wavefunctions considered above are solutions for the Schrodinger equation (4.1). Their combinations are also solutions for (4.1). The linear combinations provide a preferred direction for the wavefunction and the probability area. An example of a sum of two wavefunctions is given below:

$$\Psi_{211} + \Psi_{21-1} = C_2\rho \sin\theta e^{-\rho/2}[\exp(i\varphi) + \exp(-i\varphi)] = C_2\rho \sin\theta e^{-\rho/2} 2\cos\varphi; \tag{4.18}$$

The plot of (Eq. 4.18) wavefunction resembles dumbbells along x and y axis. The wavefunctions may be used for describing orbital magnetic momenta. The spin characteristic is the spin projection $m_s = \pm\frac{1}{2}$. In the context of a wavefunction the spin and space parts may be combined as:

$$\psi = \phi(x)\chi; \tag{4.19}$$

where $\chi = \uparrow$ and $m_s = 1/2$ or $\chi = \downarrow$ and $m_s = -\frac{1}{2}$ for a single electron.

4.3 CARBON ATOM STATES AND BONDING

The graphene has a covalent type of bonding and involves two electrons. The simplest analogy of this bonding would be that of hydrogen. For the carbon complete states, there are two types of spin exist: parallel (S = 1) and antiparallel (S = 0). For the antiparallel spin $m_s = 0$ and the parallel spin allows $m_s = 1, 0, or -1$ (spin triplet). Triplet is defined as three states with total spin angular momentum 1:

$$|1,1\rangle = \uparrow\uparrow \; for\, S = 1$$

$$|1,0\rangle = (\uparrow\downarrow + \downarrow\uparrow)/\sqrt{2} \; for\; S = 1$$

$$|1,-1\rangle = \downarrow\downarrow \; for\; S = 1$$

Using the accepted notation:

$$x_{1,1} = \uparrow_1\uparrow_2, x_{1,-1} = \downarrow_1\downarrow_2; \text{ and for } x_{1,0} = \uparrow_1\downarrow_2 + \downarrow_1\uparrow_2; \tag{4.20}$$

For $S = 0$ (spin singlet):

$$x_{0,0} = \uparrow_1\downarrow_2 - \downarrow_1\uparrow_2; \tag{4.21}$$

For the parallel spins ($S = 1$), the states are symmetric as in (4.20) and for the antiparallel spins, the states are antisymmetric as in (4.21). The wavefunctions for the symmetric and antisymmetric cases are as follows:

$$\Psi_A(1,2) = \phi_{sym}(1,2)\chi_{anti}(1,2) \text{ for antiparallel spins} \tag{4.22}$$

and

$$\Psi_A(1,2) = \phi_{anti}(1,2)\chi_{sym}(1,2) = \phi_{anti}(1,2) \text{ for parallel spins } (S = 1) \tag{4.23}$$

In a carbon atom, the nucleus is Z = 6. Since carbon has a multi-state atom, the Pauli Exclusion Principle dictates that only one electron is allowed per fully specified state. Fig. 4.1 shows the filled carbon atomic levels[1]. The levels are designated as $1s^2\, 2s^2\, 2p^2$. The successive ionization energy level per carbon are 11.26, 24.38, 47.88, 64.49, 392.1 and 490.0[2]. The energy level values may be explained approximately by the Bohr atom calculations. The level separation may be explained by electron electrostatic repulsion where the Bohr's radius is $r_n = (n^2/z)a_0$. It is the other electrons that determine graphene's hybridization by a linear combination of 2s and 2p states. The fifths electron seems to overlap/penetrate the space occupied by the first four electrons closest to the core which results in a greater effective charge $Z^* = 2.67$ instead Z = 2. In graphene, the outer three electrons form a trigonal sp^2. The remaining $2p_z$ electrons can go to the delocalized extended states. These electron states constitute the valence and conduction bands making graphene a material with superior conducting characteristics.

4.4 FORMATION OF CRYSTALLINE CARBON

The formation of crystalline (or it may be molecular) takes place with combinations of wavefunctions in a particular direction. Usually, the combinations are trigonal sp^3. The combinations imply combining of $\vec{s}$ states ($l = 0$) with $\vec{p}$ states ($l = 1$) and that is why the word "hybrid" is used. The states have different energies and the nuclei that form the bond are separate.

The carbon molecule contains the atom of carbon and hydrogen with the latter bound by an electrostatic attraction. The mutual electron interaction is a covalent bond. It can involve several electrons.

4.4.1 Quantum Mechanics of One-Electron Covalent Bond

The effect of H_2^* electron interaction is quantum mechanical by nature. There are two wavefunctions $\psi_a(x_1)$ and $\psi_b(x_2)$ for the system of two protons and one electron. The both wavefunctions describe the behaviors of the two electrons (each function for each electron). An instability of close-positioned states cause tunneling from one state to another. The ground state's solution in a general form is:

$$\Psi' = A\psi_a(x_1) + B\psi_b(x_2); \tag{4.24}$$

where $A^2 + B^2 = 1$.

An electron tunnels from $\psi_a(x_1)$ to $\psi_a(x_1)$ with frequency f. The temporal stability may be in the symmetric form:

$$\Psi_S = 2^{-1/2}[\psi_a(x_1) + \psi_b(x_2)]; \tag{4.25}$$

and the antisymmetric form:

$$\Psi_A = 2^{-1/2}[\psi_a(x_1) - \psi_b(x_2)]; \tag{4.26}$$

The symmetric and antisymmetric states provide equal probability for an electron's presence in either state or what accounts for the system's stability. This system has two levels that may be utilized in a quantum computer. From the energy perspective the point in the middle of the distance connecting both states is the most favorable because of the electron being attracted from both states. The symmetric and antisymmetric states differ in energy by:

$$\Delta E = hf; \tag{4.27}$$

The tunneling of the electron may be described as resonant since the energy difference ΔE quantized (ΔE = 2.65 eV = twice the H_2^+ binding energy)[3]. The tunneling "hopping" takes place 1.28 x 10^{15} times per second (1/s). The electron resides at one of each state for 0.778 fs. The Schrodinger equation for this case is:

$$[-(\hbar^2/2m)\nabla^2 + U(r)]\psi = \mathrm{H}\psi = E\psi; \tag{4.28}$$

where r is the position of the electron.

The solutions are in the form (with no interactions):

$$\psi = \psi_a(x_1) = \psi = \psi_b(x_2); \tag{4.29}$$

The electron is traveling ("hopping") between two protons. $E = E_0$ when there is no interaction. The difference of energies for the interaction is:

$$E - E_0 = (k_c e^2 / D a_0) + (J + K)/(1 + \Delta); \tag{4.30}$$

The interaction of the electron and two protons may be also written as:

$$\Delta E = -k_c e^2 (1/r_{a,2} + 1/R) \text{ for } \Psi_S \tag{4.31}$$

where R is the distance between two protons.

And
$$E - E_0 = k_C e^2 (1/r_{a,2} + 1/R) \text{ for } \Psi_A; \tag{4.32}$$

where
$$K = \iint \psi_b^*(x_2)[-k_C e^2 (1/r_{a,2})]\psi_a(x_1) d^3x_1 d^3x_2 = -(k_C e^2/a_0) e^{-D}(1 + D); \tag{4.33}$$

And
$$J = \iint \psi_a^*(x_1)[-k_C e^2 (1/r_{a,2})]\psi_a(x_1) d^3x_1 d^3x_2 =$$

$$= (k_C e^2/a_0)[-D^{-1} + e^{-2D}(1 + D^{-1})]; \tag{4.34}$$

And
$$\Delta = \iint \psi_b^*(x_2)\psi_a(x_1) d^3x_1 d^3x_2 = e^{-D}(1 + D + D2/3); \tag{4.35}$$

where $D = R/a_0$.

Eq. (4.33) is the integral that evaluates the rate with which electrons move to the neighboring positions. The K-integral is a function of the atomic forces and distances (the latter is an exponential term). The first time ($-(k_C e^2/a_0)$) is equal to $-2E_0 = -27$ eV. The resonance K – integral determines the energy difference between states in (4.32). The energy difference between the symmetric and antisymmetric case for H_2^+ is 2 x 2.65 eV = 5.3 eV.

In molecular hydrogen, the single covalent bond joints two 1s states where $n = 1$ and $l = 0$.

$$[-(\hbar^2/2m)(\nabla_1^2 + \nabla_1^2) + U(r_1) + U(r_2)]\psi = (\mathrm{H}_1 + \mathrm{H}_2)\psi = E\psi; \tag{4.36}$$

The solution to (4.36) is in the form:

$$\psi = \psi_a(x_1)\psi_b(x_2) \text{ or } \psi = \psi_a(x_2)\psi_b(x_1); \tag{4.37}$$

where a and b are two protons at the distance R.

The interaction between the two protons is repulsive which is $k_C e^2/r_{1,2}$, where $r_{1,2}$ is the distance between the atoms. The attractive force exists between each electron and the opposite proton which is $-k_C e^2(1/r_{a,2}+1/r_{b,1})$. On the whole, taking into account the attraction and repulsive forces, the summary interaction is:

$$\mathrm{H}_{\text{int}} = k_C e^2[1/R + 1/r_{1,2} - 1/r_{a,2} - 1/r_{b,1}]; \tag{4.38}$$

The energy expectation value of the considered molecular system is[3]:

$$\langle E_{\text{int}} \rangle = \int \psi^* \mathrm{H}_{\text{int}} \psi dq; \tag{4.39}$$

From (4.20), (4.21) and having antisymmetric waveforms for the antisymmetric singlet spin state ($\delta = 0$):

$$\langle E_{\text{int}} \rangle = A^2(K_{1,2} + J_{1,2}); \tag{4.40}$$

And the symmetric triplet spin state:

$$\langle E_{\text{int}} \rangle = B^2(K_{1,2} - J_{1,2}); \tag{4.41}$$

With $$K_{1,2}(R) = \iint \phi_a^*(x_1)\phi_b^*(x_2)\mathrm{H}_{\text{int}}\phi_b(x_2)\phi_a(x_1)d^3x_1d^3x_2; \tag{4.42}$$

And $$J_{1,2}(R) = \iint \phi_a^*(x_1)\phi_b^*(x_2)\mathrm{H}_{\text{int}}\phi_a(x_2)\phi_b(x_1)d^3x_1d^3x_2; \tag{4.43}$$

For the hydrogen molecule, the covalent bonding requires the spin singlet case (the spins are antiparallel). The orbital wavefunction is, however, symmetric. The charge distribution varies from case to case with the spin-singlet case having more electrons in-between the two protons. The electrostatic energy (as well as the charge) depends on the orientation of the electron magnetic moments. The exchange interaction is:

$$\mathrm{H}_e = -2J_e\vec{s}_1 \cdot \vec{s}_2; \tag{4.44}$$

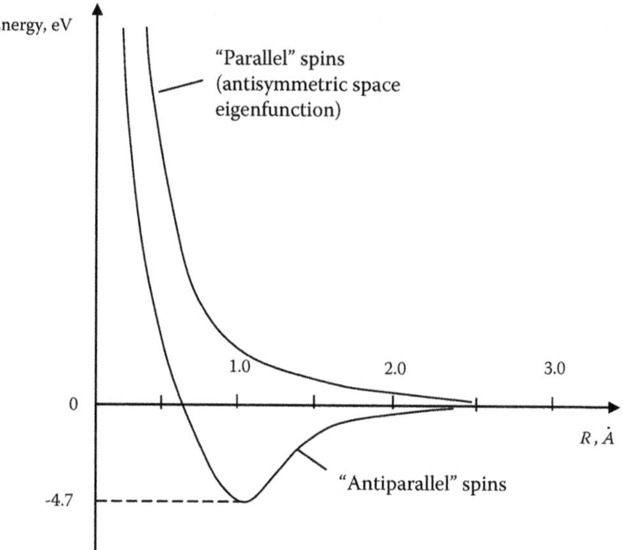

FIGURE 4.2 Energy dependence for bonding and antibonding states of the H – molecule.

The bonding state requires antiparallel spins (Fig. 4.2). The negative sign in (4.44) accounts for a negative bonding for antiparallel spins. From (4.2), the bonding energy is 4.7 eV with R = 0.074 nm.

4.4.2 Tetrahedral Bonding Example

Methane is a good example of tetrahedral bonding that involves 2s $2p^3$ which are wavefunctions. For 3D projections, $2p_x$, $2p_y$ and $2p_z$ are (1, -1, 1), (-1, 1, 1), (1, -1, 1), (1, 1, -1), and (-1, -1, -1). In diamond and in silicon crystals, like in methane, the bonds are tetrahedral sp^3 type. The sigma bond in particular has a rotational symmetry with respect to the axis connecting two atoms. Ethane, C_2H_6 also has sigma bonds: C_4H_{10} are also based on sp^3 sigma bond type. The significance of this atomic description comes from methane use for the chemical vapor deposition (CVD) of graphene. The current substrates include metals such as carbon.

4.4.3 Planar sp^2 and π-Bonding

Another kind of carbon bonding is a planar bonding of trigonal symmetry in such carbon substances as benzene, graphite and graphene. The sp^2 -bonding is schematically drawn in Fig. 4.3.

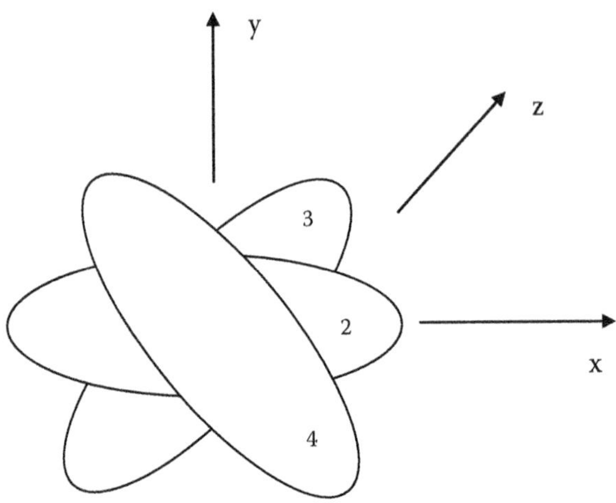

FIGURE 4.3 Planar sp^2 – bonding. The z-direction of "z" in out of the page: 2, 3, 4 are the linear combinations for the single bonds in-plane.

The additional p_z orbitals form a double bond or a π –bond. Using the bracket notation:

$$|1\rangle = |p_z\rangle \tag{4.45}$$

$$|2\rangle = \sqrt{1/3}|S\rangle + \sqrt{2/3}|p_x\rangle \tag{4.46}$$

$$|3\rangle = \sqrt{1/3}|S\rangle - \sqrt{1/6}|p_x\rangle + \sqrt{1/2}|p_y\rangle \tag{4.47}$$

$$|4\rangle = \sqrt{1/3}|S\rangle - \sqrt{1/6}|p_x\rangle - \sqrt{1/2}|p_y\rangle \tag{4.48}$$

It is worthwhile to linger on the benzene molecular-atomic structure since benzene can be regarded as the starting point of developing graphene structurally. Benzene, C_6H_6, has a hexagonal planar ring where each atom contributes three out of four valence electrons to sigma bonds and to one hydrogen atom of the carbon molecule. $2p_z$ valence electrons, positioned perpendicular to the carbon-carbon axis, create a π – bond located between the both carbon atoms. The created structure is planar because of the mentioned p_z electrons: orbitals must be parallel to form a π-bond. The sigma and π-bonds form a carbon-carbon double bond. The hexagonal benzene ring has six carbon atoms divided by equal distances. The C_6H_6 anomality allows one π –bond move from side to side. Similar ring arrangement happens continuously in organic compounds and, importantly for our discussion, it exists in graphite, graphene, and nanotubes.

Some significance for understanding of the graphene bonding has a triple covalent bond. *sp* hybrids form 180^0 angles among the bonds. Such a triple covalent bond is the basis of acetylene C_2H_2. Each carbon atom gives out one 2s electron to bond a hydrogen atom. The two carbon atoms bond each other by three 2s2p2 electrons. In total, six electrons are engaged in π and σ-bonds. The distances within carbon molecules are in the range of 100 – 150 *nm*. The atomic distances for the hydrogen are within 2-3 angstroms.

The transfer of electrons in closed benzene rings has been described as "aromacity"[4]. F. A. Kekule did his famous work on the structure of benzene. He suggested that the benzene structure contained a six-part ring of carbon atoms with alternating single and double bonds. The empirical formula for benzene had been known before Kekule but its saturated structure had not been unidentified. Kekule's proposed structure implied the number of isomers known for benzene derivatives. For each mono-derivative of benzene only one isomer existed, i.e. all six carbon atoms were equivalent.

The symmetry of the benzene rings is broken by substituted atoms and a single "Kekule" structure is formed. If, instead of a substituted benzene ring, we have a symmetric ring, the delocalization takes place: the π-electrons containing in a "box" with the typical carbon-carbon double-band spacing (L = 134 *pm*), the confinement energy for L is $h^2 / 8mL^2 = 20.9\,eV$. The delocalization is critical for understanding of the behavior of large bonding networks in graphite, graphene and carbon nanotubes. Planar arrays that benzene rings can form include graphene sheets, where there are no hydrogen atoms and every 2p electron in every atom is delocalized (there is one hydrogen atom for each carbon one).

The benzene ring anormality is featured by its diamagnetism. An electric current goes through covalent bonds. The diamagnetic effect is generation of a small counteracting field B' response to an applied magnetic field $\boldsymbol{B}$:

$$\vec{B}' = \chi_m \mu_0 B; \tag{4.49}$$

where μ_0 = the permeability of free space and χ_m = the magnetic susceptibility. The possibility of using the covalent bonds to conduct electric current is one of the applications for molecular electronics. Thus, we have one more important property of benzene rings. For conductance to occur, there are available P_z electrons in the benzene ring. Traditionally, the benzene ring is considered to be composed of linear combinations of the wave functions of the carbon's six atoms. A perpendicular magnetic $\boldsymbol{B}$ is applied to an electron circling a closed orbit. The diamagnetic susceptibility is a measure of how strong the influence of the field on the electron is. Let us consider the steps that lead to measuring of this influence and calculating the diamagnetic susceptibility. The physics of the diamagnetic effect are as follows[5].

For a charge q at a distance r (orbit radius), the encircling area is πr^2. The charge plane, a perpendicular magnetic field, $\vec{B}(t)$ is applied which intensity increases from zero. The change in the magnetic field causes an electric field E to appear (the Faraday's law). The resulting voltage in the loop is equal $d\Phi_m/dt$ or:

$$V = -\pi r^2 dB/dt = 2\pi r E; \tag{4.50}$$

From (4.50)
$$E = \frac{V}{2\pi r} = -\frac{rdB}{2dt}; \tag{4.51}$$

The angular momentum on the charge moving around a circular orbit changes with time constituting a force, $\tau = qrE$:

$$\tau = dL/dt; \tag{4.52}$$

$$\frac{dL}{dt} = -\frac{qr^2}{2}\frac{dB}{dt}; \tag{4.53}$$

After some time, Δt, the corresponding change in the angular momentum is ΔL:

$$\Delta L = -\frac{qr^2}{2}B; \tag{4.54}$$

The magnetic field changes also the magnetic momentum, $\Delta\mu$ which is $\mu = (q/2m)L$. Inserting (4.54) for ΔL:

$$\Delta\mu = \frac{\Delta L q}{2m} = -\frac{q^2 r^2}{4m}B; \tag{4.55}$$

For the fixed area of the orbiting charge, A, the current ΔI:

$$\Delta I = \frac{\Delta\mu}{\pi r^2}; \tag{4.56}$$

For Z electrons:

$$\Delta I = -\frac{Ze^2}{4\pi m}B; \tag{4.57}$$

Eq. (4.57) may be modified for the case of a spherical atom with an average r of $r_{av}^2 = x^2 + y^2 + z^2$, then μ becomes:

$$\mu = -\frac{Ze^2}{6m}r_{av}^2 B; \tag{4.58}$$

The current ΔI is on the order of nanoamperes. r_{av}^2 includes orbits with different orientations with respect to B. For N induced magnetic moments, the diamagnetic susceptibility is:

$$\chi_m = -\mu_0 N(Ze^2/6m)r_{av}^2; \tag{4.59}$$

where, by definition, $\chi_m = \mu_0 M/B$.
In a planar benzene ring, the angle θ between the z-axis and the normal to the plane of the ring, change from 0 to 180^0. If the area of the plane is A, the magnetic flux equals $BA\cos\theta$. For the susceptibility calculations:

$$\Delta\mu_z = -(Ze^2r^2/4m)B\cos^2\theta; \tag{4.60}$$

A somewhat simplified version of (4.60) for $\Delta\mu_z$;

$$\Delta\mu_z = -(Ze^2r^2/8m)B; \tag{4.61}$$

For N benzene rings per unit volume the magnetic moment per unit volume M, is (4.61) times N.

A careful study of diamagnetic susceptibility of benzene rings provides an experimental value for ring contribution to excess diamagnetism:

$$\chi_m = \frac{M}{B} = -N(Z_ne^2r^2/8m_e^*c^2); \tag{4.62}$$

This is for planar module, where m_e^*- effective mass of the electrons in the ring. For example, C_6H_6 has two π electrons with effective mass $m_e^*/m_e = 1.14$. The density distribution in the benzene ring is shown in Fig. 4.4. The radius of the carbon ring is 139.9 *nm*. The *C-H* bond is 110 *pm* long. The diamagnetic carbon susceptibility is 72.2 *pm*. To make sure that the current path lies within the radius, we can take half the length of the *C-H* bond which is 55 *pm* (110 pm/2). The screening current for two electrons from (4.56):

$$\Delta I = \Delta\mu/\pi r^2 = -Z(e^2/4\pi m)B; \tag{4.63}$$

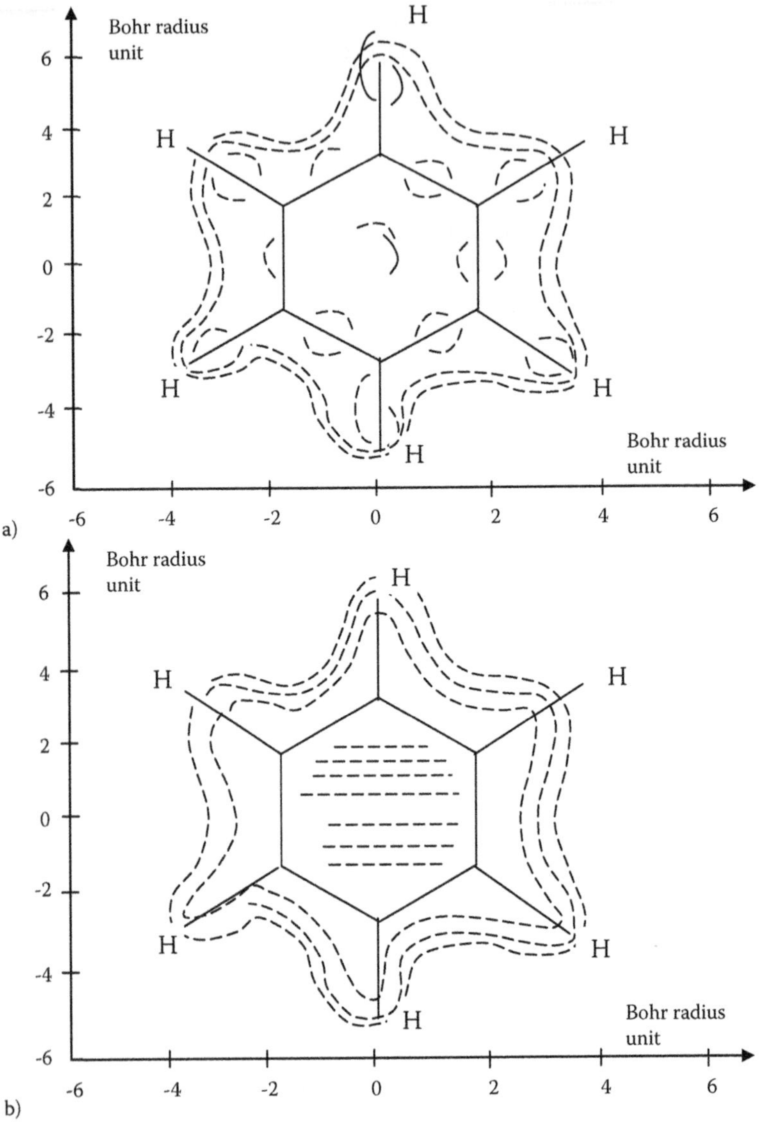

FIGURE 4.4 The induced current density distribution in the benzene ring a) Experimental current density distribution; b) Calculated current density distribution.

Calculations of the reduced benzene ring currents have been performed[7]. One is called "Hontree-Fock self-consistent-field methods". In Fig. 4.4, a magnetic field is applied perpendicular to the ring's plane. The magnetic lines in Fig. 4.4 show the diamagnetic response. Fig. 4.4 (a) shows a simulation of calculated in the benzene ring plane.

Z-direction is absent and paramagnetic currents circulate inside the ring. The main induced current is in the p_z. The calculated current distribution extends beyond the middle of *C-H* bonds. The whirls of line currents are seen around the midpoints of *C-C* and *C-H* bonds. It has distance of 170 *pm* along the direction perpendicular to a *C-C* bond[8]. A larger part of the screening current flows in the sigma bonds that results in the total diamagnetic screening current to the value of 11.8 A/T for benzene. This idea that the trigonal sigma-bond network contributes to the magnetic susceptibility is opposed by the fact that in graphene only p_z electrons are necessary for an accurate description of diamagnetic susceptibility and electrical conduction.

The Schrodinger equation treatment of the bonding electrons in the benzene ring confines the motion to the *x-y* plane. The azimuthal angle, φ, is the only variable. The potential energy, although dependable on the angle φ, may be considered as constant for simplicity.

The Schrodinger equation for the electron in the ring:

$$\frac{\hbar^2}{2mr^2}\delta^2\psi/\delta\varphi^2 + U(\varphi)\psi = E(\psi); \tag{4.64}$$

The solution should be in the form:

$$\psi = A\exp(im\varphi); \tag{4.65}$$

where $m' = 0, \pm 1, \pm 2...$

Using the solution for (4.64):

$$\frac{\hbar^2}{2mr^2}m'^2 = E - U; \tag{4.66}$$

where $E > U$. For $U = 0$, then

$$E = m'^2 E_1; \tag{4.67}$$

where $E_1 = \dfrac{\hbar}{2mr^2} \approx 1.05eV$ and $m' = 0, \pm 1, \pm 2, ...$

The state $m' = 0$ is occupied by two electrons (one spin up and one spin down). If a magnetic field is turned on, the two delocalized states associated with two electrons can develop a rotation.

The solution for (4.67) are $\exp(im\varphi)$ and $\exp(-im\varphi)$. Each state in the solution may be occupied by two electrons with spins up and down ("up" for one state and "down" for the other). Another solution for (4.67) is $A\exp[im(\varphi - \varphi_0)]$ that has the same energy for φ_0.

$A\cos[m(\varphi - \varphi_0)]$ and $B\sin[m(\varphi - \varphi_0)]$ are the linear combinations of the above solutions (i.e. $\exp(im\varphi)$ and $\exp(-im\varphi)$. The corresponding probability distributions of the solutions are $\cos^2(m\varphi)$ and $\sin^2(m\varphi)$. The potential energy $U(\varphi)$ from (4.64) represents the interaction of the ions. The six ions are positioned by angles $\Delta\varphi = 60^0$. The six-fold potential has a Fourier form of $U_6 = -U_6\cos^2(3\varphi)$ with one of the ions positioned at $\varphi = 0$. The electrons experience bonding with the probability distribution of $\cos^2(3\varphi)$ (bonding) and $\sin^2(3\varphi)$ (antibonding).

Two equivalent dimerized states (each state consists of three dimers) leave the other three states (having three double bonds) with a wider spaced carbon on the other three positions. Under these conditions (with each carbon atom dimerizing) the corresponding potential is then:

$$U_3 = -U_{03}\cos^2(3\varphi/2); \tag{4.68}$$

A solution to (4.68) appears in the form:

$$\psi = A\cos(3\varphi/2) \pm iB\sin(3\varphi/2); \tag{4.69}$$

There are two wavefunctions and $\varphi = 0$ is at a dimer position. The origin φ is shifted in the middle of the distance between two carbon atoms by the angle of 30^0. A "dimer" is oligomer that consists of two monomers similar in structure. A dimer can be weak, strong, covalent or intermolecular. Oligomer is term that describes a molecular complex that consists of a few monomers. Oligomer is a counterpart of a polymer, for which the number of monomers (as constituents) is not limited.

The probability distribution is:

$$P(\varphi) = A^2\cos^2(3\varphi/2) + B^2\sin^2(3\varphi/2); \tag{4.70}$$

The distribution function's first part relates to atoms *1 – 2, 3 – 4,* and *5 – 6* (the extrema are at *0^0*, *120^0*, and *240^0*) and the second part corresponds to atoms *2 – 3, 4 – 5,* and *6 – 1* (the extrema are *60^0*, *180^0*, and *300^0*). The dimer location shifts one bond length, i.e. from one side of the atom to another. The distribution function (4.70) assumes that the four electrons are uniformly distributed between the corresponding atoms. The uniformity of double bonds distribution is also implied, in which case $A = B$ in (4.70). The above wavefuctions suggest four available electrons for two Kekule states which are represented by (4.70). These four electrons do not influence the magnetic susceptibility. These electrons belong to the resonating bond that does not have the angular momentum ±1. The momentum is cancelled by the linear combination from $\sin(k_x)$ and $\cos(k_x)$.

In addition to the model described above, there are others that help with the understanding of the diamagnetic susceptibility and chemical changes caused by the remaining two free electrons in addition to the bond assumed to be equal in the benzene ring. One of the alternative theories holds that six π – electrons occupy three bonding states having two electrons each and one nodal plane for each electron. The states are called *"HOMO"* (*Highest Occupied Molecular Orbital*). A perpendicular magnetic field is reportedly screened by π – electron clouds one on each side of the lowest model plane[9]. The absence of electron circulation in the benzene ring the *HOMO* model is explained by the presence of two degenerate states with a higher energy and nodal planes that intersect the nodal planes at the right angle. The other two *HOMO* states are reported to distribute the charge at different angles φ which contradicts to the Kekule states that are located at *120^0* angle. The contradiction may not be as crucial as it seems since the molecular arrangement does not seem to be unique. Thus, going back to our earlier discourse, a linear combination of Kekule states is as good a choice as are any other existing theories.

On a more fundamental level, a set of linear combinations of atomic orbitals with the $2p_z$ atomic orbital function as a basis we can write[1]:

$$\Phi_\mu(\vec{r}) = 6^{-1/2}\Sigma_{(j=1\ to\ j=6)}\exp(i\pi\mu j/3)\cdot u_{2pz}(\vec{r}-\vec{R}_j); \tag{4.71}$$

where $\mu = 0, \pm 1, \pm 2, 3$.
The suggested approximations are given for the states $\mu = \pm 1$:

$$E_\mu = \langle \Phi_\mu | H_e | \Phi_\mu \rangle = \alpha + 2\beta\cos(\pi\mu/3); \tag{4.72}$$

It is implied that the probability density is $\Phi^*_\mu(\vec{r})\Phi_\mu(\vec{r})$ is invariable with rotation by any multiple of the angle of 60^0. The lowest state ($\mu = 0$) corresponds to $m' = 0$ in (4.67). The coefficients in (3.57) are approximately, $\alpha = -7\,eV$and $\beta = -24\,eV$. Since they are both negative, the lowest state where the wavefunctions add in phase is $\mu = 0$. The two degenerate states are $E = \alpha + 2\beta\cos(\pi/3)$ at $\mu = \pm 1$ with $E = -31$ eV or 24 eV. The states are occupied when $\mu = 0, 1, -1$; and $\mu = 0$ has circulation of charge with an applied magnetic field that provides the diamagnetic screening. The probability densities for $\mu = \pm 1$ may be connected by the following equation:

$$\psi_\pm = 2^{-1/2}[\Phi_1(\vec{r}) \pm \vec{i}\,\Phi_{-1}(\vec{r})]; \tag{4.73}$$

4.4.4 Molecular Carbon Variations: Fullerene C_{60}, Graphene and Fluorographene

The Fullerene C_{60} molecule may be presented as a sheet of graphene that is installed into a sphere. Being a close relative of graphene, fullerene C_{60}'s structure deserves our attention. The molecules comprising the material are spheres containing 60 atoms of carbon and arranged in five and six-fold rings. There are 32 rings total that have 10 hexagons and 12 pentagons. Incidentally, the name *"Fullerene"* was adopted from the architect R. Buckminster Fuller, whose icosahedral (a polyhedron with 20 faces) geodesic domes resemble the Fullerene molecular structure. Elser and Haddon suggested that the bonds of the molecule have resonance ("hopping") integrals of approximately 2.44 eV[10]. Each hexagonal benzene ring's neighbors are three hexagonal and three pentagonal rings.

Fullerene molecules exist in nature and spontaneously form at high temperatures in the media containing free atoms of carbon and no oxygen. Crystalline graphene ($(CH)_N$ or $(C_6H_6)_N$) can be made from an exfoliated graphene crystal in an atmosphere of cold discharge of atomic hydrogen. The bonds change from sp^2 to sp^3 (Fig. 4.5).

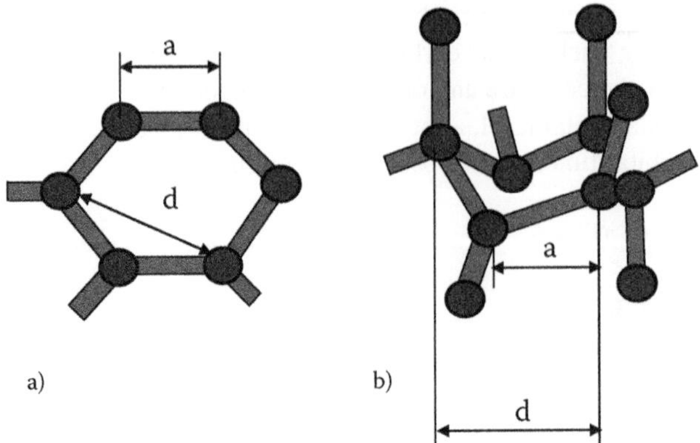

FIGURE 4.5 Comparison of graphene a) and graphane b) molecule arrangement.

Figure 4.5 illustrates a transformation from metallic to semiconductor bonds. Graphane is a two-dimensional polymer that consists of carbon and hydrogen. The formula of the molecule is $(CH)_n$ and n is large. Graphane usually alternates the hydrogen atoms in A and B sublattices sides, we receive a diamond-like structure. The graphane lattice constant, a = 0.244 *nm* (graphene a = 0.242 *nm*). The crystalline structure of graphene may be easily transformed into that of graphene by the process of annealing in the atmosphere of Ar at 450^{0}C for 24 hours. Another version of graphene was suggested in 2010[11]. Fluorographene is a derivative of graphene which is a fluorocarbon in chemical terms. It is a two-dimensional carbon structure that consists of sp^3 hybridized carbons where every carbon atom is bound with one fluorine atom. The chemical formula, $(CF)_n$, is close to that of Teflon (chemical formula $(CF_2)_n$) which has carbon chains: each carbon is connected with two fluorines. In contrast, graphene is unsaturated (sp^2 hybridized). Fluographene has a hexagonal lattice with one fluorine at each carbon atom and lattice constant of 0.248 *nm*. Fluographene's mechanical properties are similar to those of graphene. The bandgap of 3 eV modes fluorographene makes it a good isolator. Fluorine (F) is a chemical element which is the lightest halogen with the atomic number 9. Fluorine is the most electronegative element and is extremely reactive forming bonds almost with all other elements with the exception of noble gases. Fluorographene can be produced from graphite monofluoride $(CF)_n$ that has layers of weakly band stacks of fluorographene. Its most stable formation consists of an infinite array of trans-linked cyclohexane chains that have covalent $C-F$ bonds in an AB-stacking sequence.

4.4.5 Formation of Graphite

Graphite is formed in the earth as the end result of several steps from coal that includes, among others, bituminous and anthracite. The high pressure makes graphite a denser material than coal: e.g. the density of graphite 2.27 g/cc versus 1.3 – 1.4 g/cc for anthracite. In addition to pressure, heating plays a significant role in forming of graphite. During the process of carbonization, the temperature of higher than 2000^{0}C releases hydrogen (with no ambient oxygen). As a result a planar graphite structure is produced. Graphite may exist in nature as thin pieces of < 0.1 mm thick and several millimeters long[12]. In general, graphite is mined for industrial applications. A higher temperature of 3000^{0}C is used to make graphite electrodes for furnaces using electric discharge for steel production.

In addition to conventional mixing methods, there have been new synthetic techniques developed during the past several decades. One of them provides pieces (flakes) of graphite mainly for scientific purposes, a by-product of steel making, the so-called "kish graphite" which is a eutectic mixture of iron and carbon with the 4.3% content of graphite that exists at T = 1147^{0}C. The dimensions are bigger than with the already described graphite flakes, with one possible slope of a circle with a diameter of approximately 1 *mm*[12].

Synthetically, graphite has been produced from carbonaceous materials at high temperatures and pressure. The process is called *HOPG* (Highly Oriented Pyrolytic Graphite) which means that pyrolytic graphite is received by thermal decomposition of hydrocarbon gas on a heated substrate. Pressure is also applied in order to improve the quality. The subsequent annealing under compression gives *HOPG*. The typical temperatures range from 2800^{0}C to 3500^{0}C and pressures are in the range of 4000 to 5000 *psi*. Pyrolytic graphite (carbon) is a material which is close to graphite that has covalent bonding between the layers as a result of defects in its production. The typical production process includes heating of hydrocarbon almost to its temperature of decomposition and then permitting the graphite to crystallize. The angular misalignment of the crystal is improved by annealing of the graphite at temperature of 3300^{0}C. As a result, we have a specimen about 1 *mm* long (0.1 *μm* in c-direction). Graphite in general has a lamellar structure, i.e. a microstructure that is composed of thin, alternating layers of various materials which exist in the form of lamellae. Similar to other layered materials, it consists of stacked planes. The forces within the lateral planes are much stronger than between the planes. Because of this, *HOPG* cleaves like mica. In an atomic resolution scanning tunneling microscopy there are several typical images: one is a close-packed array where each atom is surrounded by six nearest neighbors. The distance between them is 0.246 *nm*. The hexagonal rings have the center to center distance of 0.1415 *nm* (see Fig. 4.6).

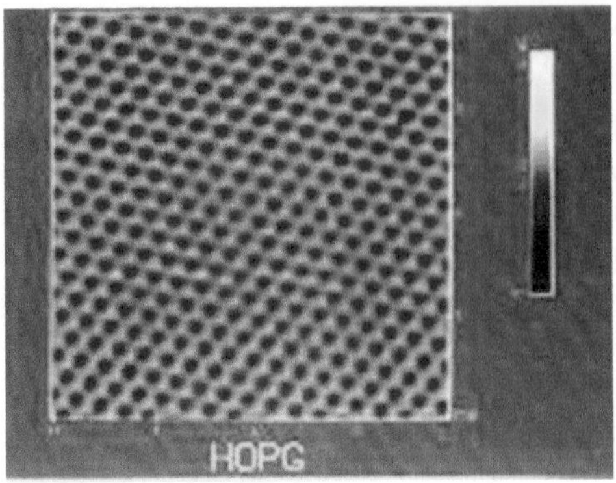

FIGURE 4.6 The hexagonal structure of HOPG.

The image in Fig. 4.6 is a close-packed array obtained from the basal plane of HOPG which has an *AB* packing of adjacent layers (Fig. 4.7).

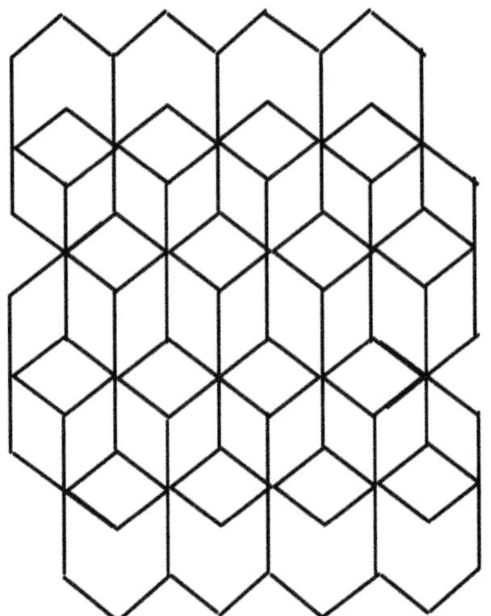

FIGURE 4.7 Graphite *AB* stacking.

The rotation of stacking planes stands for the graphite lubricating feature. It is also a source of stacking faults. The layers become incommensurate destroying the crystalline order of graphite and making it grease-like. The *HOPG* thickness can reach 40 layers with a lateral grain size up to 10 *mm*. Graphite sheets are used as materials that can stand high temperatures, suitable for *PC* heat sinks, for example, for moderators in nuclear reactors.

REFERENCES

1. Bransden, B. H., Joachain, C. J., *"Physics of atoms and molecules"*, Prentice Hall (2003)
2. Lide, D.R. (ed), *"Handbook of chemistry and physics"* 84 ed. CRC (2004)
3. Pauling, L., Wilson, E. B., *"Introduction to quantum mechanics"*, McGraw-Hill Book Company (1935)
4. Von Schlayer, P.R., and Jiao, H., "What is aromaticity", *Pure and Applied Chemistry,* **68**, Iss. 2 pp. 209-218
5. Feynman, R., Leighton, R., Sands, M., "The Feynman Lectures on Physics", Caltech (1964)
6. Jiao, H., and von Schlayer, P.R., "Aromaticity of pericyclic reaction transition structures: magnetic evidence", *Journal of Physical Organic Chemistry,* Vol. 11, pp. 655 – 662 (1998)
7. Stamenov, P., Coey, J.M.D. "Magnetic susceptibility of carbon-experiment and theory", *Journal of Magnetism and Magnetic Materials,* Vol. 290-291, Part 1, pp. 279-285 (2005)
8. Fliegl, H., Sundholm, D., Taubert, S., Juselius, J., Klopper, W., "Magnetically induced current densities in aromatic, antiaromatic, homoaromatic, and nonaromatic hydrocarbons", *J. Phys. Chem.* A **113**, pp. 8668-8676 (2009)
9. Carey, F., Giuliano, R., *"Organic Chemistry",* 8th ed. McGraw-Hill (2011)
10. Elser, V., Haddon, R.C., "Icosahedral C_{60}: an aromatic molecule with a vanishingly small ring current magnetic susceptibility", *Nature,* **325**, pp. 792-794 (2002)
11. Robinson, J.T., Burgess, J.S., Junkemeier, C.E., Badescu, S.C., Reinecke, T.L., Perkin, F.K., Zalalutdniov, M.K., Baldwin, J.W., Culbertson, J.C., Sheehan, P.E., and Snow E.S., "Properties of fluorinated graphene films", *Nanoletters* **10** (8), pp. 3001 – 3005 (2010)
12. Dresselhaus, M.S., Dresselhaus, G., Eklund, P.C., *"Science of fullerenes and carbon nanotubes".* Academic, San Diego (1996)

5 Properties of Electrons in Graphene

The Schrodinger's equations were used to develop the band theory of semiconductors. Graphite's (the graphene's precursor) electronic band structure was worked out to demonstrate the capabilities of solid state physics[1]. More recently, the Dirac approach (utilizing the symmetry of graphene) has explained (at least to some extent) the graphene's unique electronic properties, such as high electron mobility among others.

5.1 BASIC ELECTRON BANDS OF GRAPHENE

The electron bands in graphene are based on the honeycomb lattice (Fig. 5.1).

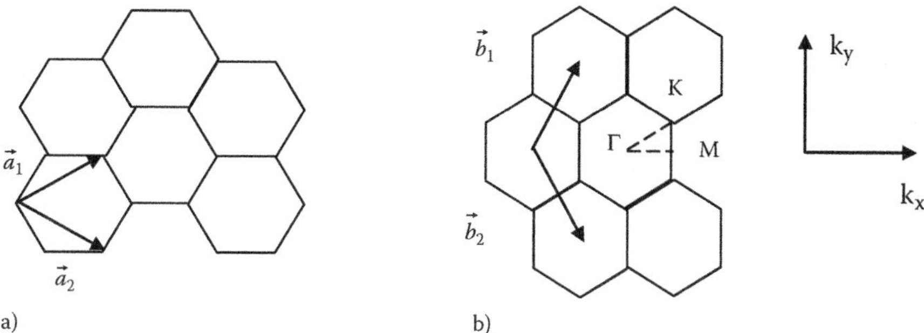

FIGURE 5.1 Schematic representation of graphene crystal structure: a) the translation by unit vectors $\vec{a}_1$ and $\vec{a}_2$ by which the whole crystal can be formed; b) the first Brillouin zone for graphene (the center in Γ).

The tight-binding theory indicates that the energy bands are formed by the $2p_z$ electrons. In graphene, the linearity of $E(k)$ close to E_F gives the carrier speed and the effective carrier mass close to zero. The distance between the graphite planes is 0.337 *nm* while the distance in the hexagonal structure is 142 *pm* which is less than half the inter plane distance. The fact that can be sufficient to neglect the inter plane interaction. Thus, graphene may be treated as a single hexagonal layer crystal. Three out of four valence electrons are involved in sp^2 trigonal bonding. The fourth electron, on the other hand, is a conduction electron. The Bloch wavefunction is given[1]:

$$\psi_k(r) = N^{-1}\sum_j \exp(i\vec{k}\cdot\vec{r})\varphi(\vec{r}-\vec{r}_j); \tag{5.1}$$

The Hamiltonian for the tight-binding model:

$$H = -t\sum(a_i^* b_j + b_i^* a_j) - t'\sum[a_i^* a_j + b_i^* b_j) + H.C.]; \tag{5.2}$$

where $H.C.$ is Hamiltonian conjugate and $a_i^* a_j$, $b_i^* b_j$ for $t' \sim 0.1\ eV$.

The interactions ("hops") of $t \sim 2.8\ eV$ are included in the form $a_i^* b_j$. When electrons move from lattice to lattice, the term $a_i^* b_j$ annihilates an electron on site j on the lattice B while creating an electron on site I of the lattice A.

$$\langle \vec{k}|H|\vec{k}\rangle = \sum_m \exp(-i\vec{k}\cdot\vec{\rho}_m)\int dV\varphi^*(\vec{r}-\vec{\rho}_m)H\varphi(\vec{r}); \tag{5.3}$$

Eq. (5.3) is determined by finding its diagonal elements.

Please note: $\vec{\rho}_m = \vec{r}_m - r_j$ in Eq. (5.3).

For the nearest sites to be separated by ρ such that there are two coefficients α and t, we have the first order energy for the bond:

$$\langle \vec{k}|H|\vec{k}\rangle = -\alpha - t\sum_m \exp(-i\vec{k}\cdot\rho_m) = E_k; \tag{5.4}$$

where $\alpha = -\int dV\varphi^*(\vec{r})H\varphi(r)$ and $t = -\int dV\varphi^*(\vec{r}-\vec{\rho})H\varphi(\vec{r})$;

The energy necessary to "hop" from one nearest neighbor to another t, is:

$$t(R_y) = 2(1+\rho/a_0)\exp(-\rho/a_0); \tag{5.5}$$

Eq. (5.5) is based on two hydrogen atom calculations and expressed in Rydberg units, $R_y = me^4/2\hbar^2 \approx 13.6\ eV$.

The energy t exponentially diminishes with distance ρ. In Fig. 5.1 b) for the honeycomb lattice[1]:

$$E_k = E_{\pm}(\vec{k}) = \pm t[3+f(\vec{k})]^{1/2} - t'f(\vec{k}); \tag{5.6}$$

where the indices of energy stand: "-" for the lower (π) one-electron band and "+" stands for the upper (π^*) one-electron band associated with $2p_z$ state for carbonite *f(k)* is calculated as:

$$f(\vec{k}) = 2\cos(\sqrt{3k_y}a) + 4\cos(\sqrt{3k_y}a/2)\cos(3k_x a/2); \tag{5.7}$$

where a is separation between carbon atoms (1.42 A) and $\sqrt{3a}$ is the lattice constant.

For the first Brillouin zone, the lattice constant is connected with the wavevector k:

$$-\pi/a < k < \pi/a; \tag{5.8}$$

The wavevectors are pointed from the center of the first Brillouin zone, Γ. The honeycomb lattice of graphene may be achieved by vector translations $\vec{a}_1$ and $\vec{a}_2$ (Fig. 5.1). The points in the first Brillouin zone, Γ, K and M represent important directions for electron motion in the graphene lattice. The point K is where the conduction and valence bands come together. In case, when a graphene layer is shifted so that the inherent inversion symmetry is not valid, $\vec{K}' = -\vec{K}$. In other words, the point K does not coincide with the point $-K$ and is denoted K' to underline the difference. The full energy dispersion is for crossing points $\vec{K}$ and $\vec{K}'$(dispersion is expanded for $\vec{k} = \vec{K} + \vec{q}$):

$$E_{\pm}(q) \approx 3t' \pm \upsilon_F(q) - [9t'a^2/4 \pm 3ta^2/8\sin(3\theta_q)|q|^2]; \tag{5.9}$$

where for $\hbar = 1$, $\upsilon_F = 3ta/2$ and $\theta_q = \tan^{-1}(q_x/q_y)$; t = the lopping energy and t' is the overlap energy (offset). If t' is negligible then:

$$E_{\pm}(q) \cong \pm\upsilon_F |q| + O(q/K)^2; \tag{5.10}$$

The band structure surfaces are conical for the energy distribution along k-vectors. One of the outcomes of the above conical structure is that the electron cyclotron mass close to the Dirac points in graphene is anomalous.

$$m^* = (2\pi)^{-1}[\delta A(E)/\delta E] \text{ at } \mathrm{E} = \mathrm{E_F} \tag{5.11}$$

The area of conical energy distribution in q – space:

$$A(E) = \pi q(E)^2 = \pi E^2/E_F^2; \tag{5.12}$$

Thus, the effective mass is:

$$m^* = E_F/\upsilon_F^2 = k_F/\upsilon_F; \tag{5.13}$$

Using the electronic density $n = k_F^2/\pi$, we have:

$$m^* = (\sqrt{\pi}/\upsilon_F)\sqrt{n}; \tag{5.14}$$

The effective mass is zero at the Dirac points. The density of the state close to the Dirac point is:

$$g(E) = 2A_c|E|/\pi\upsilon_F^2; \tag{5.15}$$

where $A_c = 3\sqrt{3}a^2/2$.

5.1.1 Dual-Lattice Aspect

Taking into account the dual-lattice, we can write in the Dirac form in the vicinity of $\vec{K}$ and $\vec{K}'$:

$$\Psi(\vec{k},\vec{r}) = f_1(\vec{\mathrm{K}})\exp(i\vec{\mathrm{K}}\cdot\vec{r})\Psi_1^S(\vec{\mathrm{K}},\vec{r}) + f_2(\vec{\mathrm{K}})\exp(i\mathrm{K}\cdot\vec{r})\Psi_2^S(\vec{\mathrm{K}}\cdot\vec{r}); \tag{5.16}$$

where $\vec{k} = \vec{\mathrm{K}} + \vec{\kappa}$ and the effective mass theory is applied including $\vec{k}\cdot\vec{p}$ function[2].

Taking (5.2) and taking the Fourier transform of the operators a^*a around the Brillouin zone points $\vec{K}$, $\vec{K}'$, we have:

$$a_n = (N_c)^{-1/2} \sum \exp(-i\vec{k} \cdot \vec{R}_n) a(\vec{k}); \tag{5.17}$$

where R_n – lattice locations, N_c – number of cells[3].

Now, the operators become:

$$a_n = \exp(-i\vec{K} \cdot \vec{R}_n) a_{1,n} + \exp(-i\vec{K}^1 \cdot \vec{R}_n) a_{2,n}; \tag{5.18}$$

$$b_n = \exp(-i\vec{K} \cdot \vec{R}_n) b_{1,n} + \exp(-i\vec{K}^1 \cdot \vec{R}_n) b_{2,n}; \tag{5.19}$$

The subscripts *1* and *2* refer to the points $\vec{K}$ and $\vec{K}'$. The next step is to substitute a_n and b_n into (5.2) which gives:

$$H = -i\hbar \upsilon_F \iint dxdy[\Psi_1^*(\vec{r})\vec{\sigma} \cdot grad\Psi_1(\vec{r}) + \Psi_2^*(\vec{r})\sigma^* \cdot grad\Psi_2(\vec{r})]; \tag{5.20}$$

where $\upsilon_F = 3ta / 2\hbar e$, $\vec{\sigma} = (\sigma_x, -\sigma_y)$ are Pauli matrices, and $\Psi_i^*(\vec{r}) = (a_i^*, b_i^*)$, $i = (1, 2)$.

Eq. (5.20) is only valid close to $\vec{K}'$.

A similar approach to the Hamiltonian-Dirac formalism may be found in earlier works Keldysh described as "Landau-Zener" tunneling that takes place between the electron bands. Semenoff and Mele in 1984 also contributed theoretically. However, only the experimental studies by Novoselov et al have classified the above-mentioned theories. The presence of *A* and *B* sublattices, in particular, gives coupling of orbital motions.

From the chemical point of view, graphene may be described as a molecule with alternative double bonds (Fig. 5.2). This approach dates back to 1930's and emphasizes bonds among carbon atoms with their four valence electrons per atom to form bonds with each carbon atom's three neighbors. This approach, however, does not explain the graphene dual lattice symmetry that does not allow carrier backscattering.

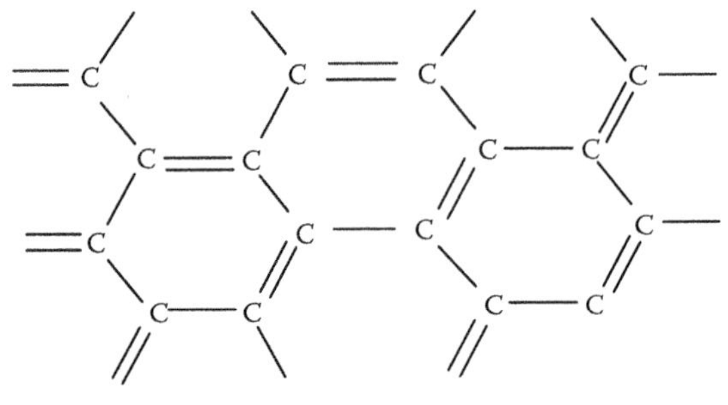

FIGURE 5.2 Graphene structure in its chemical-bond representation.

5.2 BILAYER GRAPHENE

Bilayer graphene can be easily extracted and can be a basis for electronic devices. Bilayer graphene has two layers stacked in the $A - B$ Bernal stacking described earlier. The production process of the bilayer graphene involves extracting a layer from a single graphite crystal or its production by epitaxial or catalytic growth. The process of graphene growth uses SiC single crystal (which is bH polytype with (0001)-orientation) as the substrate. SiC in vacuum is heated to T = 1400^{0}C so that silicon atoms are removed. As a result, we have a layer of carbon on the same crystal lattice substrate - large epitaxial domains of bilayer graphene[4]. The electron mobility may achieve 2.75 m^2/Vs at room temperature and the current goes up to 1 nA per a carbon atom. For practically available specimen of the dimension on the order of 10 mm, the current may be approximately half an Ampere.

Applying voltage between two graphene sheets, a bandgap of about 200 meV is formed by shifting the volume and conduction bands. Thus, a semiconductor rather than a semimetal is built. Potassium atoms were deposited into the graphene bilayer. Potassium provided its single valence electron for the necessary free carrier concentration. The crystal structure of bilayer graphene is given in Fig. 5.3. In Fig. 5.3 a) a unit cell of its lattice structure, in particular, is depicted. In Fig. 5.3 b) the graphene bilayer has two conical surfaces that represent energy in the conduction and valence bands. The applied voltage creates an energy gap of 200 meV. The applied voltage increases the carrier concentration in bilayer graphene. For example, the calculations show that values up to 120 meV are achieved as the carrier density reaches 10^{13} cm^2. Si has an energy gap of approximately 1 eV which corresponds to a local density of about 1 x 10^{11} cm^{-2}.

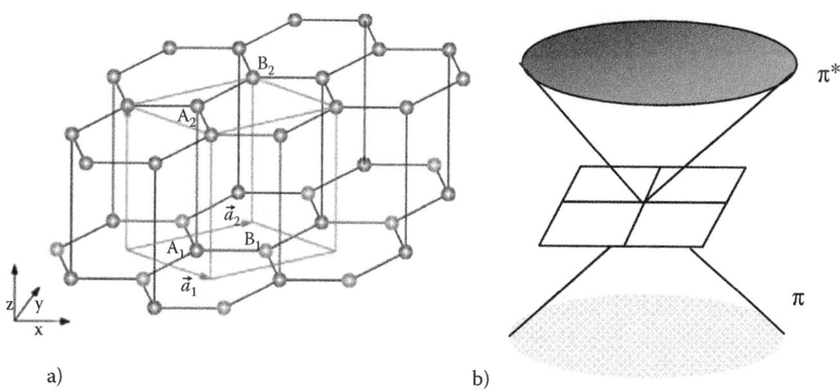

FIGURE 5.3 a) Crystal structure of bilayer graphene. Voltage is applied to the layers: t (the hopping integral) is ~ 3.1 eV (within a layer), and $t_p \approx 0.22$ eV (between layers), b) The energy bands near K. The parabolic bands are characteristic for a semimetal[4].

The gap can be introduced if the symmetry between the graphene layers is shifted (see Fig. 5.4)[5]. It is a possible application for semiconductor devices with variable parameters which is advantageous for switching transistor modes. In particular, a bigger bandgap increases the voltage difference between "on" and "off" states. At lower carrier density, the density of states has four minima. A different band structure means that parabolic minima are replaced by minima with linear dispersion. It may be surmised that a high mobility that ensues is linked with the features of the above band structure differences. Currently, the carrier mobility of 10^6 cm^2/Vs. Graphene samples are cleaved and etched in buffered hydrofluoric acid[7]. The doping is ~ 10^8 cm^2. Silicon wafers are used as substrates. Constant current of about 1 mA/μm is used to heat the bilayer graphene. This process sets a high resistance value if the applied voltage is zero. The conductivity increases almost sinusoidally and proportionally to the carrier density, N (Fig. 5.5).

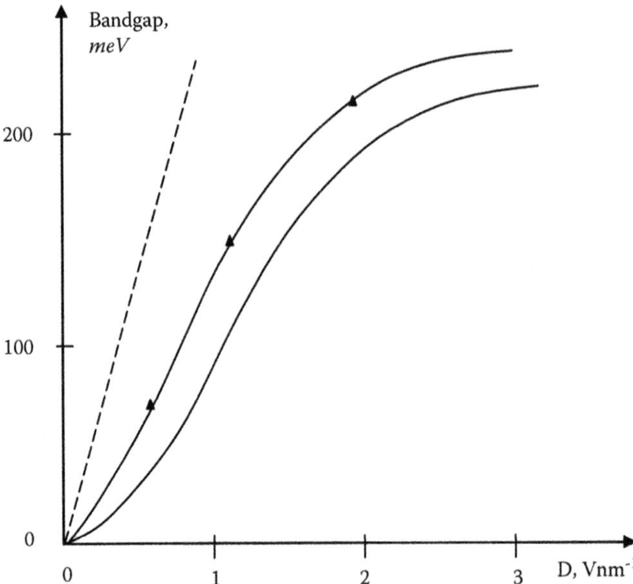

FIGURE 5.4 Energy-gap dependence in graphene on electric field. The upper solid line represents theoretical prediction (self-consistent tight binding). The lower solid line is ab initio density functional (DFT) (based on tight-binding calculations). The dashed line represents unscrewed tight-binding calculations[6].

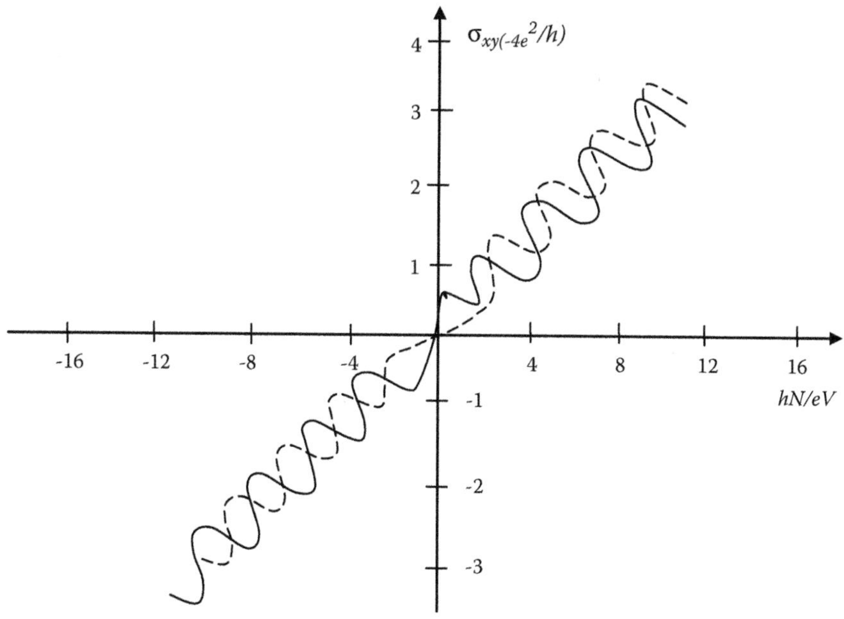

FIGURE 5.5 Dependence of conductivity on carrier density. The solid line corresponds to a bilayer and the dashed line stands for a mono layer graphene.

REFERENCES

1. Wallace, P.R. "The Band Theory of Graphene", *Physical Review*, **71** (9): 622 (1947)
2. DiVincenzo, D.P., Mele, E.J., "Self-consistent effective-mass theory for intralayer screening in graphite intercalation compounds", *Phys. Rev.* B, Vol. 29, Iss. 4 (1984)
3. Castro Neto, A.H., Guinea, F., Peres, N.M.R., Novoselov, K.S., and Geim, A.K., "The electronic properties of graphene", *Rev. Mod. Phys.,* **81** (2009)
4. Ohta, T., Bostwick, Seyller, T., Horn, K., "Controlling the electronic structure of bilayer graphene", *Science*, **18**, Vol. 313, pp. 951 – 954 (2006)
5. McCann, E., "Asymmetry gap in the electronic band structure of bilayer graphene", *Phys. Rev.* B **74** (2006)
6. Zhang, Y., Tang, T., Girit, C., Hao, Z., Martin, M.C., Zettl, A., Crommie, M., Ron Shen, Y. and Wang, F., "Direct observation of a widely tunable bandgap in bilayer graphene", *Nature* **459** pp. 820-823 (2009)
7. Mayorov, A.S., Elias, D.C., Mucha-Kruczynski, Gorbachev, R.V., Tudorovskiy, T., Zhukov, A., Morozov, S.V., Katsnelson, M.I., Fal'ko, V.I., Geim, A.K., Novoselov, K.S., "Interaction-driven spectrum reconstruction in bilayer graphene", *Science*, **333** (2011)

6 Producing Graphene
Methods and Sources

The basic method of obtaining graphene suggested by Geim and Novoselov includes mono- and a few-layer samples. The usual procedure is cleaving crystalline graphite, exfoliation, or a direct synthesis by vapor deposition from gaseous carbon (e.g. methane or acetylene). The state-of-the art method of obtaining graphene has been suggested by the Nobel laureates, Geim and Novoselov in 2011 which implied micromechanical cleaving producing separate graphene plates. This simple approach has off-set more traditional techniques, such as graphite exfoliation resulting in low quality samples. Another advantage of mechanical cleaving is that a high temperature is not necessary. While the mechanical cleaving may be done at room temperature, exfoliation and HOPG (highly-oriented pyrolytic graphite) requires temperatures of several thousand degrees. Kish graphite is made at $T = 1100^0C$ and HOPG at 3000 K (*Kish* is a mixture of graphite and slag).

Early attempts of mechanical cleaving produced not less than 10 layer-samples while Novoselov et al succeeded in making one layer graphene sheets. The top layer of a graphite crystal is removed by the means of an adhesive type and then placed on a substrate. It is important that the adhesion force between the substrate and the graphene layer is stronger than between the graphene layers. In this case, a layer of graphene can be detached from the graphene stack (crystal), thus providing a quality specimen. The graphene specimen characterization may be carried out by the means of usual optical microscopy[1]. In particular, Novoselov et al found that optical contrast may be helpful in locating a single graphene layer if the layer has a certain substrate (e.g. Si/SiO_2 where the silicon oxide is 300 nm thick). Another substrate thickness may be 100 nm[2]. Other substrate (such as h – BN (hexagonal boron nitride)) did not offer the advantage of locating a single graphene layer as its thickness could not be determined by the above optical means[3]. However, Novoselov et al have suggested further improvements of their single graphene layer methods to produce single layer compounds such as BN and some others, such as MoS_2, $Bi_2Sr_2CaCu_2O_{5+\delta}$ (BSCCO), and $NbSe_2$. Further investigations of substrates used for graphene monolayers led to finding of fluctuations of the Fermi energy ("puddles") that affected locally the carrier mobility. As it was indicated earlier, h-BN substrate is superior to Si/SiO_2 where the density of impurities is higher. It has been noticed that the quality of substrate surface also substantially influences the monolayer's quality (such as mechanical roughness).

The influence of a substrate on the graphene monolayer is illustrated in Fig. 6.1. Making measurement similar to those shown in Fig. 6.1 helps in choosing the most suitable substrate. In particular, roughness from Si comes from its irregularities and trapped charge in the amorphous of the grown oxide. Different methods, including annealing can be used to improve the substrate quality.

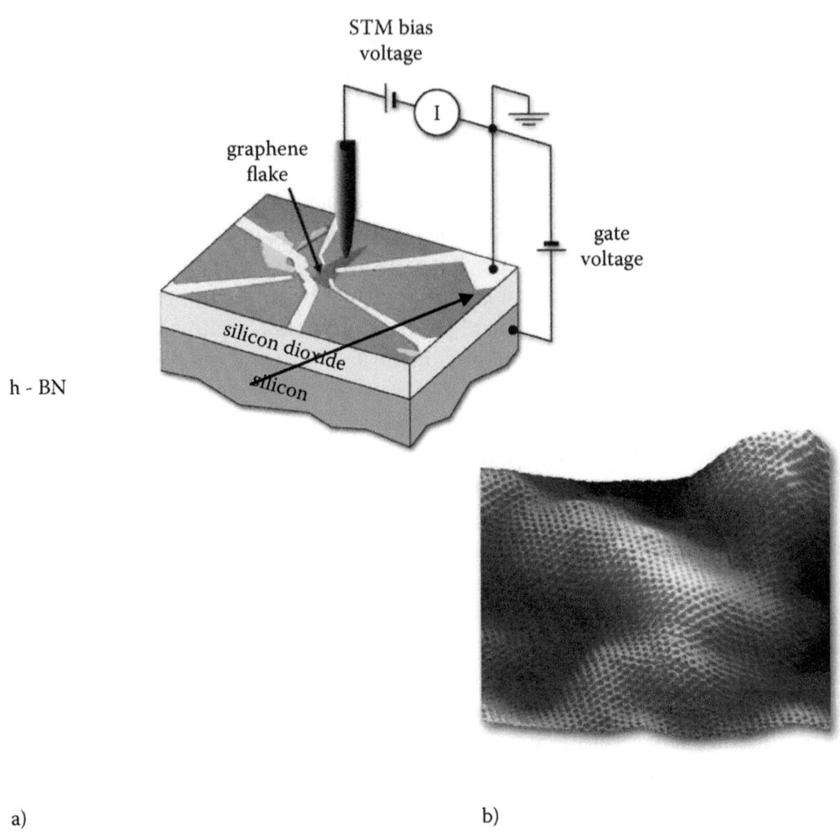

FIGURE 6.1 a) A graphene monolayer on a substrate; b) Surface non-uniformity because of $h - BN$ surface corrugations. *(Continued)*

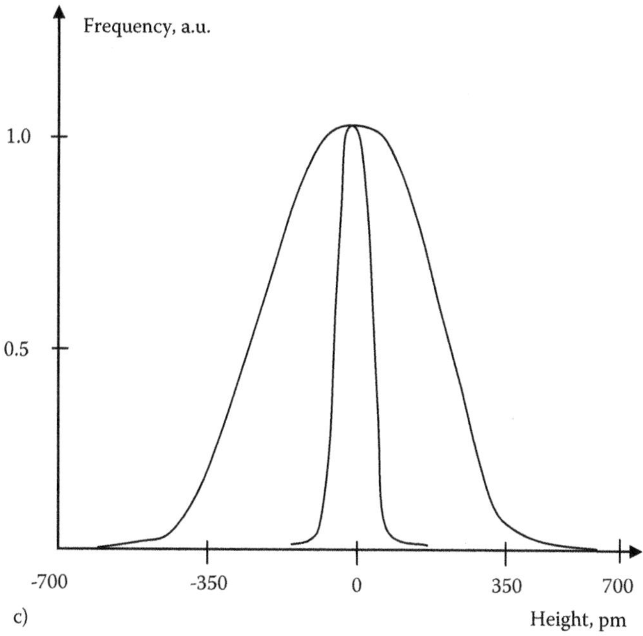

FIGURE 6.1 (Continued) c) Graphene height distribution on SiO_2 substrate (the outer curve) on h – BN (the inner curve).

6.1 CHEMICAL METHODS OF PRODUCING GRAPHENE LAYERS

Chemical methods of making sheets, flakes of platelets of different thickness out of graphite have existed for a long time. A number of such methods have been developed comparatively recently: individual graphene layers are released from highly oriented pyrolytic graphite (HOPG)[4]. Schematically, the process is shown in Fig. 6.2. The process includes: intercalation (i.e. a reversible inclusion of a molecule into layered structures) of graphite with potassium (KC8) and graphite heating in the presence of *K*. Potassium Graphite is otherwise known as graphite intercalation compound (GIC) in which potassium atoms are located between layers of graphitic carbon. Potassium Graphite is a very strong reducing agent used as a catalyst. The next step is growth of linear polymers in the presence of styrene or butadiene vapors. The last step depicts sheets of graphene. The growing of linear polymer in Step 3 forces the graphite planes to separate. The subsequent heating of elastic polymer at temperatures greater than 400^{0}C results in producing completely exfoliated graphic carbon. Similar work was performed by B.C. Brodie in 1859[5]. The graphite was subjected to acids. The resulting graphite oxide was called "graphone". Graphone is a new material that adds magnetism to the graphene properties. The electrical and structural properties are retained, for the most part. Since magnetism and electronic spin are closely related, graphone has new qualities and paves the way to applications in spintronics. This is a low quality graphene according to Novoselov, et al (2011)[6]. The authors regard the above method as one not providing high quality graphene sheets since the graphite under consideration is intercalated with oxygen and hydroxyl groups.

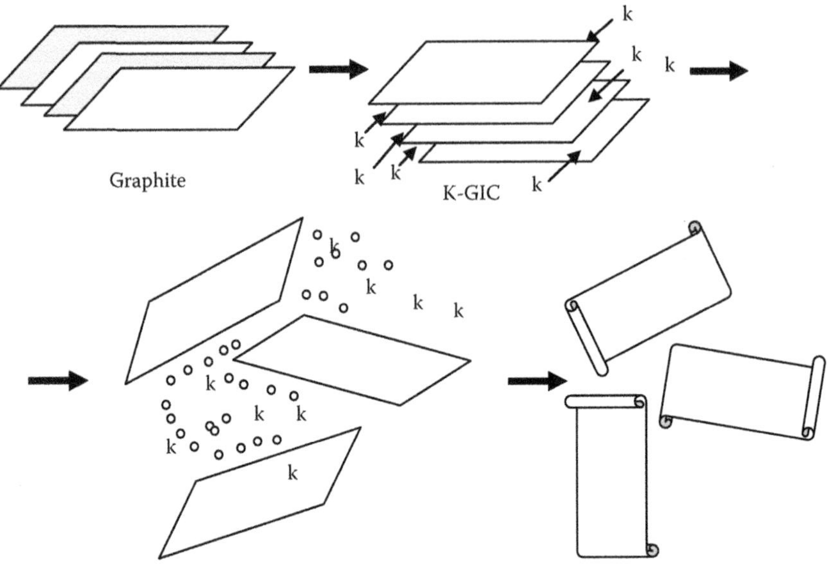

FIGURE 6.2 Chemical process of graphene exfoliation[4].

As such, the groups imply an easy water penetration into the specimens. Measurements confirm that mobility in chemically processed flakes is two to three orders of magnitude lower than in mechanically processed (exfoliated) graphene flakes. The chemical approach is, however, more practical than the mechanical one. Thus, there is a future for the chemical techniques for e.g. solar cells, LEDs and some other applications where an enhanced electron mobility is important (such as transparent conductors)[7].

Another alternative for producing single graphene sheets was suggested by Schniepp et al[8]. The suggested method belongs to the chemical graphite processing. The authors described the received single graphene sheets as "functionalized". The structural defects such as defects caused by oxidizing and reduction reactors and level structural defects are characteristic of the process. The final product includes also specific chemical defects, such as impurities (e.g. L – O – L (epoxy)). C – OH groups may exist between the graphene planes and at the edges – C – OH and - COOH[8].

First, the graphene flakes were placed in an oxidizing solution of nitric, sulfuric acids and potassium chlorate for 96 hours, the time which is necessary to remove the interplannar space characteristic of graphite (0.34 nm wide) and to change the space to 0.654 – 0.75 nm which what the solid graphite has. Thermal exfoliation that uses preheated environment (T = 1050^{0}C) goes to a much higher temperature at a speed of more than 2000^{0}C per minute releasing single graphene sheets with the forced release of CO_2 gas. A careful study of the graphene sheets, however, gives evidence of defects both at the edges as well as at the base of the specimens. The defects are easily identified with Atomic Force Microscopy (AFM) and are obstacles for high quality of electronic devices but may be useful additives for composite materials.

6.1.1 Bulk Exfoliation

Bulk exfoliation of graphene has been one of the effective, low-cost methods for graphene nano-sheet (suspensions) that have superior electrical conductivity. Mixed colloidal suspensions exist for reduced graphene oxide and layered metal oxide nano-sheets. In order to reduce the number of defects in the graphene sheet, the exfoliation starts at the edges. The graphite is milled to produce edge-carbonated graphite (ECG) sheets, the size of which is in the range of 100 – 500 nm[9]. This process gives better quality graphene samples than by the means of chemical exfoliation since only outer surfaces are affected – the internal hexagonal structure remains intact. The carboxylated edges make the graphene samples dispersible in water which separates the graphene sheets in the sample. Subsequently, the edges are removed by heating the sample. The goal is to obtain transparent high conductivity films on substrates (e.g. on Si) for series production. The ECG process is depicted schematically in Fig. 6.3.

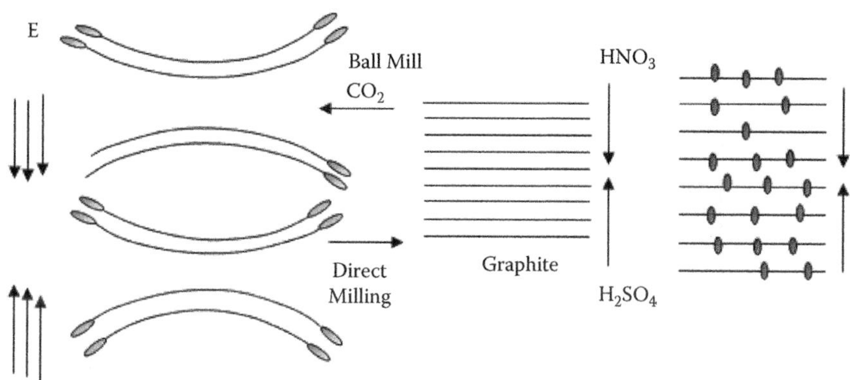

FIGURE 6.3 ECG (Edge – Corboxylated Graphite) process.

In Fig. 6.3 graphite oxide (GO) is depicted on the right-hand side. Direct annealing gives pure graphite. Exfoliation, in contrast, is applied to GO, producing graphene with plane defects. Exfoliated graphene has been produced on a large scale and its quality has been improved along the way[10].

Another method of exfoliation is used for "expandable graphite"[11]. The exfoliation of graphite takes place at $T = 1000^0C$ for 60 sec in the atmosphere of argon + 3% hydrogen. The heating causes intercalants formation of volatile gaseous species. As a result, graphite expands and we receive a few layers of graphene sheets. The thermal exfoliation allows to control its parameters in order to reduce the number of graphene sheets to one or two. Subsequently, exfoliated graphene is submerged in a 1,2 dichloroethane (DCE) of poly (*m*-phenylenevinylene-co-2.5-dioctoxy-*p*-phenylevinylene). 30-minute sonication produces a homogeneous suspension. Sonication is used to weaken the interlayer stresses. A further centrifugation removes redundant large pieces leaving planes and ribbons of graphene. The PmPV polymer is used to make assist the sonication process to stabilize the process. Later, the PmPV is removed by calcination at 400⁰C. Obviously, the polymer binds the graphene by van der Waals forces preventing graphene from dispersing in the organic solvent. The sonication time may be optimized in order to control the process. In particular, to make nanoribbons the sonication time is reduced. AFM (Atomic Force Microscopy) may be employed for the product characterization that allows seeing ribbons and graphene sheets.

A possible application for nanoribbons is for field-effect transistors[11]. An oxidized $p^{++}Si$ wafer served as the back-gate of the FET. The ribbons with widths of 10 – 55 nm were placed on the wafer. The source and drain were formed by Pd (palladium) contacts that were attached to the ribbons. The described design of the FET allows widening of the difference between "OFF" and "ON" states at the ribbon width of less than 10 nm. The data is given in Fig. 6.4.

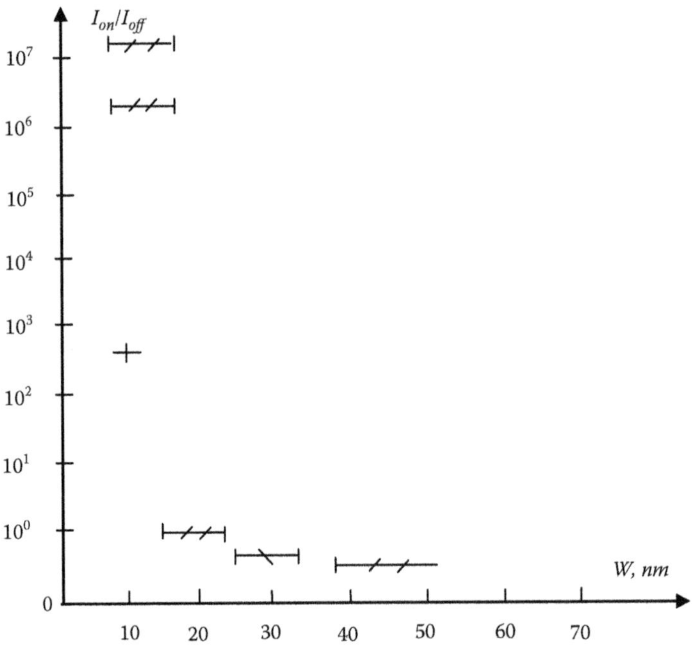

FIGURE 6.4 The range of achievable rations of ON/OFF states with respect to the ribbon width.

Empirically, the band gap in the nanoribbon depends on the ribbon width, as E_g, ≈ 0.8 eV. Nanotubes form a p – type semiconductor with the palladium contacts that act like a Schottky barrier since the Pd work function is high. Taking this into account, the ON/OFF ratio becomes:

$$I_{ON} / I_{OFF} \sim exp\,(E_g / k_B T); \tag{6.1}$$

It has been reported that the energy gaps are caused by electron confinement that is restricted by the width W and specific details of the edges of the nanoribbon's configuration.

Experimentally, it has been found that graphene nanoribbons (GNR) always have a band gap. Nanotubes have fewer defects than nanoribbons which may suggest making nanotubes out of nanoribbons that, in its turn, may lead to better defined configurations of better quality[12]. Solution-produced nanoribbons give hole mobility on the order of 200 cm^2/Vs. The fact that the nanoribbons always have bandgaps makes them well-suited for electronic devices.

An alternative to chemical exfoliation has been reported by Hernandez et al[13]. It is liquid phase exfoliation which implies a direct dispersion of graphite. Fig. 6.5 shows the usage of solvent N – methylpyrrolidone (NMP) for liquid-phase exfoliation. The counts of dispersion show the amount of dispersion vs. the number of layers. The exfoliation implies solving of powered graphite in the 0.01 mg/ml solvent, NMP.

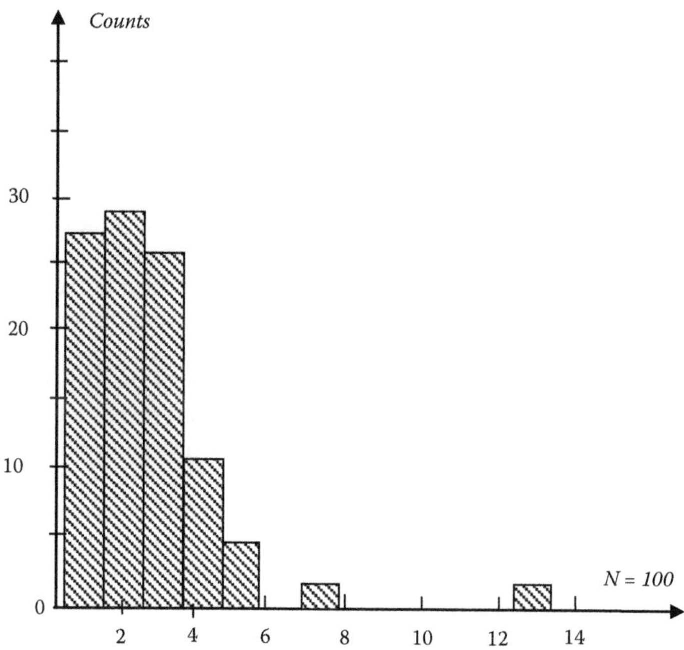

FIGURE 6.5 Counts of dispersion vs number of layers per sheet. Direct liquid phase exfoliation with solvent NMP characterized by TEM[13].

The dispersed graphene was deposited on a carbon grid and then dried. The subsequent TEM characterization showed that there were no oxidation defects. About 70% of the yield specimen contained 3 or less graphene layers (see Fig. 6.5).

6.2 EPITAXIAL METHODS OF PRODUCING GRAPHENE LAYERS

An important direction of graphene technology development is creation of a manufacturable process, i.e. a process that leads to series or large-scale (mass) production. Micromechanical cleavage, for example, is not a manufacturable process. An epitaxial growth has a potential of becoming such a process. The present manufacturable processes are chemical vapor deposition, atomic layer deposition and photolithography-based processes. A classical substrate for epitaxial growth is SiC and metals (such as Ni or Cu)[14]. The disadvantage is the absence of crystalline compatibility between graphene and the substrate. An ultra- high vacuum growth of graphite carbon on Si (111) surface by e-beam deposition was reported[15]. First, a thin amorphous carbon layer was grown on a silicon surface at 560^{0}C and then graphitic carbon was grown at T = 830^{0}C. It is advantageous that SiC is insulating since electrical current cannot flow through the substrate. The disadvantage lies in the inferior quality of graphene grown on SiC substrates. The hexagonal substrate BN described earlier (see Fig. 6.1) yields fewer defects and provides a higher carrier mobility for micro-mechanically cleaved graphite but, on the other hand, methods of direct deposition on BN substrates are still to be developed.

6.2.1 Carbide Substrates for Epitaxial Growth

SiC remains the most widespread substrate for graphene growing. The carbon crystallography for the substrate surface is given as [000(-1)] and the silicon's surface is "0001" for hexagonal "polytypes" 4H and 6H. Rotation along the covalent bond, during the growth of SiC generates alternate stacking layers of Si and C. The stacking has a different structural sequence which may be identified in the $(11\underline{2}0)$ plane, however, not along the basal plane. The constant growth energy is not constant and two or more SiC polytypes may form at the same time. The process is called "polytypism". One of the methods to detect polytypes, is Raman spectroscopy. SiC can be α or β –type. The α-type has a hexagonal and the β-type has a cubic symmetry. The latter are 4 and 6 hexagonally stacked bilayers (per unit cell). The SiC can be considered a composite material that consists of SiC layers, and there are two types of the layer stacking: cubic and hexagonal. The SiC compound retains the covalent bonding of both Si and C. The epitaxial process using heating has been known for more than four decades[16]: SiC is heated in vacuum and a graphene layer is formed out of C atoms while Si atoms leave the SiC sample's surface. A growth of a bilayer was reported and characterized by ARPES (Angle-Resolved Photoemission Spectroscopy)[17]. Another characterization method is X-ray (SXRD) that involves diffraction from the sample surface. The present graphene growing process produces specimens size of a wafer at atmospheric pressure[18]. One of the challenges is an elimination of steps of SiC surface. The steps change the resistance of the epitaxially grown layer resulting in a lower quality.

One of the characterization methods employs Scanning Tunneling Potentiometry. The method provides the dependence of the local potential upon the induced current flow (Fig. 6.6).

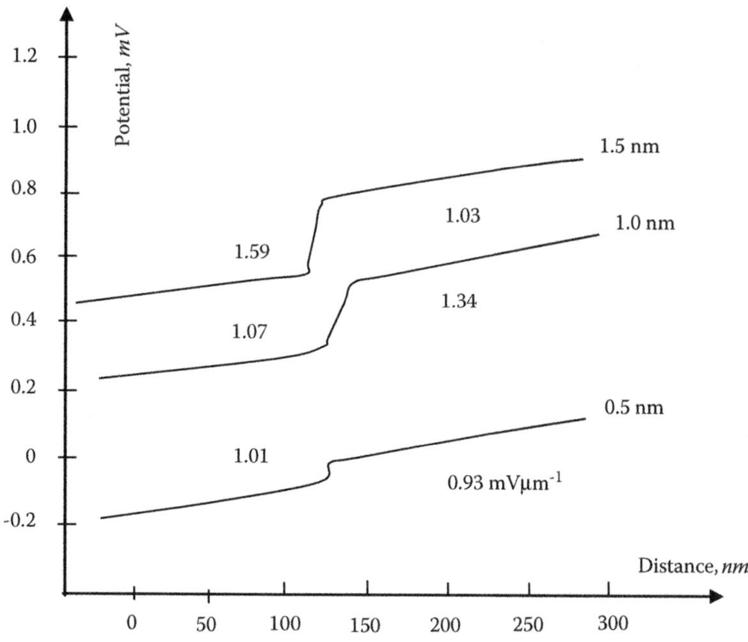

FIGURE 6.6 The identified steps on a single graphene layer on SiC. The steps have heights of 0.5 nm, 1.0 nm, and 1.5 nm (from the bottom to the top curves). The two numbers before and after the slope give the change of the measured voltage vs the height increase[14].

The measurements in Fig. 6.6 were done at T = 72 K. The carrier mobility in the graphene layer was estimated as 0.3 m^2/Vs and carrier density as 10^{13} cm^{-2} for the given specimen[14]. Every step adds to the total resistance across a sample. More specifically, the "step-resistance effect" gives changes in the carrier density of graphene of graphene layers. This effect contributes to the lower electron mobility for epitaxial films on SiC with respect to mechanically exfoliated graphene which is placed on oxidized Si. One of the hypothesis suggests that the mobility reduction is due to the electrical coupling between the graphene layer and the substrate. The resistance varies in the vicinity of the steps due to the variations in coupling. SiC substrate is supposedly dopes the graphene[19].

6.2.2 Graphene Growth from Carbon-Containing Gases

Graphene can be grown from carbon-containing gases on metal substrates that can catalyze the growth. Cu, for one, is a weakly binding metal while Ru is strong enough to provide epitaxial growth of a layer. On the other hand, the restricted solubility of graphene in copper makes it possible to produce only single layers of graphene, a very much desirable result. Low-pressure CVD on Cu foil reportedly gave single crystal domain, up to 0.5 mm with large graphene grains only weakly attached to the substrate[20]. However, the gain boundaries are not desirable and a work is being done to improve the CVD accordingly. Polymethyl methacylate (PMMA) membrane may be detached from the substrate with graphite or graphene sheets attached to it.

Subsequently, PMMA may be released from the graphene sheets and then dissolved in acetone. At the moment graphene monolayers can be grown on transition metal substrates, such as Cu, Ni, Ru, Co, Ir, Pt. Transition metals, like all metals, are ductile and malleable. They conduct heat and electricity. Transition metal's feature is the fact that their valence electrons are present in more than one shell. And that is why these metals exhibit several common oxidation states. Methane, as a carbon source, is used to deposit atoms of carbon (CVD) on copper foil (25 μm thick) at T = 1000⁰C[21]. A layer of graphene was grown in 30 minutes. The self-limiting procedures ensures that only 5% of the grown surface has more than one graphene layer. No discontinuity defects were reported in the graphene grown on a copper substrate with mobility of 0.4 m^2/Vs. The mobility may decrease somewhat (to about 0.3 m^2/Vs) with larger films (as large as 4 x 4 inch)[22].

The above techniques to produce large-area graphene are promising for advanced electronic devices. The mechanism of growth on different substrates is different: e.g. graphene layer growth on Ni takes place by segregation of Cu by surface roughness. The physical properties of SiC and SiO_2/Si are different and, thus, the growth on SiC substrates may be useful only for SiO_2 devices since SiO_2/Si have not been used so far. Growth of graphene layers on polycrystalline Ni and Cu substrates by Chemical Vapor Deposition (CVD) is advantageous for large-area continuous films. The graphene layers on Nifoil/film have still a wide variation of thickness over the surface. Recently, it has been shown that Cu is going to be an excellent material for making large-area uniform thickness since C is not dissolved substantially in Cu[22].

Large 4 x 4 inch film layers on Cu (later transferred to oxidized silicon) were reported. The carrier mobility was 0.3 m^2/Vs with quantum Hall effect. Similar results were received on Ni deposited on a silicon wafer. Graphene was grown by CVD process. Co substrate was found to be suitable as well[23].

There are number of other alternative methods of graphene growth have been proposed in the last decade that include a substrate-free plasma enhanced (PECVD) growth and PECVD with substrates applied to carbon nanostructures. Some of the techniques rely on Cu and Ni. In particular, large-area graphene layers (single or multilayers) may be deposited on Ni substrates by radio-frequency plasma-enhanced CVD (RF-PECVD) at T = 650⁰C (which is comparatively low)[24]. During the RF – PECVD process CH_4 gas at 2 – 8 sccm (sccm means standard cubic centimeter per minute at STP). Carbon atoms from the CH_4 diffuse into the Ni film. Then methane (CH_4) segregates at the *Ni* surface. The number of graphene layers may be increased with the methane gas flow. The large area of graphene layers and low cost of the CVD process is favorable for graphene manufacturing. The application of the CVD is also advantageous because of the lower cost of the process in comparison to large-area graphene epitaxially grown on single crystal SiC. SiC substrates are expensive and graphene exfoliation from SiC to another substrate poses substantial difficulties.

Moving toward large-scale graphene monolayer production, a roll-to-roll process has been described in a number of publications. CVD growth on *Cu* foil is a well-elaborated technique presently available. The foil is wrapped around a cylinder that serves as a substrate which is placed in a CVD reactor. The cylinder is a 7.5 inch in diameter quartz tube on which the copper foil is rolled around. The tube length is 39 inches with the foil of up to 30 inches in the diagonal direction. The copper foil is preliminary annealed in H_2 at $T = 1000^0C$ in order to increase the size of the copper grain from several to a hundred micrometers. Methane is added after the annealing is finished. The gas flows for 30 minutes at 460 mTorr (Torr is a unit of pressure defined as a fraction of a standard atmosphere (pressure). One Torr is 101,325/760 pascals (~133.3) PA. The flow rate is 24 cc/min for methane and 8 cc/min for hydrogen. Afterwards the furnace is cooled in hydrogen at the rate of $10^0C/s$ and pressure of 90 mTorr. At the next step, the copper foil with a graphene layer on it is attached to a "thermal release tape". The tape is a thin polymer sheet. The attachment is achieved by pressing the tape between two running soft rollers at about 2 MPa pressure[25]. The next step is copper removal. The tape that consists now of graphene on polymer on a copper sheet, is pulled through a basin thus removing the copper. After that the graphene layer is attached to an appropriate substrate. The usual substrate choice is 188 μm-thick polyethylene terephthalate (PET). Then, the graphene layer on the substrate is subjected to a pull between rollers at T ~ 100^0C. The procedure may be repeated to receive a more complicated structure (usually four graphene layers)[25]. The manufactured graphene layer may be doped and has $\rho = 30\,\Omega/\,square$. The chemical doping is done with nitric acid HNO_3 that creates a strong $p-$type conductor. This flexible transparent (~ 90% optical transmission) conductor has improved characteristics in comparison with, e.g. indium tin oxide (ITO) and carbon nano-tubes[26]. One of the applications of the above transparent conductors is touch - screens of superior flexibility for electronic devices.

The flexible conductors have been improved by reduction of their resistivity to the values of around $\rho = 10\,\Omega\,/square$. The transmission, however, has decreased to 85%. The resistance was lower by stacking four monolayers (an increased number of layers) and by chemical doping[25]. The resistance could be decreased even further (e.g. to $\rho = 8.8\,\Omega\,/square$) but the transparence also decreases in this case. The monolayers were intercalated with $FeCl_3$ to obtain the above results[27]. This method of achieving lower resistivity, however, is not for series production since it is based on mechanically exfoliated graphene.

6.2.3 Quality Control of CVD Graphene

Practical electronic applications of graphene bring out the necessity of its quality control. One of the concerns is electrical effects of grain boundaries. The work has been done on clarifying what those effects are and how to control them. The CVD process produces polycrystalline graphene. The surface grains range from 1 to 50 μm for the fast and the low growth rates respectively. The grain effect is shown in Fig. 6.7.

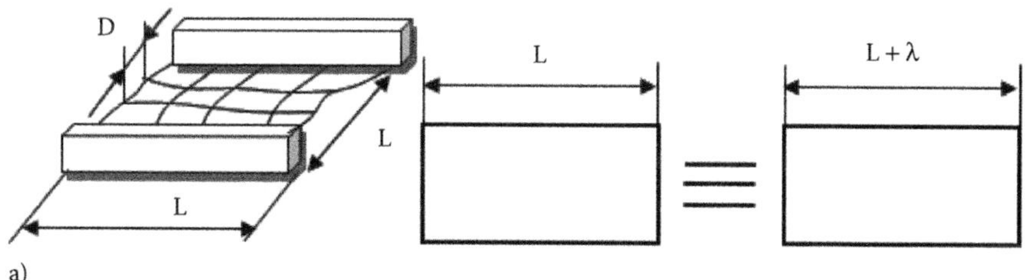

FIGURE 6.7 a) Grain boundaries dimensions. *(Continued)*

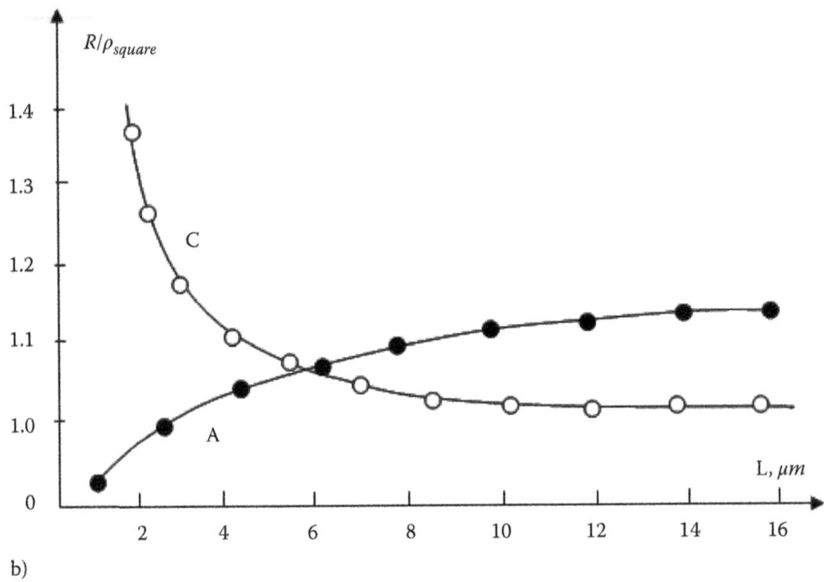

FIGURE 6.7 (Continued) b) Dependence of device resistance vs device size. The "A" – curve is a fast method that gives smaller grain size and "C" – curve represents a slow process with large grain size[28].

The grain size for the fast process, $D \approx 1\,\mu m$ and the grain size for the slow process, $D \approx 50\,\mu m$. The smaller grain size implies a better contact among grains. The parameter λ is associated with the grain size and contact (Fig. 6.7 a)). The influence of the grain structure is modelled by introducing the parameter λ or $L + \lambda$ (Fig. 6.7 a)). λ increases with the presence of defects that separate grains. The empirical variation of resistivity vs the film dimension was measured with or without a grain boundary (Fig. 6.7 b)): $R = (\rho / square)(L / W)$, where L and W are the film dimensions. R' is the change of R with the presence of a grain boundary:

$$R' = (\rho / square)(L / W) + \rho_{GB}(L / W) = (\rho / square)(L + \lambda)^{29}; \tag{6.2}$$

$$\text{where } \lambda = \rho_{GB} / (\rho / square); \tag{6.3}$$

λ ranges from 200 nm to 1.8 μm. The deviation of λ may exceed the dimensions of an electronic device itself. In this case, the following expression takes into account the grain size D as $n = L/D$ and

$$R = (\rho / square)(L + n\lambda) / W; \tag{6.4}$$

For large n, the above expression reduces to:

$$R = (\rho / square)(1 + \lambda / D); \tag{6.5}$$

The grain size, therefore, determines the minimal size of a device. At the linear density of $2\times10^{7}/cm$, the critical defect size is about 2 nm. Practically, in order for defects not to influence a device performance, the device size should not exceed 5 μm per side. Overlapping of grains during the growth of a layer may improve the quality.

One of the concrete improving methods to reduce the grain negative effect on the grown graphene properties was to selectively oxidize the copper substrate by UV illumination producing O and OH radicals[30]. The boundaries were defined by continuous exposure that widens the copper regions. These boundaries determined the graphene boundaries. The subsequent annealing helped restore the graphene, low resistivity. The oxidation process required moist atmosphere assisted by UV illumination. An increase in temperature could decrease the graphene sheet resistance, e.g. at T = 1060^{0}C, the graphene size was 72 μm^2. The size-resistance was[30]:

$$R=(\rho/square)_0[1+(A/A_c)^{-n}]; \tag{6.6}$$

where A = average grain size; n = 4 and A_c – fitting parameter. Empirically, $(\rho/square)_0 = 230\,\Omega/square$. The higher than the ideal resistivity apparently is a result of defects. Also, the boundary conditions of the substrate contribute to scattering and higher resistivity. The above study suggests a resistivity of $57.5\,\Omega/square$ for a four-layer stack that was heavily doped with HNO_3 making it n – type[25]. The discussed grain boundaries cause a lower mobility than in mechanically-cleaved graphene. Growing graphene without grain boundaries may be a major success. One of the theoretical studies considers different Cu surface orientation. As the typical surface orientation is (100), an alternative may be (111)[31]. The Cu surface has, in this, a hexagonal symmetry that matches the graphene's symmetry of a benzene ring. Cu (111) may be produced from the Cu evaporation. At the beginning of the growth, defects are more numerous. They cannot be rectified with islands enlargement, the process that still keeps the grain boundaries. The benzene ring symmetry may be stabilized by adding Mn to the Cu. Mn – Cu (111) surface is thus formed and the hexagonal symmetry maintained[32].

6.2.4 Chemically Modified Graphene and Graphene-Like Structures

Boron-nitride and graphene have similar crystal structures. The crystal constants for both are almost the same. Boron-nitride (BN) is, however, almost an insulator unlike graphene. The bandgap of BN, nevertheless, can be different being in the form of $(BN)_n(C_2)_m$, where n, m = 1, 2, 3.... (Fig. 6.8).

The grown structure is not homogeneous, though. There are NB_3 and BN_3 groupings. The bandgaps differ from one grouping to another. There is no uniform composition $(BN)_n(C_2)_m$. The manufacturing process of producing BCN films by CVD is established although with the above non-uniformity drawback[33]. One particular process involves a CVD grown h – BNC film. A gas mixture of equally mixed B and N (methane) and NH_3 – BH_3 (ammonia borane). The carbon ingredient is between 10 and 100 %. Several layers of h – BCN are deposited and may be transferred from Cu to other substrates. Lithography can be used to produce different patterns in oxygen plasma. All of the above means the feasibility of a manufacturing of electronic devices.

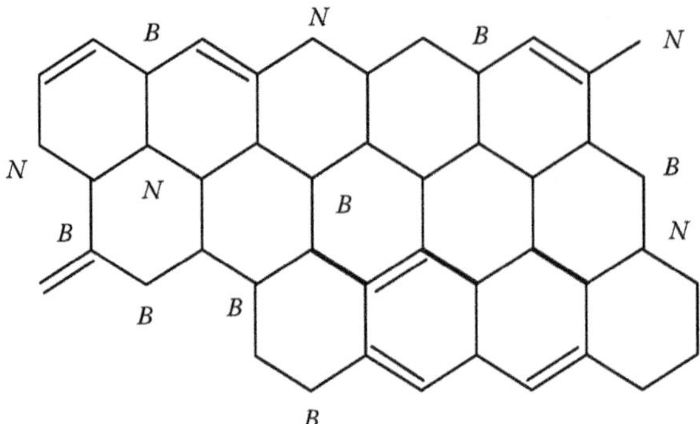

FIGURE 6.8 *BCN* crystal structure. The unspecified crystal vertices have *C* atoms.

Chemically separated reactions, such as *A B* binary reaction may be in the form of Eq.(6.7) and Eq.(6.8):

$$HfCl4 + 2H_2O \rightarrow HfO2 + 4HCl \tag{6.7}$$

$$2Al(CH_3)_3 + 3H_2O \rightarrow 3Al_2O_3 + 6CH_4 \tag{6.8}$$

The oxide HfO_2 has a high kappa; $\kappa = 20 \div 22$ as opposed to $\kappa = 3.9$ for conventional insulator SiO_2[34]. The usage of HfO_2 prevents the gate current leakage which has to be so thin that the unavoidable leakage occurs. The gate insulator has to have a substantial capacitance to introduce into the conductance channel to provide an adequate density of carriers when the bias is applied. The species A are grown on the substrate so that only one monolayer is produced. Then, the species B are applied to the layer A and species B react with available A sites, forming one epitaxial layer. The compounds in the form of A_nB_m, band on self-limiting surface chemistry, help solving problems connected with the evaporated metal film breaking into separate regions that restrict miniaturization of electronic devices. Attempts to eliminate these irregularities are made to achieve that exceptional carrier mobility that one-layer graphene has. The quality of electronic devices requires perfection of graphene layers, their continuity that minimizes scattering. The CVD method uses inert chamber atmosphere as opposed to the molecular beam epitaxy (MBE)'s high vacuum. The described process is supposed to surpass the conventional MBE in order to produce high quality graphene layers.

The oxide layers may be deposited also by atomic layer deposition (ALD). In particular, sapphire (Al_2O_3) is deposited first by evaporating and oxidizing a thin layer of Al. ALD is also used to grow HfO_2, mentioned earlier. Field effect transistors and spintronic devices are among the most popular applications of graphene technology. These applications require very thin dielectric layers for gates and tunnel barriers. ALD is one of the promising methods to produce this kind of dielectrics. The practical application of ALD, in this case, is a very complex task since there are no reactive adsorption sites, at least, not on the perfect graphene surface.

The more popular CVD does not give the necessary uniformity. In particular, the boundary regions contain graphene flakes close to step edges. CVD also has basal plane defects and impurities from the lift-off procedure that can further contaminate the edges of the graphene layer. The described drawbacks are exacerbated by very low sticking probability of defect-free areas. As a result, direct deposition of Al_2O_3 oxide is difficult given selective growth of the dielectric.

6.3 GRAPHENE NANORIBBONS

The physical properties of a graphene ribbon are different from those of an infinite plane of graphene. Now, we have a finite width L, and we have conditions connected with boundaries. The electromagnetic wave propagates with the wavefunction:

$$\psi(x,y) \sim e^{ikx}\sin(n\pi y/L); \tag{6.9}$$

where $k_y = n\pi L$ for the propagation from $y = 0$ to $y = L$.

The confinement energy equation:

$$(\hbar k_y)^2/2m = \hbar^2/8mL^2; \tag{6.10}$$

In a graphene monolayer, in a 2D space, the energy E_n of the n – subband is:

$$E_n = \upsilon_F|\rho| = \hbar\upsilon_F(k_x^2 + k_y^2)^{1/2} = [E_x^2 + n^2(\Delta E)^2]^{1/2}; \tag{6.11}$$

where υ_F = the Fermi velocity and $n = 1, 2, 3...$

Depending on the nanoribbon width, L, the energy gap is:

$$\Delta E = \pi\hbar\upsilon_F/L \sim (eV - nm)/L; \tag{6.12}$$

With nanoribbon width on the order of tens of a *nm*, the energy gap is on the order of one hundred meV[35]. The dependence of energy gap vs width is shown in Fig. 6.9 a) (the width is *W*, not *L* in this case).

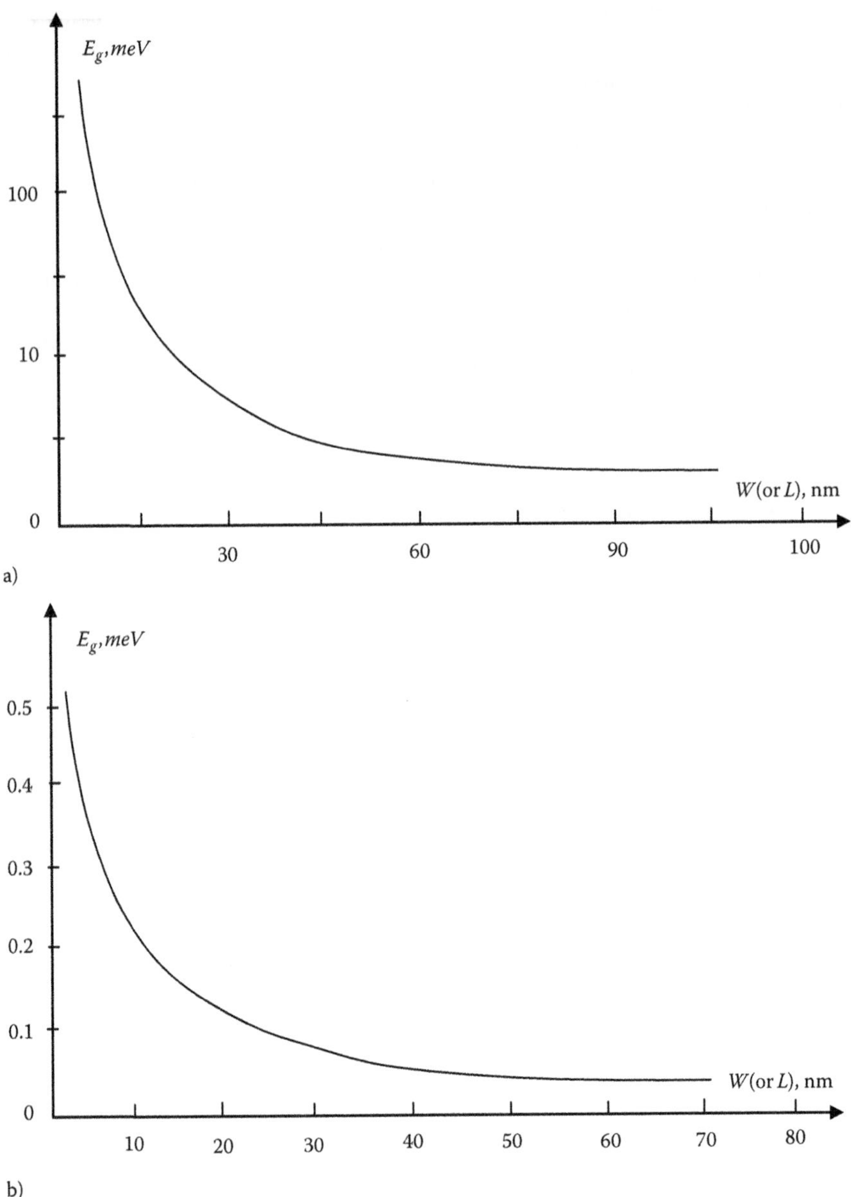

FIGURE 6.9 a) Dependence of the energy gap, E_g vs the width W for mechanically pulled graphene on oxidized Si[35] and b) The graph is inferred from ON/OFF current rations based on estimates[36].

In the manufacturing, it is difficult to achieve perfect boundary conditions resulting in the size-and boundary effects. The edge roughness causes shorter mean free paths and the appearance of half-metallic properties. 2D graphene introduces new qualities of the material, e.g. quantum confinement effect (QCE). Normally, graphene nanoribbons (GNR) may be divided into two types: Armchair and Zigzag (Fig. 6.10).

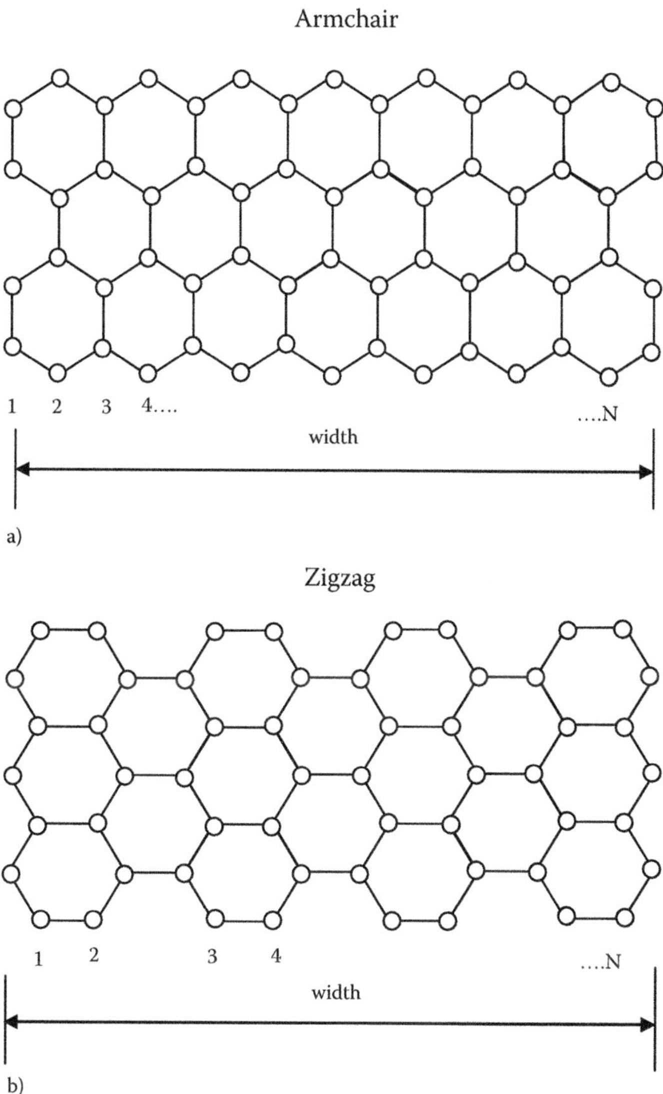

FIGURE 6.10 a) Atomic structure of an "Armchair" GNR and b) Atomic structure of a "Zigzag" GNR.

Edge atoms are not saturated since GNR are stripes of graphene. The edge states are important for the edge conditions. For the "Armchair" GNR there are no restrictions.

The properties of graphene ribbons depend on their boundaries, in particular, on their atomic structure. The ability to control the structure in many ways determines the ribbons' properties. The TEM analysis is used to study the changes in the boundary conditions. The conventional technique to bring about the above changes is to heat a graphene ribbon. Graphene sublimates at T = 3900 K and subsequent annealing reconstructs the boundaries. The annealing process, however, does not allow to achieve straight boundaries and alternatives to extreme heating are considered. One of the approaches to achieve regular boundary geometry is reactive-ion-etching but alignment with the underlying graphene monolayers' crystalline areas is very difficult – one of the principal difficulties of producing graphene nanoribbons, although some applications do not require perfect edges. Some FET-transistors, interconnects and flash memory devices can tolerate the edge's imperfection. At the moment, the edge quality is the main restriction factor for graphene nanoribbons wider range of applications. Some publications raise doubts about the above conclusions. Ultrasmooth graphene nanoribbon edges were reported[37].

6.3.1 Graphene Nanoribbons' Band Gaps

The dependence of the energy band gap and the width W is shown in Fig. 6.9 a). The data was received from micro-mechanically-cleaved graphite. Fig. 6.9 b) shows for comparison data for chemically-exfoliated ribbons. For the chemical exfoliation the ribbons were taken from a suspension. The AFM identified the ribbons with widths between 10 and 55 nm[37]. These results had been predicted for the gap of E_geV ~0.8/W, nm[36]. The prediction stated that GNR (graphene nanoribbons) with homogeneous edges ("Armchair" and "Zigzag") have bandgaps decreasing with an increase of the width of a ribbon. The decrease is useful for transistor switching applications as the ratio ON/OFF increases. The hole mobility in nanoribbons reaches 200 cm^2/Vs[35].

Carbon nanotubes can exhibit semiconductor or metallic qualities depending on the structure parameters, i.e. the diameter of the nanotube or the direction in which the axes bend while forming a nanotube from graphene. A criterion for the existence of metallic qualities of a nanotube is the circumference length that contains an integer number of Fermi wavelengths: $\pi D = n\lambda_F = n3\sqrt{3}a/2$ or $D = n3\sqrt{3}a/2\pi$, where n is an integer. The integer n and m relate to the diameter D_t as:

$$D_t = \left(\sqrt{3}a/\pi\right)\left(n^2 + m^2 + nm\right)^{1/2}; \tag{6.13}$$

where a is the closest distance between neighboring atoms.

The nanotubes whose edges have an "Armchair" configurations (with n = m) are metallic and $D_t = m(3a/\pi)$. The circumference length of a nanotube has an integer number of Fermi wavelengths. In a crystal, the energy associated with the highest occupied crystal orbital is called the Fermi energy (or Fermi level). The wavelength that corresponds to this Fermi energy is called the Femi wavelength. The Fermi velocity corresponds to the kinetic energy of the above-mentioned Fermi energy.

At present, it seems that the boundary conditions determine whether the GNR is metallic or semiconductor. In practice, 1/3 of possible nanotubes exhibit metallic qualities. Some researchers believe that using the Schrodinger equation in simple tight-binding approximation or the Dirac equation using conical bands and the Fermi velocity constant, metallic or semiconductor properties depend on the width *L*. Specific boundary conditions may specify the ribbon properties. For example, zigzag boundary ribbons are all metallic (not depending on the width *L*)[36]. The same way, deviations from a boundary regular shape may lead to semiconducting qualities. In this case, the atomic spacing change produces variable bandgaps. The graph in Fig. 6.9 b) gives the bandgap of 0.4 eV for the smallest ribbon width. A small gap gives quasi-metallic properties, marginally suitable for a minimal ON/OFF ratio in transistor logic but may be suitable for interconnects. One idea is to use nanoribbons with metallic and semiconductor behavior for interconnect and device purposes based on their width, *L*. In particular, the wide nanoribbons may serve for interconnects and the narrow ones being semiconductors. The advantage of almost identical electron and hole masses provide applications for complementary logic devices over analogous silicon logic.

6.3.2 Nanoribbon Manufacturing

One of the methods of chemical synthesis suggests using two anthracene modules. Anthracene consists of three benzene rings aligned next to each other. The ribbon's width is approximately 0.74 nm. This width is limited by the polymerization process up to 20 mm[33]. The bandgaps range from 1.12 eV to 1.6 eV and may be longer since ribbons may act as semiconductors. Long narrow nanoribbons are formed on a gold surface Au(111). Under the conditions of ultra-high vacuum 10,10' –dibromo 9;9' – bianthryl (DBD^A) evaporates at 470 K. The bianthryl molecule consists of two anthracene molecules.

The chemically synthesized nanoribbon has the width of two or three benzene rings. The edges are of "Armchair" – type with a perfect edge with current of 100 nA through a ribbon. This amount of current means the nanoribbons have metallic behavior. The described method is advantageous in producing long ribbons with a precise width and perfect edges. The highest temperature for the process does not exceed 400^{0}C as the commercial CVD used for making graphene has the processing temperature of 1000^{0}C.

The described process has 10,10^1 – dibromo -9,9^1 – bianthryl (DBDA). At the beginning. DBDA molecules are polymerized at T = 200^{0}C. The polymerization takes place on the golden surface free from atoms of Br. This process of polymerization makes acetylene oligomers long and non-planar. The next step is designed to remove the hydrogens and to make the nanoribbon rigid so it can be detached from the gold easily. The detachment may be done by the STM (Scanning Tunneling Microscopy) tip in UHV (ultra-high vacuum) (see Fig. 6.11). This step is analogous to annealing and requires T = 400^{0}C. Such method leaves few free carriers in the ribbon making it not metallic[33].

The question still exists how to use effectively the nanoribbons. Nanoribbons have a number of advantages when compared to carbon nanotubes since the former do not have to be sorted for metallic and nonmetallic and they have the bandgap necessary for device applications. Another possible advantage might be in using other materials instead of Au (111) but it is not clear whether gold is irreplaceable.

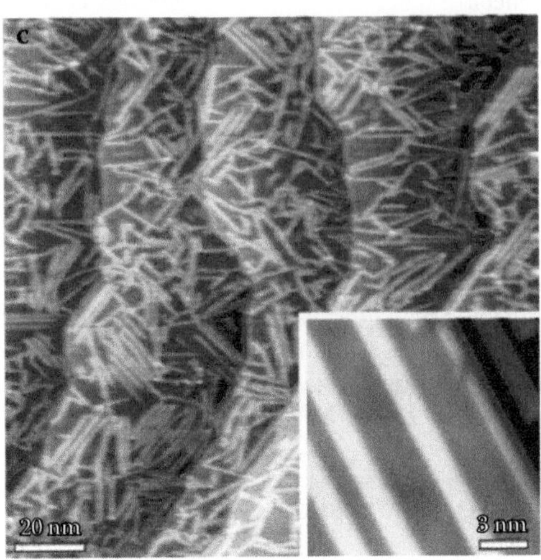

FIGURE 6.11 STM Image: AGNRs/Au(111).

REFERENCES

1. Novoselov, K.S., Geim, A.K., Morozov, S.V., Jiang, D., Zhang, Y., Dubonos, S.V., "Electric field effect in atomically thin carbon films", *Science* (2004)
2. Abergel, D., Russell, A., and Fal'ko, V.I., "Visibility of graphene flakes on a dielectric substrate", *Appl. Phys. Lett.* **91** (2007)
3. Dean, C.R., Young, A.F., Meric, I., Lee, C., Wang, L., Sorgenfrei, S., Watanabe, K., Taniguchi, T., Kim, P., Shepard, K.L., and Hone, J., "Boron nitride substrates for high-quality graphene electronics", *Nature Nanotechnology* **5** pp. 722 – 726 (2010)
4. Shioyama, H., "Cleavage of graphite to graphene", *Journal of Materials Science Letters,* **20**, Issue 6, pp. 499 – 501 (2001)
5. Brodie, B.C., "On the atomic weight of graphite", *Philosophical transactions of the Royal Society of London*, **149**, pp. 249 – 259 (1859)
6. Novoselov, K.S., *Nobel Lecture*, "Graphene: materials in the flatland", *Rev. Mod. Phys.*, **83**, pp. 837 – 849 (2011)
7. Wu, J., Pisula, W., and Mullen, K., "Graphene as potential material for electronics", *Chem. Rev.* **107**, pp. 718 – 747 (2007)
8. Schniepp H.C., Li J.L., McAllister, M.J., Sai, H., Herrera-Alonso, M., Adamson, D.H., "Functionalized single graphene sheets derived from splitting graphite oxide", *J. Phys. Chem.* **B** (2006)
9. In-Yup Jeon, Yeon-Ran Shin, Gyung-Joo Sohn, Hyun-Jung Choi, Seo-yoon Bae, Javeed Mahmood, Sun-Min Jung, Jeong-Min Seo, Min-Jung Kim, Dong Wook Chang, Liming Dai, and Jong-Beom Baek, "Edge-carboxylated graphene nanosheets via ball milling", *Proceedings of the National Academy of Sciences* (*PNAS*) **109** # 15 (2012)
10. Segal, M., "Selling graphene by the ton", *Nat. Nanotechnol.* **4,** #10, pp. 612 – 614 (2009)

11. Li, X., Wang, X., Zhang, L., Lee, S., Dai, H., "Chemically derived, ultra-smooth graphene nanoribbon semiconductors", *Science*, **319**, 1229 (2008)
12. Kosynkin, D.V., Higginbotham, A.L., Sinitski, A., Lomeda, J.R., Dimiev, A., Price, B.K., Tover, J.M., "Longitudinal unzipping of carbon nanotubes to form graphene nanoribbons", *Nature,* **458**, pp. 872 – 876 (2009)
13. Hernandez, Y., Nicolosi, V., Lotya, M., Blighe, F.M., Sun, Z., De, S., McGovern, I.T., Holland, B., Byrne, M., Gun'ko, Y.K., Boland, J.J., Niraj, P., Duesberg, G., Krishnamurthy, S., Goodhue, R., Hutchison, J., Scardaci, V., Feirari, A.C., and Coleman, J.N., "High-yield production of graphene by liquid-phase exfoliation of graphite", *Nature Nanotechnology*, **3**, pp. 563 – 568 (2008)
14. Sheshamani, S., Amini, R., "Preparation and characterization of some graphene based nanocomposite materials", *Science Direct*, **95**, Issue 1, pp. 348 – 359 (2013)
15. Hackley, J., "Graphitic carbon growth on Si (111) using solid source molecular beam epitaxy", *Appl. Phys. Lett.*, **95** (2009)
16. Van Bommel, A.J., Crombeen, J.E., and van Tooren, A., "LEED and Auger electron observations of the SiC (0001) surface", *Surf. Sci.* **48**, pp. 463 – 472 (1975)
17. Bostwick, A., Ohta, T., McChesney, J., Emtsev, K.V., Seyller, T., Horn, K., Rotenberg, E., "Symmetry breaking in epitaxial graphene probed by ARPES", *American Physical Society (APS)* (2008)
18. Emtsev, K.V., Bostwick, A., Horn, K., Jobst, J., Kellog, G.L. Ley, L., McChesney, Ohta, T., Reshanov, S.A., Rohrl, J., Rotenberg, E., Schmid, A.K., Waldmann, D., Weber, H.,B., and Seyller, T., "Towards wafer-size graphene layers by atmospheric pressure graphitization of silicon carbide", *Nature Materials,* **8**, pp. 203-207 (2009)
19. Pearce, R., Tan, X., Wang, R., Patel, T., Gallop, J., Pollard, A., Yakimova, R., and Hao, L., "Investigations of the effect of *SiC* growth face on graphene thickness uniformity and electronic properties", *Surf. Topogr.*: Metrol. Prop. **3** (2014)
20. Liu, W., Li, H., Xu, C., Khatami, Y., Banerjee, K., "Synthesis of high-quality monolayers and bilayer graphene on copper using chemical vapor deposition", *Elsevier Carbon*, **49** pp. 4122 – 4130 (2011)
21. Ji, S.-H., Hannon, J.B., Tromp, R.M., Perebeinos, V., Tersoff, J., Ross, F.M., "Atomic-scale transport in epitaxial graphene", *Nature Matter*, **11**, pp. 114 – 119 (2012)
22. Li, X.S., "Large-area synthesis of high-quality and uniform graphene films on copper foils", *Science* **324** (2009)
23. Xu, B., Zheng, D., Jia, M., Liu, H., Cao, G., Qiao, N., Wei, Y., Yang, Y., "Nano-*CaO* template carbon by CVD: From nanosheets to nanocages", *Science Direct, Materials Letters,* **143**, pp. 159 – 162 (2015)
24. Lei, Q.J., Zhang, L.X., Cao, J., Zheng, W.T., Wang, X., and Feng, J., "Synthesis of graphene on a *Ni* film by radio-frequency plasma-enhanced chemical vapor deposition", *Chinese Science Bulletin*, **57**, #23, pp. 3040 – 3044 (2012)
25. Bae, S., Kim, H., Lee, Y., Xu, X., Park, J., Zheng, Y., Balakrishnan, J., Lei, T., Kim, H.R., Song, Y.L., Kim, Y.J., Kim, K.S., Ozyilmaz, B., Ahn, J.-H., Hong, B.H., and Lijima, S., "Roll-to-roll production of 30-inch graphene films for transparent electrodes", *Nature Nanotechnology,* **5**, pp. 574 – 578 (2010)
26. Takamatsu, S., Takahata, T., Muraki, M., Iwase, E., Matsumoto, K., and Shimoyama, I., "Transparent conductive-polymer stain sensor for touch input sheets of flexible displays", *Journal of Micromechanics and Micro-Engineering"*, **20**, #7 (2010)
27. Song, Yi, Fang, W., Hsu, A.L., and Kong J., "Iron (III) Chloride doping of CVD graphene", *Nanotechnology,* **25** #39 (2014)
28. Tsen, A.W., Brown, L., Levendorf, M.P., Ghahari, F., Hung, P.,Y., Havener, R.,W., Ruiz-Vargas, C.,S., Muller, D.A., Kim, P., Park, J., "Tailoring electrical transport across grain boundaries in polycrystalline graphene", *Science*, **336**, #6085, pp. 1143 – 1146 (2012)

29. Li, X., Magnuson, C.W., Venugopal, A., Tromp, R.M., Hannon, J.B., Vogel, E.M., Colombo, L., and Ruoff, R., S., "Large-area graphene single crystals grown by low-pressure chemical vapor deposition by methane on copper", *Journal of the American Chemical Society (JACS or J. Am. Chem. Soc.)*, **133** (9), pp. 2816 – 2819 (2011)
30. Duong, T., Goud, B., and Schaver, K., "Closed-form density-based framework for automatic detection of cellular morphology changes", *PNAS*, **109**, # 22 (2012)
31. Liu, C.-M., Lin, H.-W., Lu, C.-L. and Chen, C., "Effect of grain orientations of Cu seed layers on the growth of <111> - oriented nanotwined Cu", *Scientific Reports,* **4**, #6123 (2014)
32. Tao, C., Jiao, L., Yazyev, O.V., Chen, Y.-C., Feng, J., Zhang, X., Capaz, R. B., Tour, J.M., Zettl, A., Louie, S.G., Dai, H., and Crommie M.F., "Spatially resolving edge states of chiral graphene nanoribbons", *Nature Physics,* **7**, pp. 616 – 620 (2011).
33. Ci, Lijie, Song, Li, Jin, Chuahong, Jariwala, Deep, Wu, Dangxin, Li, Yongjie, Srivastava, Anchal, Wang, Z. F., Storr, Kevin, Balicas, Luis, Liu, Feng and Ajayan, Pulickel M., "Atomic layers of hybridized boron nitride and graphene domains", *Nat. Mater.* **9** 2010
34. Klaus, J.W., Ferro, S.J., George, S.M., "Atomic layer deposition of tungsten using sequential surface chemistry with a sacrificial stripping reaction", Thin Solid Films **360** Elsevier Science S.A. pp. 145 – 153 (2000)
35. Han, M.Y., Ozyilmaz, B., Zhang, Y., and Kim, P., "Energy band-gap engineering of graphene nanoribbons", *Phys. Rev. Lett.* **98** (2007)
36. Son, Y.W., Cohen, M.L., Louie, S.G., "Half-metallic graphene nanoribbons", *Nature* (2006)
37. Li, Xiaolin, Wang, Xinran, Li, Zhang, Lee, Sangwon, Dai, Hongjie, "Chemically derived, ultrasmooth graphene nanoribbon semiconductors", *Science*, **319** (2008)

7 Methods of Materials Characterization of Graphene

One of the more advanced method of graphene materials characterization is Angle-Resolved Photoelectron Spectroscopy (ARPES). ARPES can characterize the graphene band structure (Fig. 7.1).

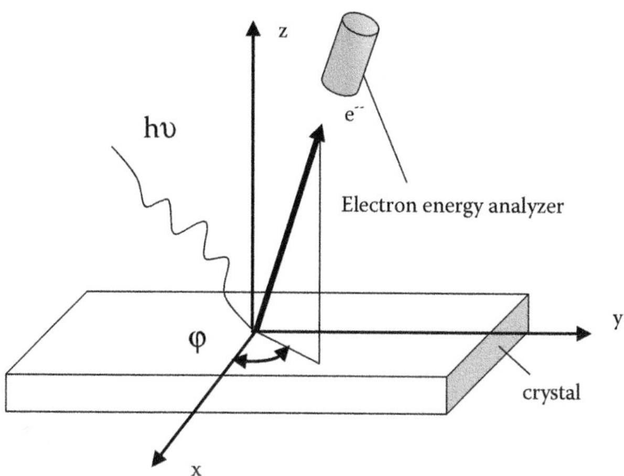

FIGURE 7.1 Energy measurement of a graphene sample by ARPES[1].

ARPES is performed at high vacuum which is necessary for momentum of excitation and energy measurement. The following equation is used for the parameters calculation:

$$E_B = h\nu - E_{kin} - \varphi, \hbar K_{11} = (2mE_{kin})^{1/2} \sin\theta; \qquad (7.1)$$

where E_B = binding energy of the material's electrons; $h\nu$ = photon energy; E_{kin} = the kinetic energy of the photo-emitted electron, k_{11} – the momentum of the material's electrons parallel to the surface and φ is the work function. $K_{11} = k_{11} + G$ is the momentum component of electrons in the sample surface. G = reciprocal lattice vectors.

The energy and momentum of the electrons in the specimen's surface (Fig. 7.1) can be received by measuring the intensity of the electrons produced by photo-emission. ARPES collects the data from a large surface but not from a separate point as it happens with scanning tunneling spectroscopy (STS).

7.1 ADDITIONAL PHYSICAL PROPERTIES OF GRAPHENE

One of the physical properties of graphene is its optical absorption (Fig. 7.2).

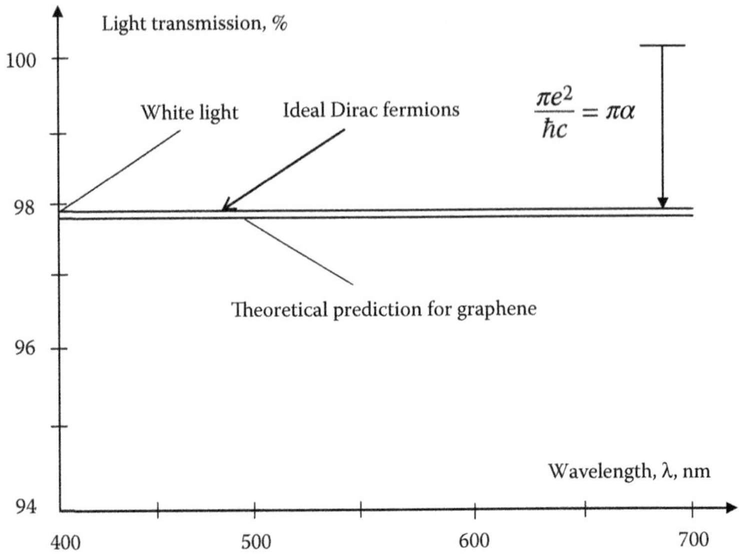

FIGURE 7.2 Optical absorption in graphene, where $\alpha = 1/137$ is the structure constant and $\pi\alpha = 2.3$ %.

The absorption is hole-dependent and depends on the magnitude of the magnetic field ***B***. The absorption is strong from atom to atom but the total absorption does not exceed 2 %. This quality of graphene makes it difficult to use for device making. However, in combination, graphene may be broadband, fast and produce gain.

The electric structure of graphene can be revealed by the Raman spectroscopy (Fig. 7.3). Usually, a sample is illuminated by a laser. The reflected electromagnetic radiation is focused by lens and directed into a monochromator. The scattered radiation is filtered out and the rest of the collected radiation is directed by a notch or a band pass filter into a detector. The collected elastic radiation corresponds to the wavelength of the laser line.

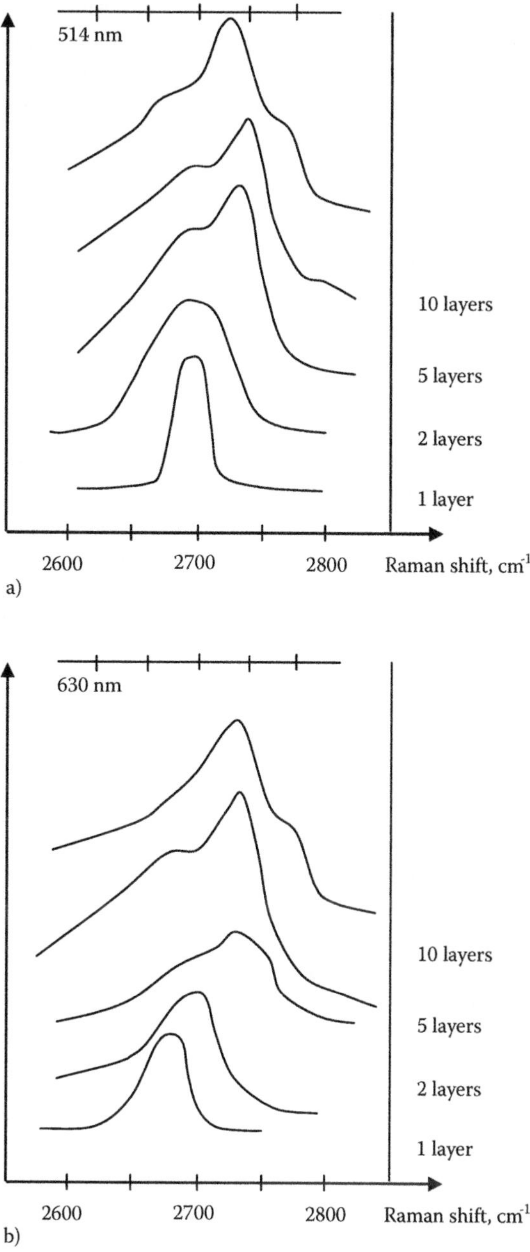

FIGURE 7.3 a) The Raman shift vs. thickness of the sample for 514 nm^2 and b) The Raman shift vs. thickness of the sample for 623 nm^2.

The Raman spectra similar for graphene and graphite. The peak at about 2700 cm^{-1} is called "2D peak". The phonons near K – point in the Brillouin zone are responsible for the peak. A shift of the energy of two phonons is caused by a photon absorption at 514 nm of the laser emitting wave[2]. The phonon energy is equal 2 x 8065 cm^{-1}/eV, (the energy near the peak of 2700 cm^{-1}).

Raman spectroscopy allows seeing the graphene electronic structure that evolves depending on the number of the layers. Electron-proton interactions provide non-destructive characterization of the material. In this aspect, Raman spectroscopy has advantages over TEM (Transmission Electron Microscopy) and AFM (Atomic Force Microscopy). In particular, TEM allows a view of a graphene sample's layers only if the specimen has a fold, otherwise the situation with AFM is analogous to that of TEM.

The strong bonds among the carbon atom influence the thermal conductivity that exceeds the thermal conductivity of the other two carbon materials: diamond and graphene. Supporting materials for graphene influence its thermal conductivity. In particular, it may cause a decrease of the conductivity because of phonon scattering of the supporting material (substrate). The measurement of κ approaches 600 W/(m-k) and much greater values of 3500 – 5300 W/(m-k) have been reported[3].

The thermal conductivity is mostly due to phonons but not to electrons. Thus, we cannot apply Wiedemann-Franz law that specifies the electronic contribution to the thermal conductivity, κ. In one dimension, the thermal conductivity, considering the diagonal elements only is:

$$k_{zz} = \Sigma c \upsilon_z^2 \tau; \tag{7.2}$$

where c – the specific heat, υ_z – the group velocity, τ – the relaxation time.

At low temperatures, impurities, defects, boundary irregularities determine the relaxation time. In anisotropic materials, the low-temperature thermal conductivity does not have the same temperature dependence as the specific heat, unlike it is the case with ordinary materials. In graphite, in particular, the thermal conductivity is weakly dependent on the interlayer phonons. In low-dimensional system, such as graphene, the thermal conductivity is substantially influenced by the adjacent layers or substrates. However, the interlayer interactions in graphite diminish the thermal conductivity by almost one order of magnitude.

7.2 SPECTROSCOPIC METHODS OF GRAPHENE CHARACTERIZATION

In the previous chapters, we considered Landau levels and the quantum Hall effect in graphene. In order to use the above phenomena for graphene characterization the scanning tunneling microscopy (STM) may be used. STM's ability to resolve on the atomic scale is due to the rapid decrease of the electron density outside the scanning tip, the acting part of which may constitute a single atomic orbit. In this case, the tomography resolution may be on the order of the Bohr's radii. In STM, the applied voltage deforms piezoelectric elements that deform the tip on an Angstrom scale. The applied voltage is controlled by the feedback electronics. The difference between the set tunnel current and the actual current that changes proportionally the piezo z-axis movement is noteworthy. The current's magnitude is in a picoamps to nanoamp range. The electron hopping frequency, $f = 10^{-9}/(1.6\text{x}10^{-19}) = 6.25$ GHz and one atom can give a current of one nA, so the set-current should be somewhat less than one nanoampere.

The tip is not smooth on an atomic scale, so in essence just one of the atoms of the tip's surface contributes to the current. The tip is usually made of *PtIr* wire and is cleaned from contamination - the process which is important for the tip's normal functioning. STM provides a dependence of density of states at the Fermi energy on a found vacancy in a graphene layer. A vacancy results in a sharp peak of density of states. Since the current in STM depends on the density of states, the density of states can be measured. The energy resolution is about $5k_BT$. STM is also able to determine Landau levels in graphene. This method provides a simpler localization inference from the STM spectroscopy than in the case of the Hall effect and acquires longitudinal and transverse conductivities that are used to determine the level positions. Fig. 7.4 the Landau levels are found based on the voltage bias applied to the sample.

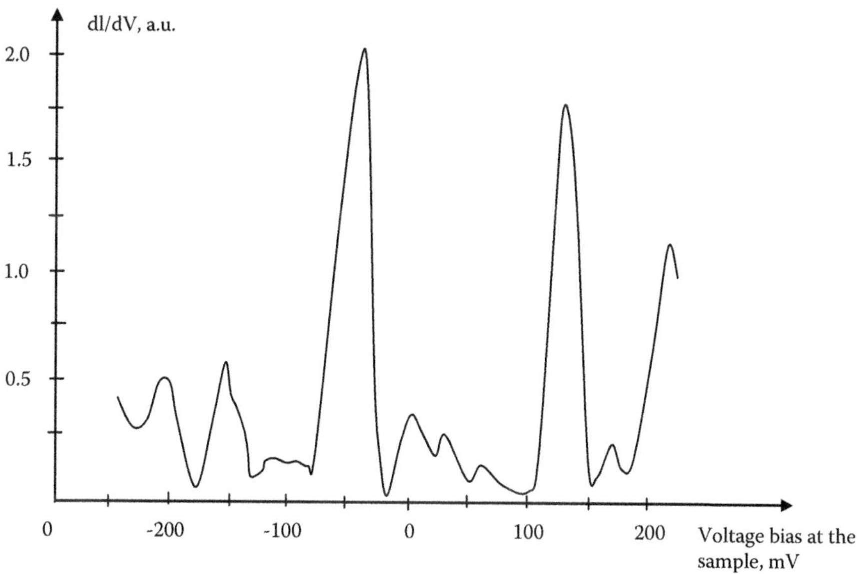

FIGURE 7.4 Landau level determination with a voltage bias applied to the sample.

7.2.1 Quantum Capacitance

Using the usual arrangement of metal-insulator-graphene-insulator-metal, we have a sheet of graphene on a metal substrate. Since graphene conducts electricity, we have two capacitors in sequence. This capacitance depends on the graphene density of states. This physical structure is called "quantum capacitance". The dependence of this capacitance on the density of states in graphene is given[4]:

$$C_q = C^2 D(E)\ ; \tag{7.3}$$

where $D(E) = dn/dE$ at E_F (the Fermi energy).

The quantum capacitance structure and its equivalent circuit is given in Fig. 7.5.

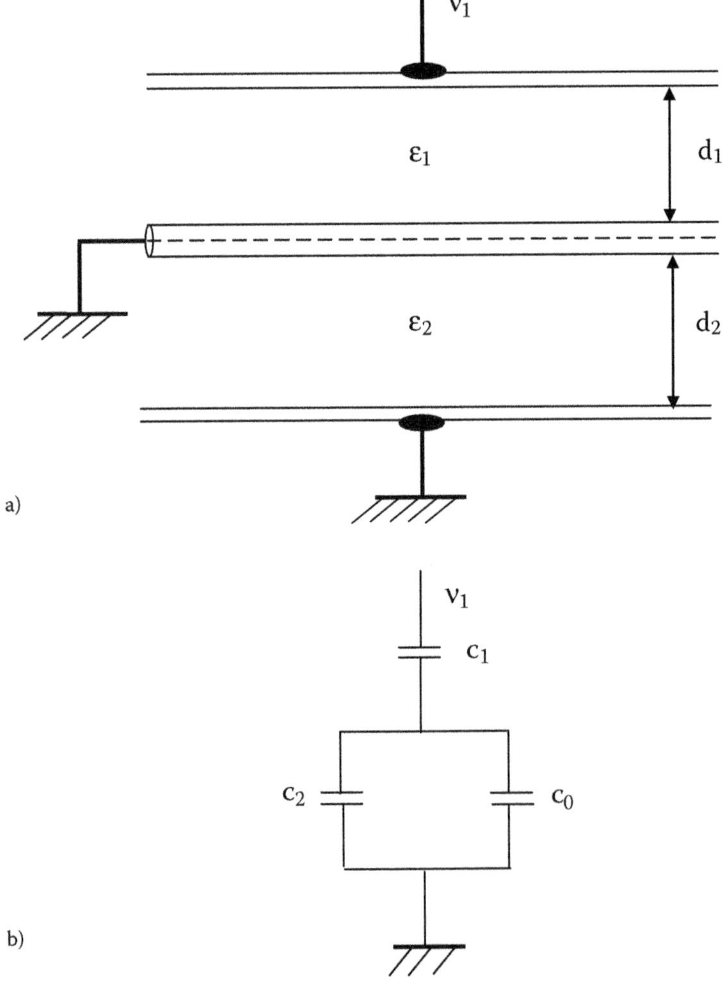

FIGURE 7.5 a) The quantum capacitance structure when the graphene acts as a two-dimensional metal and b) an equivalent circuit for the quantum capacitance structure[4].

The plate in the middle has the bottom and the top location of the electric field. Both locations influence each other. The graphene plate in the middle has a 2D structure that has an extra filling electron energy, the presence of which comes from the Pauli principle and the graphene conical density of states, *g(E)* for the *Q* of electrons and holes of the quantum capacitance[5]:

$$Q = q/g(E)[f(E+E_G/2+qV_a) - f(E+E_G/2-qV_a)]dE; \tag{7.4}$$

where *f(E)* is the Fermi function. The integration over the density of states, *g(E)* and the Fermi energy function, *f(E)* is taken place from 0 to ∞. V_a is the applied voltage. The quantum capacitance is, therefore:

$$C_Q = \partial Q/\partial V_a; \tag{7.5}$$

A graphene layer was placed between two thin insulator layers of 10 nm of Al_2O_3 each[6]. The graphene monolayer was extracted by the Novoselov/Geim' method. The graphene layer parameters were measured with electrodes made of gold and titanium. The graphene retained its high electron mobility notwithstanding the oxide application. The Fermi energy had a range of ± 0.5 eV. The top electrode provides the Dirac point at the gate voltage close to zero. The carrier concentration was $n \leq 10^{12}\ cm^{-2}$. The gate voltage was in the range of $[-3 \div +3]$,V. The corresponding of the capacitance may be calculated as follows:

$$1/C = 1/C_{ox} + 1/C_Q; \tag{7.6}$$

The last component of the above equation, C_Q can be determined from the graph in Fig. 7.6. The oxide capacitance can be found using capacitance fitting curve (Fig. 7.6). The minimum value C_Q is < 1 μF/cm² from the above graph. It is also noticeable that the capacitance C_Q is proportional to the density of states, *D* (the right axis in Fig. 7.6). The data for the graph was produced by finding the carrier density, *n* for the applied voltage at the gate. The limit line at the bottom of the curve came from experimental measurements of density of states which had a limited low value.

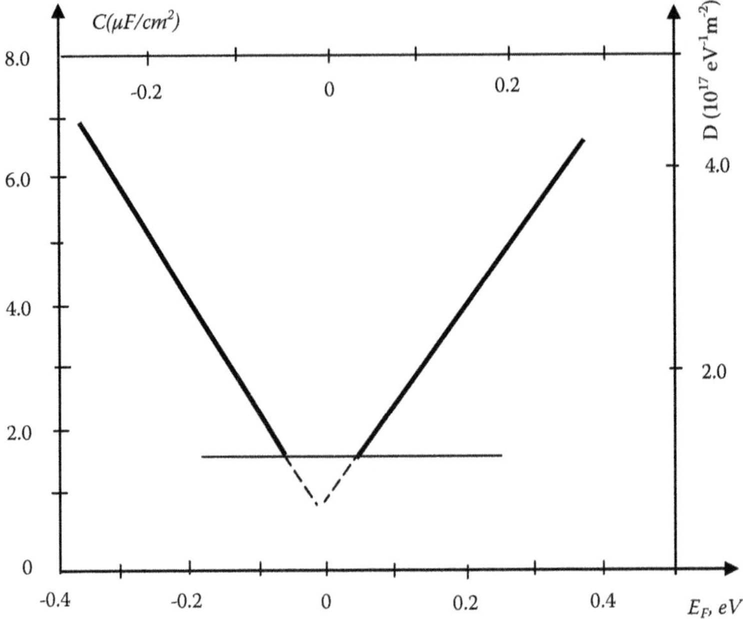

FIGURE 7.6 Dependence of C_Q on the Fermi energy, E_F and carrier concentration, *n*. The right side shows the dependence of density of states, D on the Fermi energy[6], $D = C_Q le^2$, 10^{13}/eVcm².

The minimal carrier concentration has the value of 4 x 10^{11} cm^{-2}. Using the described above the capacitance spectroscopy method[6], the Landau levels were determined (Fig. 7.7) at several temperatures within the range of [16 ÷ 250]*K*. The anomalous Landau level is shown by a dashed line.

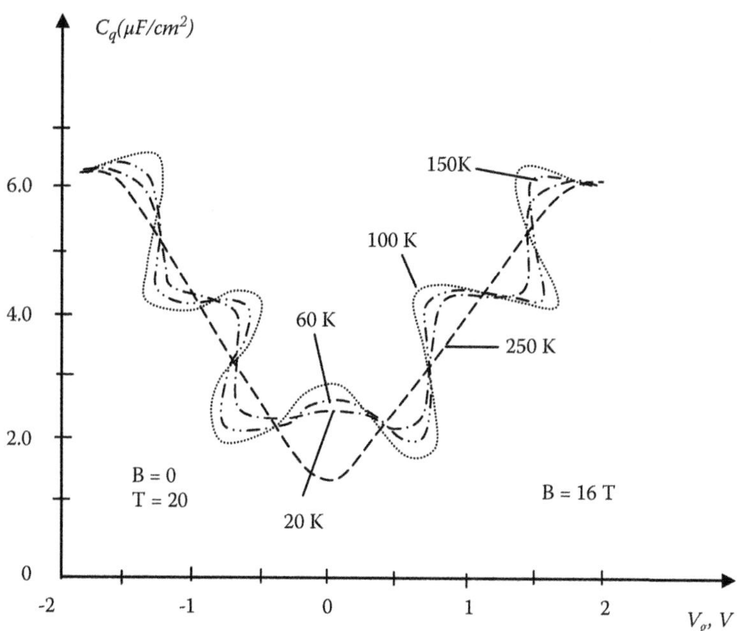

FIGURE 7.7 The Landau levels at the back gate voltage V_g and different temperatures. The data was produced capacitance spectroscopy.

7.2.2 Scanning Electron Microscopy

The scanning single-electron transistor (SSET) is one of the scanning electron methods recently suggested to measure graphene parameters. In particular, SSET provides an electric charge on the graphene surface and static electric fields with spatial resolution of 100 nm and at low temperature[7]. The tip of the SSET's device measures the local potential, V_s. V_s comes from the difference in work function between the tip and the specimen's surface, and it is analogous to the chemical potential, μ.

Similar to the field-effect transistor, the single-electron transistor, the SET has a source and a drain. In between the source and the drain, there is a sensing electrode. Because of the small size, the capacitance of the sensing electrode is very small. The single-electron charge demands an energy of $e^2/2C$ (larger than k_BT). C is the capacitance of the area between the source and the drain. The ON/OFF states of the transistor are determined by the multiplication factor of the net charge Q of the area between the source and the drain. If the factor is an odd multiple of e/2, the transistor is ON. If the multiplication factor is an integer, then the transistor is OFF. In the OFF state, a potential barrier, $e^2/2C$ exists between the source and the drain. An external field induces a net charge density $\sigma = \varepsilon_0 E, C/m^2$; where ε_0 = permittivity of space. One electron, induced under the tip produces up to 0.1 electron charge in the area between the source and the drain. The distance between the tip and the specimen's surface is 25 nm[7].

At cryogenic temperatures and at the source-drain bias = const., the source – drain current is:

$$I_{SET} = A\sin(2\pi Q/e) = A\sin[2\pi C_s(V_B + V_S)/e]; \qquad (7.7)$$

The scanning of a graphene surface takes place at a constant height. The current changes with Q with the period of one electron charge. The charge between the source and the drain is $C_S(V_B + V_S)$, where C_S is the capacitance of the tip's surface, V_B is the tip-sample bias, and V_S is the surface potential. The tip has feedback that adjusts the applied potential, V_B.

The use of sinusoidal modulation has been proposed for graphene characterization when the back-gate was employed[8]. Modulating the back potential, the graphene's chemical potential, was calculated as the ratio $d\mu/dn$, when n is the carrier concentration. A change in μ corresponds to a change in the local electrostatic potential, V_s. $d\mu/dn$ corresponds also to the local density of states. The described method allows to map the chemical potential directly. The value μ can be also calculated by integrating $d\mu/dn$. The practical application of the method is finding defective regions in graphene.

REFERENCES

1. Zhon, S.Y., Siegel, D.A., Fedorov, A.V., Gabaly, F.El., Schmid, A.K., Castro Neto, A.H., and Lanzara, A. "Origin of the energy bandgap in epitaxial graphene", *Nature Materials* **1** (2008)
2. Ferrari, A.C., Meyer, J.C., Scardaci, V., Casiraghi, C., Lazzeri, M., Mauri, F., Piscanec, S., Jiang, D., Novoselov, K.S., Roth, S., and Geim, A.K., "Raman spectrum of graphene and graphene layers", *Phys. Rev. Lett.* **97** (2006)
3. Pop, E., Mann, D.A., Goodson, K.E., Dai, H.J., "Electrical and thermal transport in metallic single-wall carbon nanotubes on insulating substrates", *Appl. Phys. Lett.* **89** (2006)
4. Eisenstein, J.P., Pfeiffer, L.N., and West K.W., "Compressibility of the two-dimensional electron gas: Measurements of the zero-field exchange energy and fractional quantum Hall gap", *Phys. Rev.* B **50** pp. 1760 – 1778 (1994)

5. John, D.L., Castro, L.C., and Pulfrey, D.L., "Quantum capacitance in nanoscale modeling", *J. of Appl. Phys.* **96** #9 (2004)
6. Ponomarenko, L.A., Yang, R., Gorbachev, R.V., Blake, P., Katsnelson, M.I., Novoselov, K.S., Geim, A.K., "Density of states and zero Landau level probed through capacitance of graphene", *Phys. Rev., Lett.* **105** (13) (2010)
7. Kalinin, S., Gruverman, A., Eds. *"Scanning probe microscopy. Electrical and electromechanical phenomenon at the nanoscale"*, Vol. 2, *Springer* (2007)
8. Martin, J., Akerman, N., Ulbricht, G., Lohmann, T., Smet, J.H., von Klitzing, K., and Yacoby, A., "Observation of electron-hole puddles in graphene using a scanning single-electron transistor", *Nature Physics* **4** 144 – 148 (2008)

8 Experimental Considerations of 2D Graphene

Graphene is unique in providing a wide spectrum of physical properties that promise superior electrical characteristics, such as effective magnet fields, elimination of charges valley degeneracy, the ability to change the local electronic potential and scattering control. Graphene has a high elastic constant, superior breaking strength and mechanical distortions, such as ripple formation that can influence the advantageous qualities. New devices based on manipulation of local strains, as one of the application examples, can be created.

Graphene is still a new material and as such it has been experimentally studied aimed to discover new qualities of the material, to confirm or refute the known observations. This chapter is devoted for discussions of the recent experimental studies and the future prospects of graphene characterization. The presented material broadens and updates the discussions presented in the previous chapter describing the physical properties of the material.

As it was discussed in the previous chapters, graphene crystals behave as an elastic material with Young's modulus, Y. Because of extreme thinness, the rigidity, $\kappa = Yt^2$ is low (we consider graphene samples with thickness < 1 nm). If a graphene sample is clamped on one side, we have a cantilever with length L and thickness t. The sample's free end frequency of oscillation is[1]:

$$\nu = 0.162(Y/\rho)^{1/2}; \tag{8.1}$$

where ρ is the mass density.

If the specimen is clamped on two sides, the factor 0.162 is increased to 1.03.

A specimen of dimensions, L, ω, t is clasped on one side and a spring constant K^* is calculated. Eq. (8.1) is rewritten then in the form:

$$\nu = (2\pi)^{-1}(K^*/m)^{1/2}; \tag{8.2}$$

where $K^* \propto (Y\omega t^3/L^3)$;

A specimen with $t = 0.34$ nm clamped on one side will be experiencing the gravity force that will cause the free specimen's side to bend downwards by L/10, where L is the specimen's length and is on the order of 50 μm. In practice, it means that a graphene electronic element must be supported. The graphene specimen deflection under a gravitational force is shown in Fig. 8.1. The assumption is the displacement[2]:

$$\xi(x) = qx^2(x^2 - 4Lx + 6L^2) / 24YI); \tag{8.3}$$

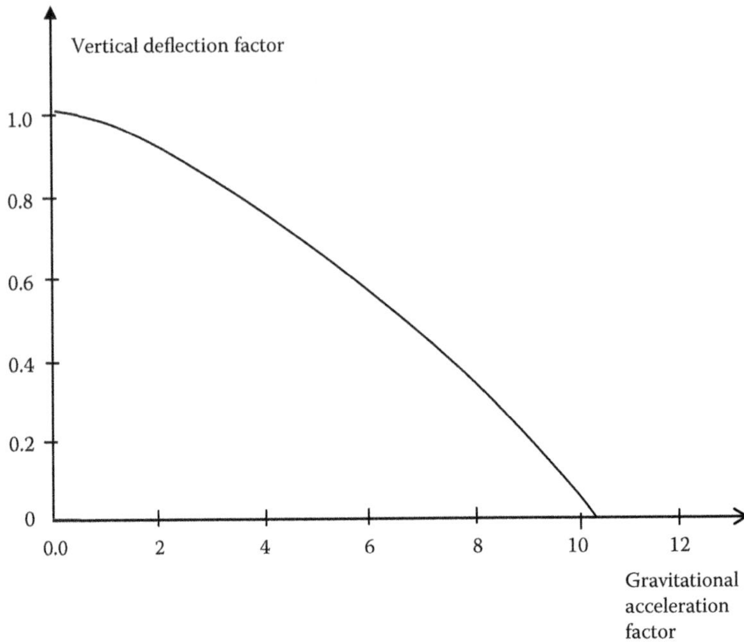

FIGURE 8.1 Vertical deflection factor multiplied by 5 of graphene cantilever 66 μm with the free end under the gravitational force acceleration (9.8 m/s²)[1].

8.1 GRAPHENE DEFORMITY UNDER A GRAVITATIONAL FORCE

The displacement of the end of a graphene specimen that acts as a horizontal cantilever of length L under a gravitational force is[2]:

$$\xi(L) = 1.5g\rho L^4/(t^2Y); \tag{8.4}$$

where g is a gravitational force, ρ – the density, Y is Young's modulus, and t is the thickness. Fig. 8.1 illustrates Eq. 8.3 displacement for $L = 66\ \mu m$. From the mechanical point of view, graphene is a very strong material. The deflection shown above was measured for the length of the sample 10^5 bigger than the thickness of 0.34 nm – an impressive ratio! An experimental set-up for measuring graphene deflection is shown in Fig. 8.2.

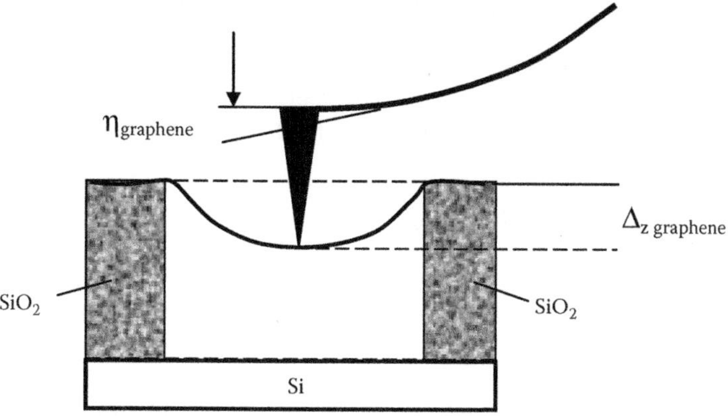

FIGURE 8.2 A graphene sheet deflection with an AFT tip[l].

$\Delta_{zgraphene}$ in Fig. 8.2 is calculated from Eq. (8.4). An experimental set-up for measuring graphene deflection is shown in Fig. 8.2. Since the fractional deformation is proportional to the cube of the sample's size, special measures need to be taken to avoid drooping of the material under a gravitational force. The measurements connected with the sample's drooping are made in vacuum to avoid the mechanical disturbance by the air and the air moisture influence.

A sheet of graphene placed on a substrate conforms the substrate surface and experiences van der Waals attraction force. The force can be estimated as $A/6\pi h^3$, where A is the Hamaker constant. The Hamaker constant A is defined from a "body-body interaction". In the van der Waals interaction:

$$A = \pi^2 \times C \times \rho_1 \times \rho_2; \tag{8.5}$$

Where ρ_1 and ρ_2 represent the number of atoms per unit volume that exist in two interacting bodies. *C* is the coefficient that comes from the interaction between a pair of particles. *A* constant allows determining *C*, the interaction parameter from the expression for van der Waals pair potential $\omega(r) = -C/r^6$. The Hamaker constant's approach ignores the medium in the particle pair. Lifshitz further developed the approach by taking into consideration the media, its dielectric properties. The van der Waals forces are suitable for distances longer than several hundred angstroms. $A \sim 10^{19}$ J, and *h* is the spacing between the graphene sheet and substrate. A gravitation force counteracts van der Waals force. If the spacing *h* is on the order of 1 μm, the van der Waals force is 1000 times stronger than the gravitational force.

The above reasoning is useful for determining the conditions for graphene growing. As it was discussed earlier, graphene may be grown on metal surfaces (such as Cu or Ni) at $T = 1000^0$ C without a 3D supporting structure.

8.2 STRUCTURAL DEFECTS UNDER APPLIED STRAIN

Under applied force graphene bends but does not fracture. Bending of carbon nanotubes allows analyzing the results of bending graphene sheets. Carbon nanotubes have diameters in the range of nanometers. The diameter of a carbon tube may be calculated:

$$d = (a/\pi)(n^2 + m^2 + nm)^{1/2}; \tag{8.6}$$

where n and m are integers, parameters of the axis around which the carbon tube is symmetrical. In this case, a graphene sheet is rolled around a cylinder axis. a is the lattice constant of graphene. $a = 246$ pm and is slightly temperature dependent because of anomalous negative expansion coefficient. As it was discussed earlier, there are "armchair" (m = n) and "zigzag" (m = 0) crystal structure configurations that give metallic ("armchair") and semiconducting ("zigzag") qualities. This pattern also can be found on the edge of a graphene/carbon sheet. Graphene is stable in most of its states. Instability manifests itself in a smooth elastic distortion. Under pressure graphene does not break but forms "buckling" shapes perpendicular to the graphene surface. In general, buckling has a sinusoidal form. It is characterized by existence of flexural phonon[3]. Single-layer graphene is characterized by existence of its flexural mode, called also "bending mode" and "out – of –plane transverse acoustic mode". Flexural mode is important for graphene thermal and mechanical qualities. Flexural mode also influence Young's modulus and nanomechanical resonance. Graphene's extraordinary thermal conductivity exists mostly because of graphene's three acoustic phonon modes at room temperature. The wavelength associated with it ranges from several nanometers to micrometers. In epitaxial growth, graphene has a ripple wavelength with 11 carbon hexagons matching the interatomic distance of the substrate[4]. The boundary conditions influence the presence of "buckling" and ripples. The wavelength influenced by the boundary conditions is the range of 0.3 – 1 nm for a free-standing bilayer graphene sample[3]. The ripples were observed by the means of TEM and they resemble similar micro- and macro-mechanical deformities. A plastic sheet if strained in outward direction will exhibit similar ripples and waves around them in a wave-like characteristic manner. The experimental work observed above shows that no spontaneous ripples appear provided there is no force applied to a graphene sample. Fig. 8.3 shows a single and multilayer exfoliated graphene forms which are put across a trench made in SiO_2/Si substrates.

Wavelength ripples are associated with distorted bonds of carbon atoms with the angles between the bonds but not the bond lengths being distorted. As it was observed previously, graphene ripples follow the classic pattern. Typical thickness, of a graphene sheet/membrane, t is up to the limit of 20 nm. The sinusoidal behavior of ripples may be described as follows:

$$\varsigma(y) = A\sin(2\pi y(\lambda); \tag{8.7}$$

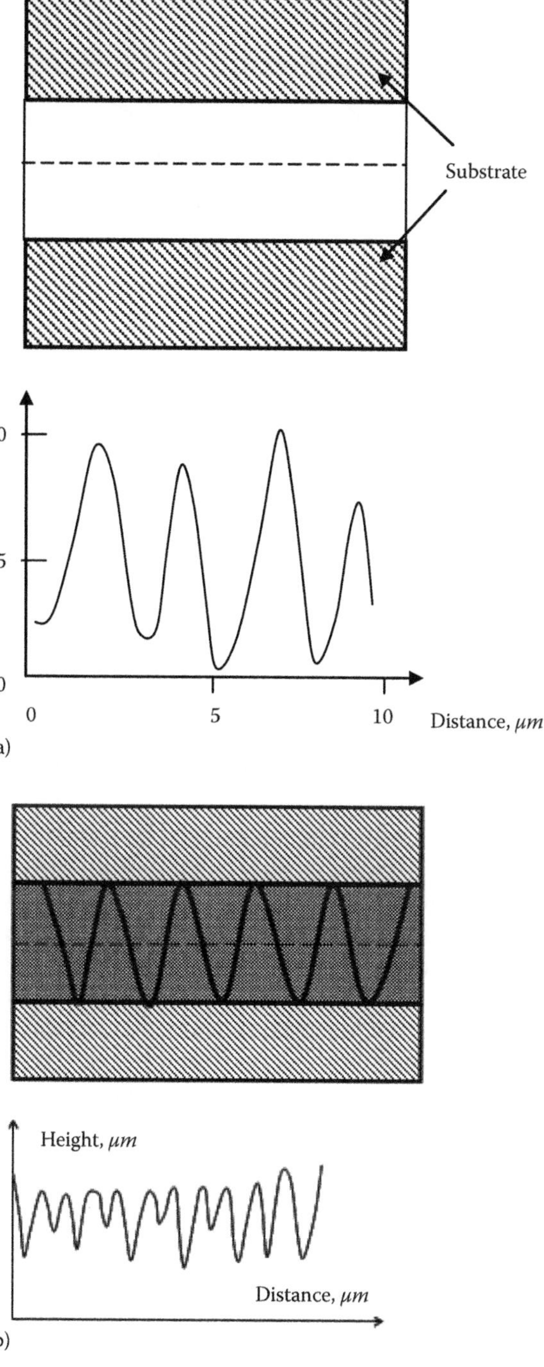

FIGURE 8.3 a) Ripple height distribution due to a strain across a trench. b) Ripple height distribution due to a strain across a trench[5].

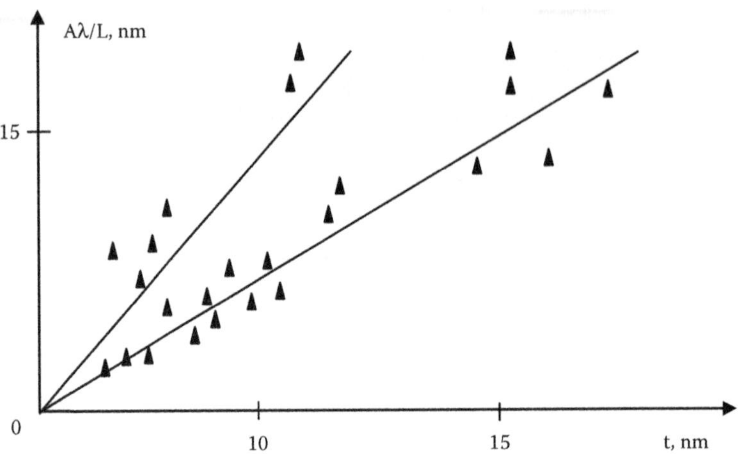

FIGURE 8.4 The appearance of ripples in response to mountain strain for a graphene sheet (layer) placed across a trench[5].

In this case, we measuring across a trench, along y-axis. The relation between the wavelength λ and Poisson ratio ν is[5]:

$$A\lambda/t = \left\{8/[3(1+\nu^2)]\right\}^{1/2}; \tag{8.8}$$

Thus, we eliminated the strain variable γ. The data in Fig. 8.4 are calculated from (8.10) for the upper line and for the lower line, assuming the tensile strain, we have:

$$A\lambda/t = \left\{8/\nu[3(1+\nu^2)]\right\}^{1/2} t; \tag{8.9}$$

The experimental results reflected in Fig. 8.4 suggest that the ripples are due to the tensile stress for specimens with thicknesses from 0.34 nm to 18 nm with ν = 0.165.

In general, the orientation of the material, wavelength or the amplitude of ripples are modified through boundary conditions or the difference in the thermal coefficients of the substrate and the sample/membrane. Of the named control factors, the thermos-mechanical manipulation is the most effective. The ripples formed on graphene membranes may be controlled by thermal manipulation. Annealing of graphene samples changes the ripple forms and their characteristics. In particular, the amplitudes undergo modification. The ability to control the substrate and the sample/membrane's thermal expansion gives the possibility to control the amplitude, wavelength and orientation of the ripples. The transverse compressive strain is:

$$\Delta \sim \sqrt{1+\frac{\zeta^2}{\lambda^2}} - 1; \tag{8.10}$$

where $A \sim \lambda\sqrt{\Delta}$ for $A << \lambda$.

Δ exists because of the difference in thermal expansion coefficient between the substrate and graphene sample.

Thermal properties of solids have contributions from phonons (such as lattice vibrations) and from electrons. The lattice heat capacity depends on the energy of phonons. The total energy of the phonons may be formulated as the total energy at temperature T:

$$E = \sum \langle n_{qp} \rangle \hbar \exp(\vec{q}); \tag{8.11}$$

where $\langle n_{qp} \rangle$ = the thermal equilibrium occupancy of phonons of wavevector $\vec{q}$ and mode $p(p= 1....3s)$, where s is the number of atoms in a unit cell.

The number of flexural mode phonons per unit area is calculated as follows[6]:

$$N_{ph} = (2\pi)^{-1} \int_0^\infty kdk[\exp(\alpha k^2) - 1]^{-1}; \tag{8.12}$$

where the integration is over wave-vector k. The exponential term in the brackets is the Planck/Bose-Einstein occupation function f_{BE}.α from Eq. (8.13) is:

$$\alpha = \hbar(\kappa/\sigma)^{1/2}/(k_B T); \tag{8.13}$$

where $\kappa = Yt^3$; and Y – Young's modulus, t = the film thickness, σ is the mass per unit area ~ ~7.5 x 10^{-7} kg/m^2.

The Planck/Bose – Einstein equation (the thermal occupancy of the mode at frequency ν):

$$f_{BE} = 1/[\exp(\hbar\nu/k_B T) - 1]; \tag{8.14}$$

Eq. (8.14) goes to $\exp(-\hbar\nu/k_B T)$ at high energy values and $k_B T/\hbar\nu$ a low energy values.

$k_B T/\hbar\nu$ is the number of phonons in the mode at frequency ν. At high temperature, the energy of phonons equals $k_B T$. Each harmonic oscillator increases the internal energy of a solid by $k_B T$. Thus, the internal energy is $3Nk_B T$ per crystal and $3k_B$ per atom.

Eq. (8.11) for a small wave-vector k, the integrand goes to $1/\alpha k$. N_{ph} in Eq. (8.11) becomes infinitely large as k comes close to zero, which means that the number of phonons per unit area goes to infinity. Phonon-phonon collisions depend on temperature or on an acquired energy by the material. At high temperature, phonon-phonon collisions are especially important (the atomic displacement are substantial). The corresponding mean free path, in this case, is inversely proportional to changes in temperature. The number of phonons increases with an increase of temperature.

Another origin of phonon scattering is impurities and crystal defects. Impurities and defects always exist in crystals. They destroy the crystal periodicity. However, at low temperature scattering from phonon-phonon and phonon-impurities is negligible. In the former case, there are few phonons and in the latter, only phonons with long wavelengths are excited. The phonon long wavelengths are much larger than the objects that cause the scattering in the first place.

Assuming a cut-off for a small k, $k_C = 2\pi/L$, where L is the sample's size:

$$N_{ph} = 2\pi / L_T^2 \ln(L/L_T); \tag{8.15}$$

where

$$L_T = 2\pi \hbar^{1/2}(\kappa/\sigma)^{1/4}/(k_B T)^{1/2}; \tag{8.16}$$

For T = 300 K $L_T \sim 0.3$ nm which is close to one lattice constant[6]. The thermal fluctuations associated with flexural phonons at room temperature are long enough to break free-floating graphene. In order for a sample to crumple, the material should experience large displacements. The crumpling is also connected with partial melting of the material, weakening the sample, membrane[7]. The coherence length associated with the size of the sample is given[8]:

$$\xi = \alpha \exp(4\pi\kappa / 3k_B T); \tag{8.17}$$

where α = lattice constant from (8.16), the density of phonons per area is 9 x 10^{20} m^{-2}. The energy range of phonons[6] starts at 1.46 x 10^{-12} eV. The phonon frequency as 2 THz at q = 4.73 x 10^9 is 355 Hz and the maximum flexural frequency is about 14 THz[9].

In Fig. 8.5 the specific heat for graphene is larger at low temperature because of a layer density of flexural phonons that exist at low frequencies[10]. The data for graphene shown by a dashed curve was received for exfoliated graphene beyond 200 K to 300 K. Since the internal energy at 300 K is based on $3k_BT$ per atom, we can calculate the energy for a platelet which can be 10s of GeV. The average thermal energy of a phonon is approximately 0.46 meV[6].

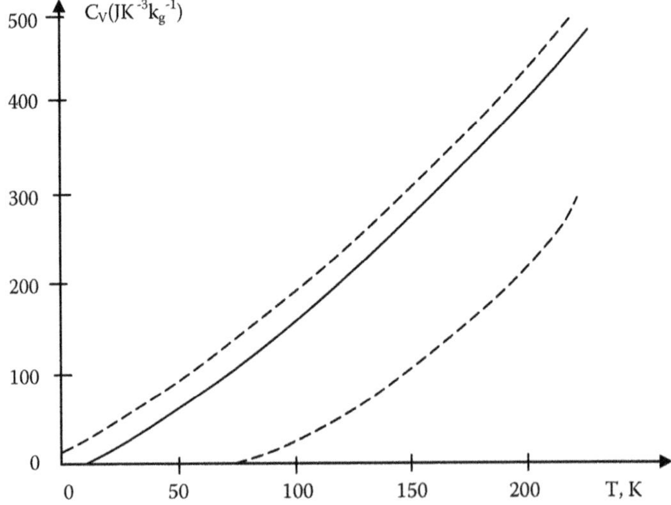

FIGURE 8.5 Comparison of specific heat calculations for graphene (dashed line) and graphite (solid line) curves.

8.3 THERMAL EXPANSION IN GRAPHENE

Graphene layers are closely separated (0.34 nm) and are bound by van der Waals forces. Graphene and graphite are similar as far as phonon dispersion is concerned. Fig. 8.6 shows phonon dispersion distribution for graphene and graphite. The following abbreviations are used:

LA = longitudinal acoustic, TA = transverse acoustic, LO = longitudinal optical, TO = transverse optical, ZA = transverse acoustical or flexural modes. The interaction between the layers in graphite may be expressed through several coefficients[11]:

$$v^2 \propto Aq^2 + Bq^4; \tag{8.18}$$

where $v \propto A|q|$, $A = C_{44}/\rho$ = sheer coefficient divided by the graphite density. $A = 0$ for one isolated layer.

A crystal even with only several atoms in a primitive cell has two kinds of phonons: acoustic and optical. A crystal with atoms $N \geq 2$ is a primitive cell that has three types of acoustic modes: two transverse acoustic and one longitudinal mode. The optical modes are calculated as $3N$ -3. The dependence of frequency on the wave-vector $\omega = \omega(k)$ is called "dispersion relation". The optical and acoustic components are shown in Fig. 8.7.

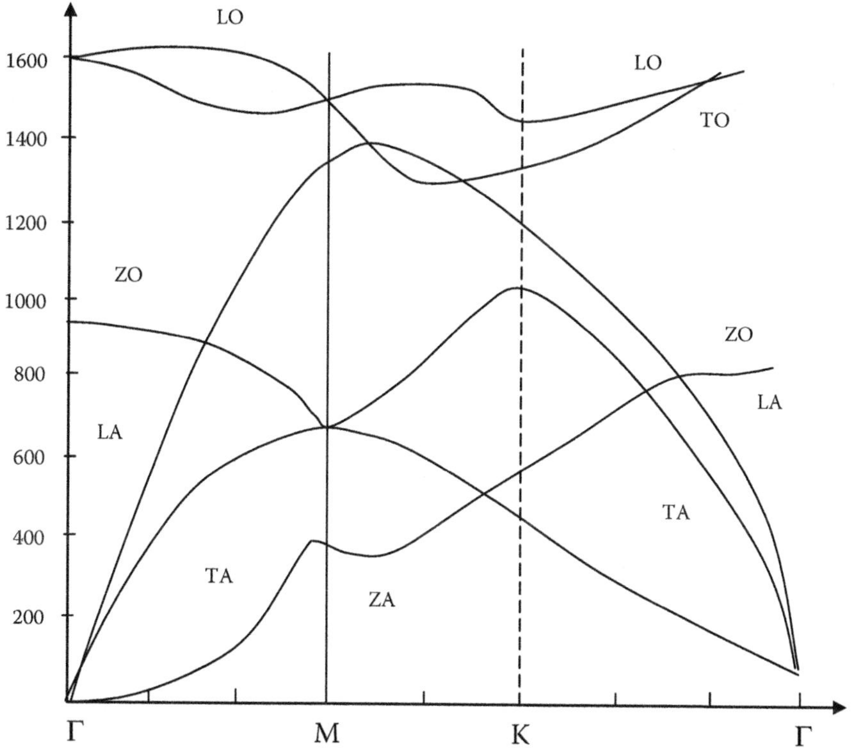

FIGURE 8.6 Comparison of graphene and graphite phonon dispersion.

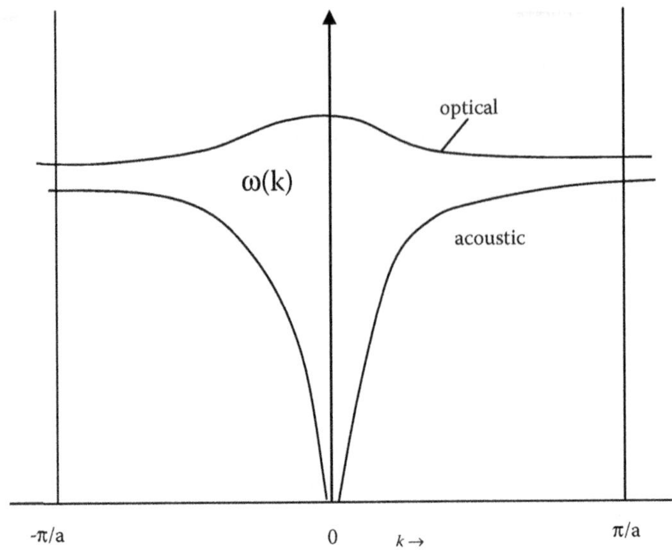

FIGURE 8.7 Acoustic and optical curves in the first Brillouin zone.

The electronic structure changes follow the lattice vibrations. The latter are small and the changes in the electronic structure take place correspondingly. The graphite expansion coefficient is negative and it reaches its maximum (the negative one) value at approximately 250 K. Graphene follows the same tendency but the coefficient is more negative, peaking up to 2300 K. At the room temperature, the graphene thermal coefficient is approximately -5.7×10^{-6} K^{-1} (see Fig. 8.8).

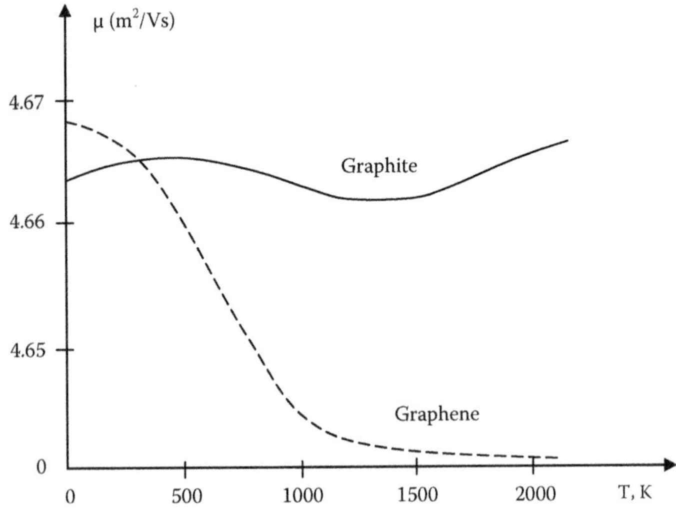

FIGURE 8.8 Graphite and graphene thermal expansion.

The growing amplitude of the out-of-phase flexural phonon is associated with graphene thermal contraction. The graphene planes bend into arcs[12]. The thermal amplitude is connected with thermal stability. The threshold of the thermal instability takes place when a thermal amplitude exceed the lattice constant. The thermal amplitude of a square sample with a side L and with the opposite sides in a fixed position is:

$$y_{th} = (kT/32Yt^3)^{1/2} L; \tag{8.19}$$

where Y is Young's modulus and t is the sample's thickness.

At room temperature the thermal amplitude is much larger than the lattice's constant and may be as large as several hundred nanometers. In this case, the lattice points are not in their stable positions, i.e. not in the positions described by the equation $\vec{R} = n\vec{a} + m\vec{b} + l\vec{c}$.

The thermal effect on the lattice is not unlike the mechanical analogy in engineering structures. The graphene thermal effects are, however, more substantial due to infinitesimally smaller case of phonon interactions.

The thermal acoustic mode (ZA) may be represented by analogy to mechanical bending:

$$h(x,t) = A\exp(-ikx + i\omega t); \tag{8.20}$$

where ω = width of the beam, A = area ($A = \omega t$) and t = thickness.

$$\omega = \sqrt{YI/\rho A}; \tag{8.21}$$

where Y = Young's modulus and $I = \omega t^3$.

A mechanical analogy including tension T gives the frequency of clamped beam as:

$$f_0 = \left\{[A(E/\rho)^{1/2} t/L^2]^2 + A^2 0.57T/\rho L^2 \omega t\right\}^{1/2}; \tag{8.22}$$

where E(or Y) is the Young's modulus, T = tension applied to the beam, N and ρ = density. L, ω and t are the dimensions of the beam[1].

The thermal change of amplitude for the beam can be found using an approximation. The beam vibration may be expressed in the form of $\omega = \sqrt{K^*/M^*}$ where the beam is clamped on both sides and the force is applied at its middle point.

$$K^* = 32Y\omega t^3/L^3; \tag{8.23}$$

At the fundamental mode, the rms thermal amplitude is given:

$$y_{rms} = (k_B T/K^*)^{1/2}; \tag{8.24}$$

The amplitude of the flexural mode is proportional to size L in case $L = \omega$:

$$y_{rms} = (k_B T / K^*)^{1/2} = (k_B T / 32Yt^3)^{1/2} L; \tag{8.25}$$

The thermal amplitude, though, does not affect a possible breaking transition of the graphene sample, since the thermal amplitude is small in comparison to the sample's dimensions:

$$y_{rms}/L = (k_B T/32Yt^3)^{1/2}; \tag{8.26}$$

Thermal vibrations encompass the range 1 MHz to 1 THz which depends on the length of the mounting, L and is restricted by the boundaries of a graphene specimen.

One important point is possibility of increasing the tensile strength of graphene by charge doping up to 17%. Incidentally, the critical tensile strain for pure graphene is only 15%. This phenomenon is explained by "stiffening" of the highest frequency mode K due to the doping (see Fig. 8.9)[13].

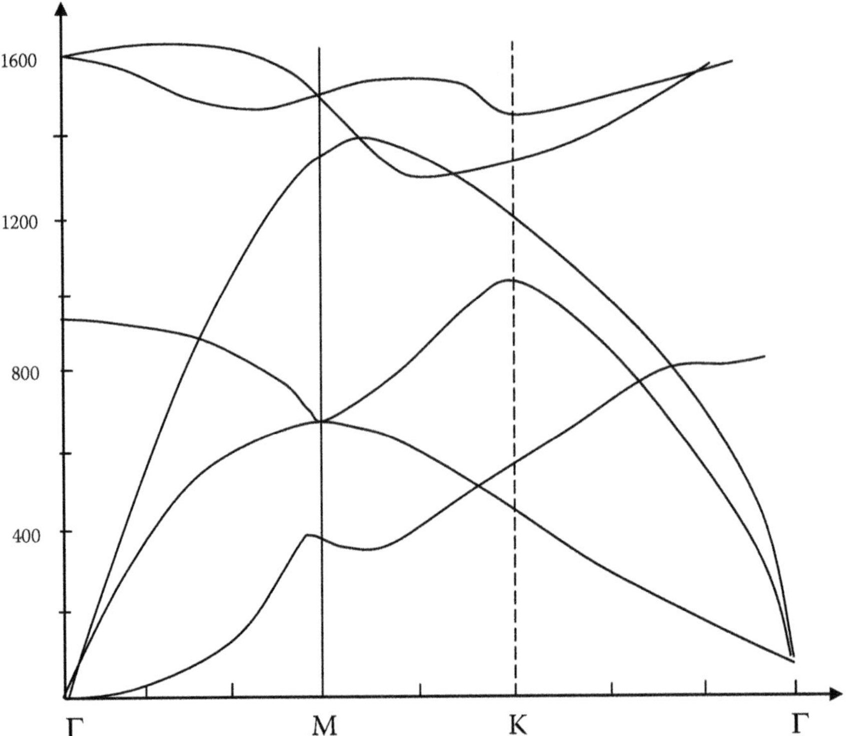

FIGURE 8.9 Phonon frequency distribution in the first Brillouin zone.

The critical wave vector q_c, has a frequency that is proportional to $q^{3/2}$ the width ω:

$$\omega = (\kappa_0 q_c/\rho)^{1/2} q^{3/2}; \tag{8.27}$$

where κ_0 = stiffness (may be on the order of 1 eV).

And the cut-off wave vector is;

$$q_c = (3K_0 k_B T / 8\pi\kappa_0^2)^{1/2}; \tag{8.28}$$

The value of q_c may be on the order of tens of a THz[14]. In Fig. 8.9 this point is between Γ and M. From Fig. 8.9 in the Brillouin zone:

$$K_0 = 4\mu(\mu+\lambda)/(2\mu+\lambda); \tag{8.29}$$

where $\mu = 4\lambda$.

The usefulness of the above theoretical description is in determining and predicting graphene's resistivity caused by electron scattering from the flexural modes. The resistivity prediction is[14]:

$$\rho \propto T^{5/2}\ln(T); \tag{8.30}$$

where $\ln(T)$ = relaxation of the flexural mode.

The tensile strain induces linear dispersion[15]:

$$\omega^2 = (\kappa_0/\rho)q^4 + \omega v_L^2 q^2; \tag{8.31}$$

where $v_L^2 = (2u+\lambda)/\rho$; where u = tensile strain.

8.4 GRAPHENE SURFACE NON-UNIFORMITY AND ELECTRON DIFFRACTION METHODS

Transmission electron microscopy (TEM) provides methods of graphene surface characterization. Different diffraction patterns give an image of the reciprocal lattice of graphene crystal structure[16]. Diffraction spots from graphene from the reciprocal lattice atoms in case of a single layer look like rods (Fig. 8.10 a) and b)). The appearance of rods comes from the two-dimensionality of graphene which may be visualized as a three-dimensional structure stretched in one direction. Thus, diffraction points of one of the 3D become a line. The Ewald sphere intersects the diffraction rods (Fig. 8.10 c)). The Ewald sphere is a geometrical figure used in crystallography investigated by X-rays, electron or neutron bombardment. The Ewald sphere allows finding wave vectors of the incident diffracted beams, diffraction angles if reflection angle is known. It also permits building of the reciprocal lattice (Fig. 8.11).

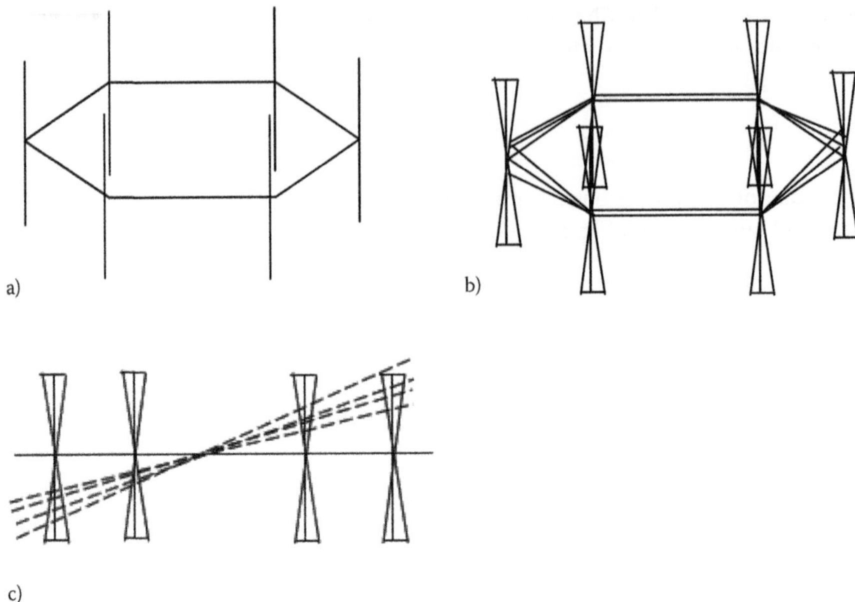

FIGURE 8.10 a) Diffraction lines for a perfect graphene crystal structure; b) Tilled diffraction lines caused by tilted graphene planes; c) Tilting of the Ewald sphere with its radius depicted by dotted lines at different tilting angles[16].

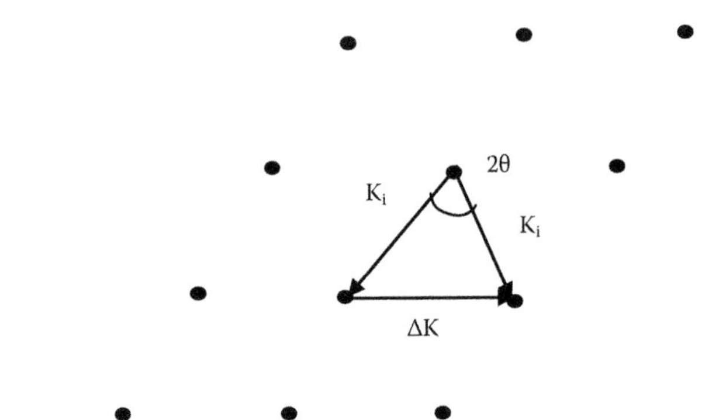

FIGURE 8.11 The Ewald sphere application to crystal characterization.

The Ewald sphere has a wave vector K_i (with the length of $2\pi/\lambda$) that is associated with the incident plane wave directed at the crystal. The diffracted plane wave acquires a wave vector K_f. K_f has the same length as K_i provided there is no energy loss in the diffraction process.

$\Delta K = K_f - K_i$ is defined as a scattering vector. Since wave vectors K_i and K_f have equal length, the scattering vector is present on the Ewald sphere with the radius $2\pi/\lambda$. The diffracted intensity *I(k)* from a graphene sample is:

$$I(k) = [\sin(\pi\omega k) / \pi k]^2; \quad (8.32)$$

where ω = sample's diameter and the full-width at half maximum is $\pi\omega k = \pi/2$. The observed width encompasses diffraction region of width $\omega \geq 5\text{nm}$. The actual width may be, however, about 50 nm. The tilting angle causes the broadening of the diffraction spots (Fig. 8.12).

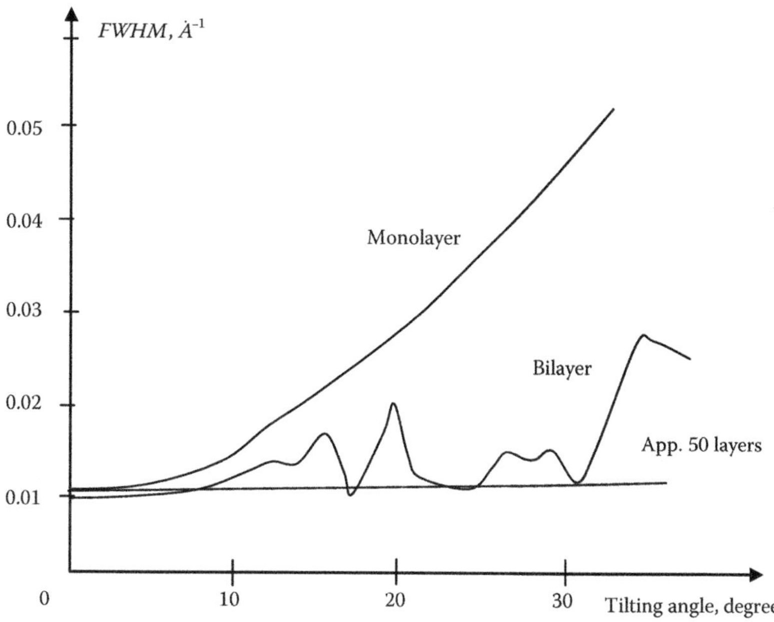

FIGURE 8.12 The dependence of the width of diffraction spots versus the angle of the tilt[17].

Another option of observing diffraction spots of graphene is explained in Fig. 8.13. The larger diameters of the diffraction spots are believed to be due to extrinsic undulations caused by the charge of the amorphous quartz layer. The layer is shown on Si substrate (see Fig. 8.14). The surface roughness is, therefore, due to the above undulations. The understanding of this effect has obvious practical applications for graphene devices.

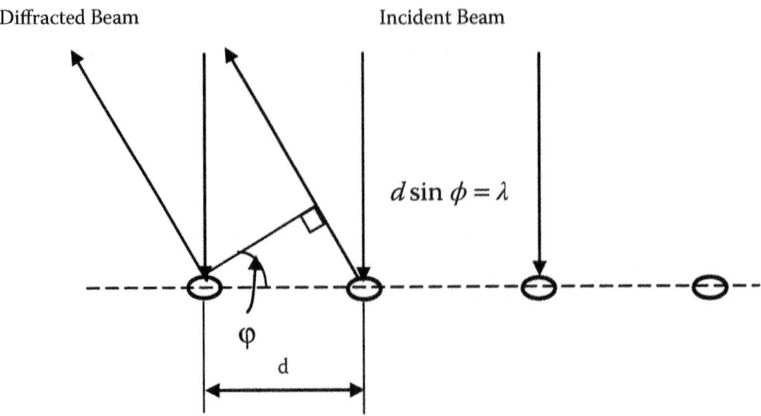

FIGURE 8.13 The diffraction condition for incident and diffracted beams.

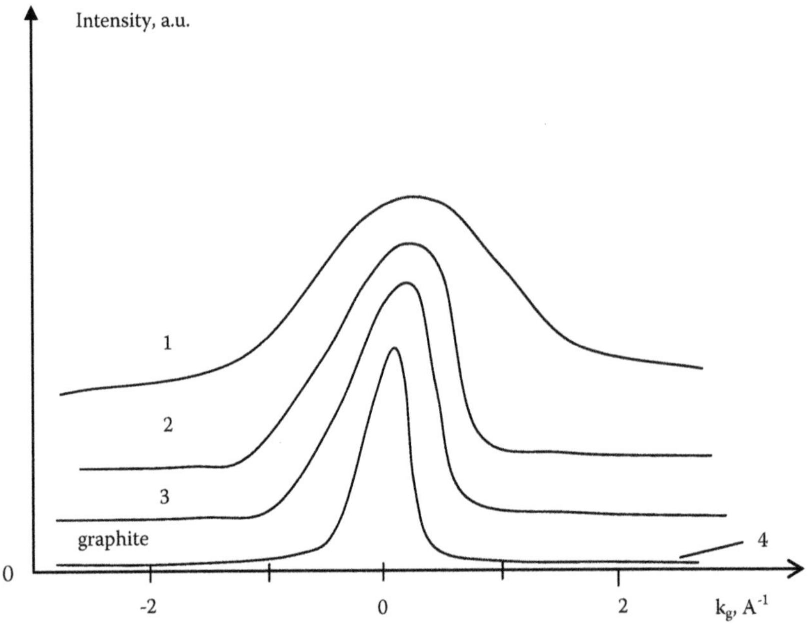

FIGURE 8.14 Intensity of low-energy electron diffraction for an exfoliated graphene monolayer (1), bilayer (2) and a trilayer (3) compared to a graphite layer (4). The substrate is a Si wafer with quartz on it.

Thermal instability of graphene also influences the roughness of the material. In particular, for suspended graphene monolayers the rms roughness reaches 1.7 A with the tilt angle of approximately 6 degrees[16]. The temperature range was from 150 K to 300 K. Also, unexpectedly the corrugation was higher at 150 K than at 300 K. The wavelength that corresponds to the lattice vibrations increases from 10 nm to 18 nm (at 500 K). The graphene samples dimensions were comparable to the width of a trench over which the sample was suspended (($0.5 \div 15\,\mu m$). The above thermal tendency does not involve the flexural phonons[16]. It seems likely that surface contamination with hydrocarbon plays the role in the graphene surface quality. Electrical measurements can indicate the degree of the contamination. On the other hand, surface absorption provides the possibility of doping the surface layer. The contamination can also result in the appearance of ripple on the surface. The compressive strain, however, does not result necessarily in graphene buckling or rippling.

The growth of graphene, in particular, on the surface (0001) of 6H –SiC results in uniform compression. The flat conformation of the graphene sample's strain is caused by van der Waals attraction to the substrate. The buckling vertical amplitude is negligible (approximately, 0.5A) with the diameter of the buckling region is 13A[18]. The buckling occurs in defect-free graphene samples. Thus, graphene has a non-zero threshold for forming of distortions, such as, e.g., buckling.

8.5 ELECTRONIC PROPERTIES OF GRAPHENE

Conductance in graphene being rather sublinear than linear is influenced by thermal processing. Annealing, in particular, decreases the minimum conductivity. In Fig. 8.15, we can see two graphene conductivities: metallic and non-metallic. Large carrier densities created metallic conductivity which decreased with increasing of temperature. Low densities produced the opposite dependence, i.e. the conductivity increased with temperature. The mobility after annealing (Fig. 8.15) was estimated as 17 m^2/Vs with the carrier density $n = 2 \times 10^{11}$ cm^{-2} at T = 40 K[19].I_m is a free path. The upper curve shows the effect of modest temperature (~ 400^0C) annealing. Here, the mean path is substantially longer (I_m ~ 1 μm). The lower curve corresponds to suspended graphene. Similar measurements for the process of annealing were performed showing a large change in conductivity (Fig. 8.16). The annealing took place in vacuum with applied high current. The temperature reached T ~ 300^0C and the Dirac point shifted from 40 V to ~ 8 V. The nanotube was mounted on a Si/SiO_2 substrate 300 nm thick[20].

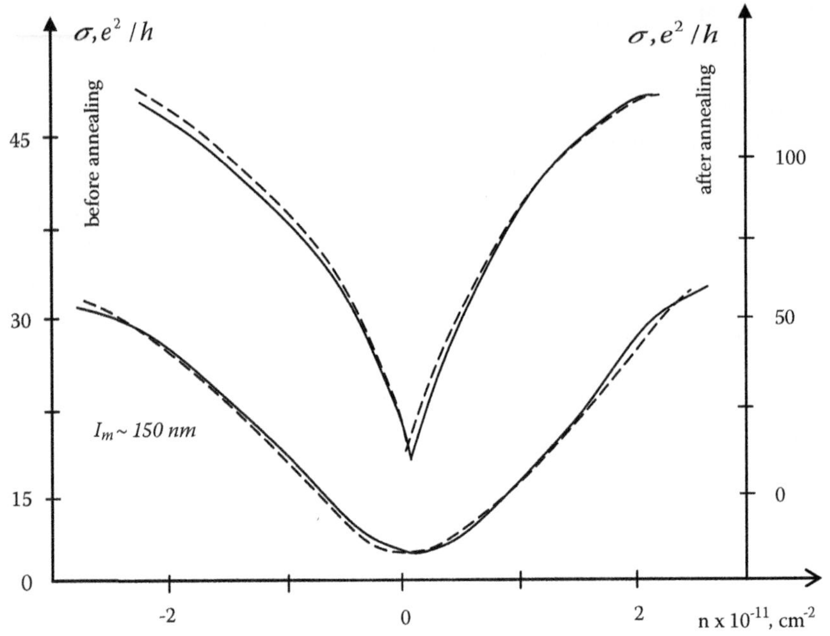

FIGURE 8.15 Conductance vs carrier density including the annealing effect.

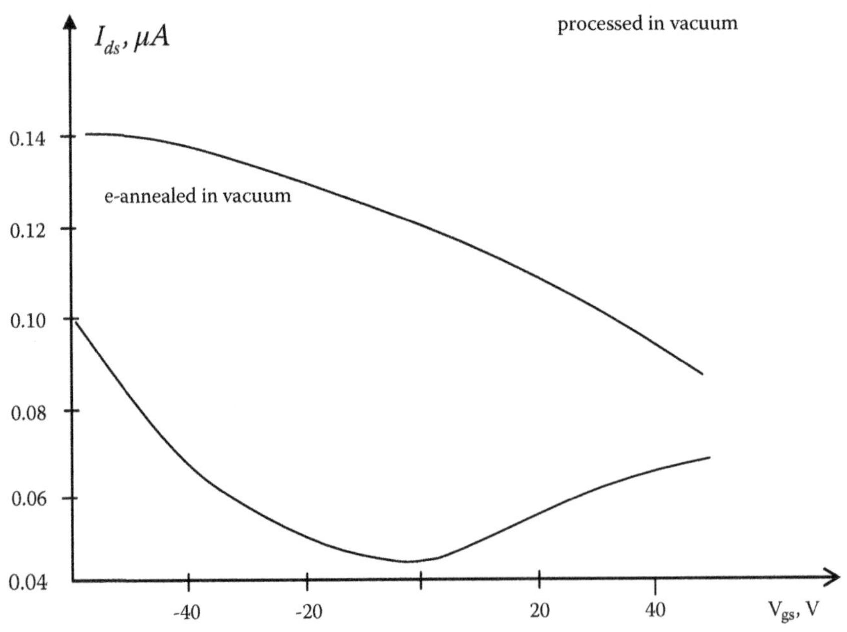

FIGURE 8.16 Effect of annealing for a wide graphene nanoribbon.

Removal of scattering centers leads to a reduction in resistivity which is, probably, related to ripple reduction. The ripple dimensions are on the scale of Angstroms. The conductivity with the metallic behavior is in the range $\mu = \sigma / en$ with n > 0.5 x 10^{11} cm^{-2}. The mobility is approximately 12 m^2/Vs at room temperature which is higher than in conventional materials. In case, of scattering from charged impurities, the conductivities is[19]:

$$\sigma = (2e^2/\hbar)k_F\upsilon_F\tau; \quad (8.33)$$

Annealing changes the dependence of σ(*n*) on *n* (carrier density) from linear to sub-linear after annealing. τ is the mean free time. The mean free path is:

$$\sigma = (2e^2/\hbar)k_F\upsilon_F\tau; \quad (8.34)$$

where $\sigma \propto n$ and τ = the mean free time.

The mean free path is:

$$l = \sigma\hbar/2e^2k_F; \quad (8.35)$$

where k_F = wave vector.

In case of substantial scattering, l may be equal to a 100 nm or more. Annealing removes scattering to a large extent, the conductivity corresponds to the dash curve in Fig. 8.15 which is ballistic in nature:

$$\sigma_{ball}(n) = 4e^2/\hbar N = 4e^2Wk_F/\pi\alpha\sqrt{n}; \quad (8.36)$$

where N is the number of longitudinal quantum channels, W = width.

The ballistic curve in Fig. 8.15 is close to the experimental data which means that at T = 40 K graphene is almost an ideal conductor. The conductivity only slightly decreases with rising temperature but the mobility remains high (12 m^2/Vs). It is higher than the highest mobility in semiconductors (7.7 m^2/Vs in InSb). The character of conductivity was also observed at room temperature[21]. The temperature dependence of mobility vs temperature is shown in Fig. 8.17.

Similar to semiconductors, a very high mobility can be measured at very low temperatures. The curves in Fig. 8.17 correspond to the measurements from three samples[6]. Measurements from a two-terminal set-up were difficult to execute creating the necessity of several experiments producing the three curves. For their analysis the following expression was used:

$$1/\mu = (1/\mu)_{T\to 0} + \gamma^{T^2}; \quad (8.37)$$

where

$$\gamma = (Dk_B/\kappa\upsilon_F)^2(64\pi e\hbar)^{-1}\ln(k_BT/\hbar\omega_c); \quad (8.38)$$

where ω_c = the cut-off frequency that corresponds to low-energy and a long wavelength changing to model the strain in the characterized samples. The bending rigidity $\kappa = Yt^2 \approx 1eV$.

$\upsilon_F \sim 10^6$ m/s, and D = deformation potential which is given:

$$D = [g^2/2 + (\beta \hbar \upsilon_F / 4\alpha)^2]^{1/2}; \tag{8.39}$$

where $g = g_0/\epsilon k_F$ = electron-phonon scattering potential which is equal approximately to 3 eV.

B is given as:

$$\beta = -\delta \log t / \delta \log \alpha; \tag{8.40}$$

where $t \sim 3$ eV.

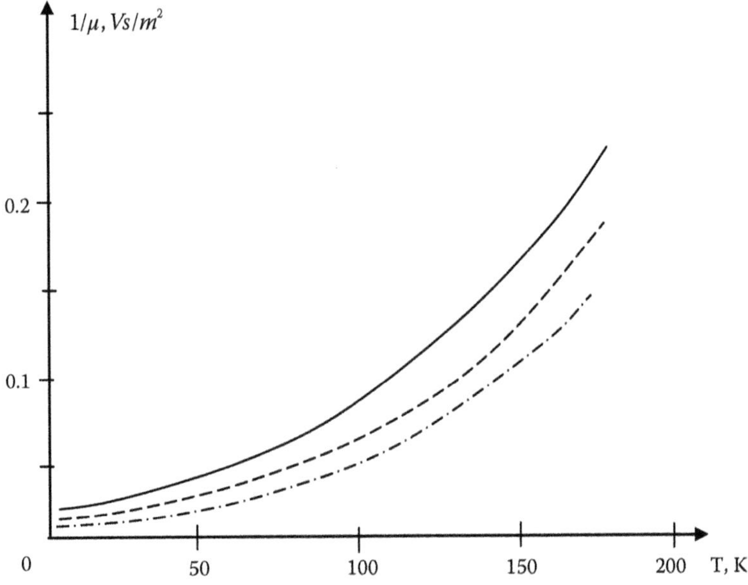

FIGURE 8.17 Dependence of inverse mobility on temperature[6].

8.6 EXTRINSIC AND INTRINSIC EFFECTS IN GRAPHENE

At the moment, it is not absolutely clear whether the nature of graphene corrugations is extrinsic or intrinsic. One version suggests that the observed effects come from mounting strain and disturbed local bonding caused by adsorbents. The thermal oscillating corrugations' mean square, h_{rms}^2 is[22]:

$$\langle h^2 \rangle = h_{rms}^2 = (k_B T / 2\pi) \int_{1/L}^{1/a} \frac{qdq}{q^2} / k(g) \approx TL^2 \varsigma; \tag{8.41}$$

where L = the length of a layer, a = lattice constant, $\zeta = 1$.

Considering coupling between in-plane and out-of-plane motions reduce ζ from 1 to 0.59[23]. There are also out-of-plane fluctuations that are in the range of the frequency of the phonons of the system. A simulation showed no defects of the hexagonal lattice and no melting up to 3500 K[24]. The heights of fluctuations varied considerably with an average of $h_{av} = 0.07$ nm at room temperature. The fluctuations had the peaks at about 80 $\dot{A}$ (in the range of 50 – 100 $\dot{A}$). The peak (height) distribution varies with the normal vector to the plane $\vec{n}$. The in-plane components of the $\vec{n}$ are given in terms of h_q, the Fourier components:

$$\langle h^2 \rangle = \sum \langle |h_q|^2 \rangle a(k_B T / \kappa) L^2; \tag{8.42}$$

where

$$\langle |h_q|^2 \rangle = (k_B T N / \kappa S_0 q^4) \tag{8.43}$$

where N = number of atoms and $S_0 = L_x L_y / N$ = the area per atom.

Further continuing the simulation[24], we use the "correlation for the normal", $G(q)$:

$$G(q) = \langle |\vec{n}_q|^2 \rangle = q^2 \langle |h_q|^2 \rangle; \tag{8.44}$$

A simplified approximation is:

$$G_0(q) N = (k_B T / \kappa S_0 q^2); \tag{8.45}$$

The angle between the local normal $\vec{n}$ and the (average) perpendicular is given as:

$$\cos\theta = 1/[1 + (\nabla h)^2] \approx 1 - \frac{1}{2}\langle \theta^2 \rangle; \tag{8.46}$$

And

$$\langle \theta^2 \rangle = \langle (\nabla h)^2 \rangle = (k_B T / 4\pi^2) \int \frac{q dq}{q^2} / \kappa(q); \tag{8.47}$$

κ must renormalized in order for the above integral to converge[25]:

$$\kappa_R(q) \approx (k_B T \kappa_0)^{1/2} q^{-1}; \tag{8.48}$$

Renormalization is a physical process based on stretching and bending interaction[26]. The height of the fluctuations reduces below $\langle h^2 \rangle \propto (k_B T) / \kappa / L^2$ because of the renormalization. Nevertheless, the fluctuations are still too high and can exceed the inter-atomic distances, at least for large samples. This phenomenon is true for 2D structures. In case of simulations lattice vibrations there exists an intrinsic tendency to forming ripples[27]. The amplitude of the transverse fluctuations is proportional to the sample size or, more precisely, to $L^{0.6}$. Thus, the size of the sample is much bigger than L and may be considered without ripples or corrugations. The calculated function $G(q)/N$ gives a peak for q. Fig. 8.18 shows reduction of fluctuation depending on renormalization. The dispersion of flexural modes as a consequence of renormalization[28] as $\omega \propto q^{1.6}$ if $\kappa_R(q) \approx q^{-\eta}$ and $\eta = 0.82$.

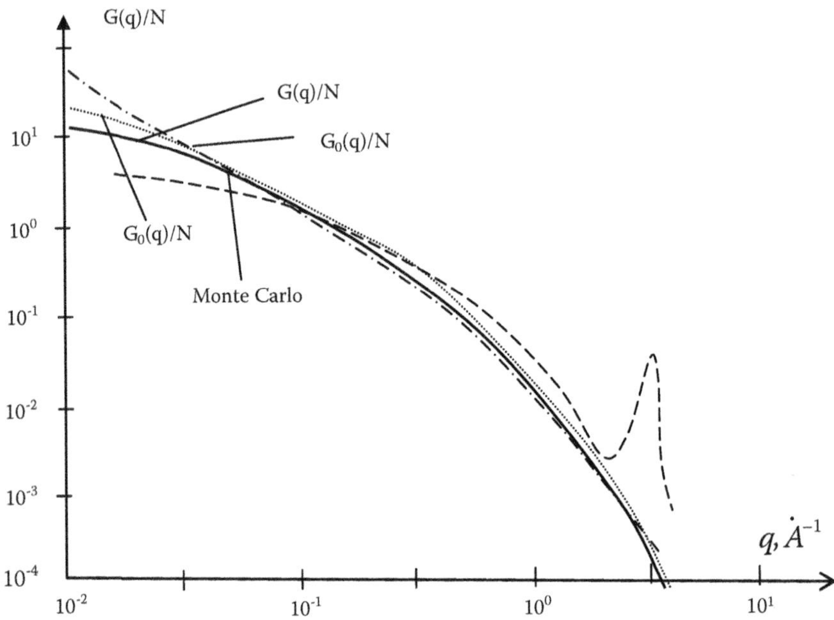

FIGURE 8.18 Comparison of different functions and Monte Carlo height-height correlation function. The two top curves are unrenormalized[28].

Radial distribution functions are given in Fig. 8.19 for two temperatures T = 300 K and T = 3500 K. The sample had 8640 atoms of carbon. The dashed line is an anomalously wide distribution of bond lengths. If centered at 0.142 nm which is a bond length. The left arrow corresponds to the double bond length of 0.131 nm and the right arrow stands for 0.154 nm band length. Similar to the benzene molecule the bonds adjust without the atoms changing their positions. The vertical motion is not likely to disturb the atoms since the extent of vertical motion (0.7 $\dot{A}$) is much less than the distance that separates graphene planes. The Debye temperature that corresponds to a crystal's highest normal mode of vibration:

$$\Theta_D = \frac{\hbar v_m}{k}; \tag{8.49}$$

where h = Planck's constant, ν_m = the Debye's frequency, k is Boltzmann's constant.

The highest temperature is achieved because of a single normal vibration. The Debye's frequency is a characteristic frequency of a crystal which is:

$$\nu_m = \left(\frac{3N}{4\pi V}\right)^{1/3} \upsilon_s; \quad (8.50)$$

where N/V = number density of atoms, υ_s = the effective speed of sound in the solid.

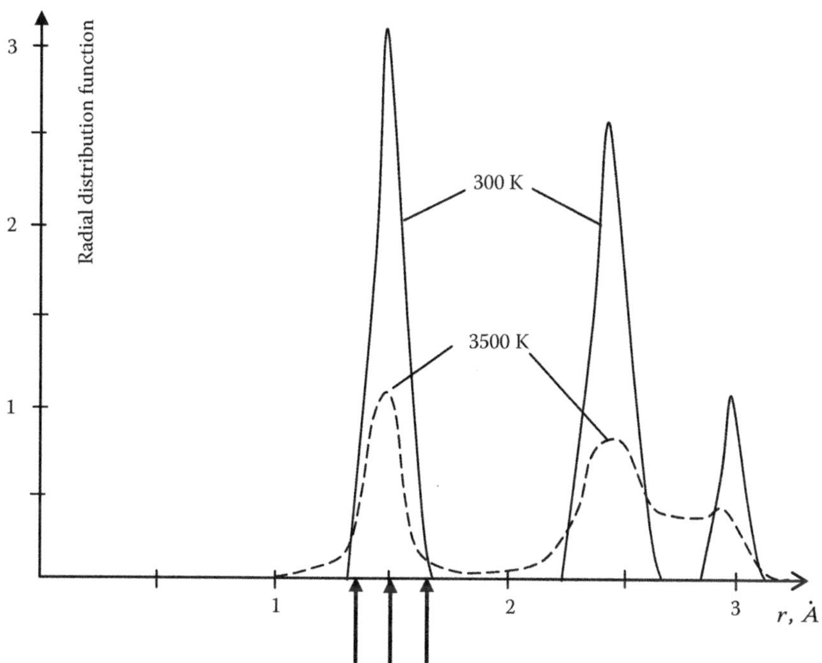

FIGURE 8.19 Radial distribution functions[27].

This atomic motion in graphene depends on the Debye temperature Q_D ~ 900 K for the motions out of plane and Q_D ~ 2500 K for motion in plane[29]. The above dependencies imply that at room temperature graphene is still in its ground state and at higher temperatures substantial distortions take place.

A number of researchers investigated defects in graphene layers and carbon tubes. Among defects, there are vacancies, topological defects, dislocations and some others.

Divacancies (Fig. 8.20) are two pentagons and an adjacent octagon ("585") and three pentagons with adjacent three heptagons ("555 777") are found in graphene and nanotubes (Fig. 8.21). The formation energy for the two above defects are 7.8 eV and 7.0 eV respectively.

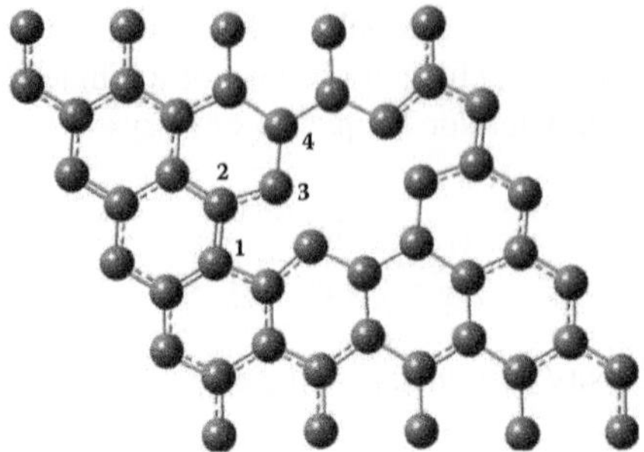

FIGURE 8.20 Forming of a divacancy in graphene when two neighboring vacancies coalesce with energy gain[30].

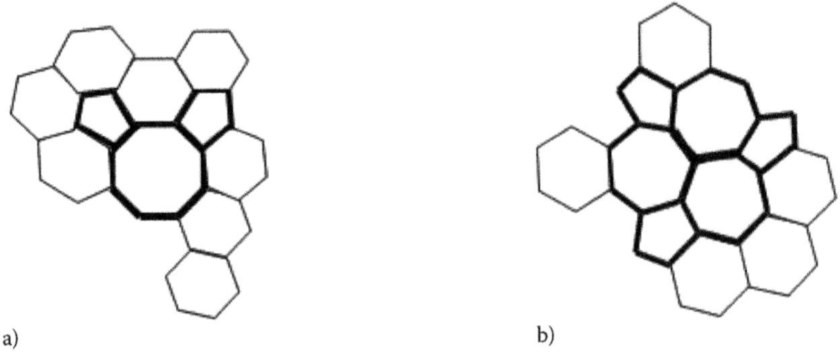

FIGURE 8.21 a) Two pentagons with an octagon ("585"); b) Three pentagons with three heptagons ("555 777")[31].

The size of a crystal when crumpling takes places is:

$$\xi_T \approx a(R_C / a)^{\beta}; \tag{8.51}$$

with $\beta = K_0 a^2 / 16\pi k_B T$; a = lattice constant; $R_C \approx 60a$; $K_0 \approx 20 eV / \dot{A}^2$; R_C = critical size of a flat plate when crumpling can take place. ξ_T is a crystalline length, beyond which the crystalline order appears fluid. In order for the above crystalline dimensions to have effect the temperature should reach 3900 K. In particular, β is ~ 100 larger approximately at 3900 K than at room temperature K = 300 K.

Defects in graphene may be difficult to localize. Two types of a dislocation are explained in Fig. 8.21. Because of heptagon presence pentagons are formed next to them.

The formation of corrugations, on the other hand, is obvious but the true reason for their formation is still being debated. One possible origin is based on adsorption[32]. A simulation of ripple appearance on a graphene sheet is shown in Fig. 8.22.

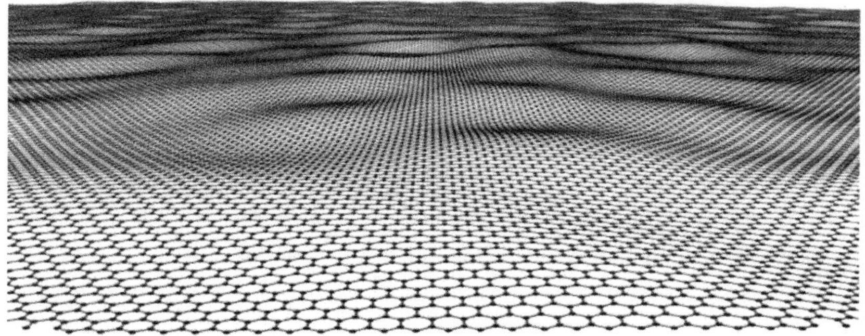

FIGURE 8.22 A simulation of a ripple formation of a graphene sample surface. Small displacements produce waves which do not penetrate the sample.

An unchanging lattice electron system will yield electron diffraction pattern. Simulations were used to specify the role of adsorbents in maintain the static distortion. Displacement of atoms on the edge of a sample produce waves without their penetration into the sample. The simulation of the above processes showed that the observed corrugations were not an inherent feature of graphene and were possibly the result of defects inside the material. A substrate can add to the presence of surface non-uniformities. An example is graphene deposited on quartz which was, in its turn, deposited on Si. Current measurement after annealing allows distinguishing between undulations caused by the substrate and by adsorbates. Adsorption is a surface phenomenon in which atoms, ions or molecules from a gas or liquid create a film on the adsorbant's surface. The subsequent decrease of undulations due to the annealing can be explained by adsorbate-induced buckling[33].

8.7 GAS PRESSURE AND GRAPHENE CRYSTAL STRUCTURE

One feature of graphene crystal structure is that it is in impervious state with respect to gases. Only large defects allow gas molecules to penetrate the crystal[34]. The energy necessary for helium penetration through a benzene (six-fold) ring is estimated to be approximately 18.8 eV through the thickness of approximately 1.43 Å.

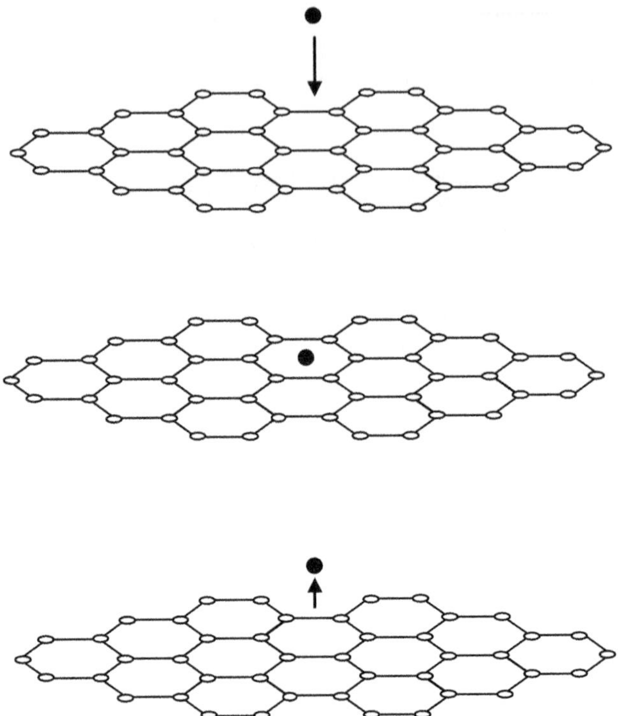

FIGURE 8.23 Reflection of a helium atom from a graphene crystal. a) A helium atom approaches a graphene crystal; b) A helium atom encounters the crystal; c) The helium atom bounces off (reflects) from the graphene surface.

The mechanism of helium penetration takes place by breaking several bonds that open a fragment of space through which atoms of helium can pass (Fig. 8.23). Without breaking bonds, however, graphene is impermeable to gases, including helium. One of the applications of this phenomenon is making devices in which gas pressure changes resonant frequencies[35].

Another possible application of graphene selective penetrating ability is blocking gases but diffusing water through graphene-based membranes. The water penetration is an array of 2D capillaries. The water goes through the network of capillaries[36]. The membranes consist of platelets that are parallel to each other with the distance of ~ 1 nm between platelets. Each membrane is about 1 μm thick.

8.8 METALLIC TRANSITIONS IN GRAPHENE

At low temperature, the Fermi level shifts to the Dirac point (Fig. 8.24).

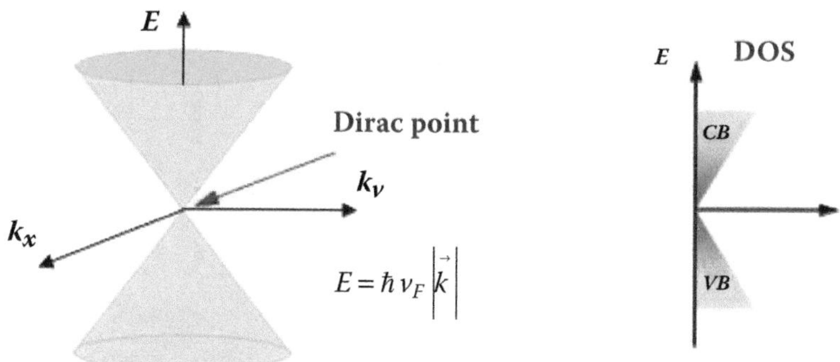

FIGURE 8.24 Dirac point in the graphene electronic structure. (Right side) density of states (conduction (CB) and valence (VB) bands). (Leibniz-Institut fur innovative Mikroelectronik).

Charge carriers linearly disperse in the vicinity of the Dirac point (the Fermi energy). The carriers are confined to a one-atom thick layer and travel at very high velocities (such as ~ 4 x 10^7 cm^2/Vs) and very high mobilities (at T = 240 K, μ ~ 120, 000 cm^2/Vs). The DOS (Density of states) at the Dirac point is equal to zero and increases linearly with respect to the energy which makes it possible for carrier modulation.

At 10 K the resistivity is 500 kΩ/square[37]. Thus, reported earlier, "minimum metallic conductivity" is a local phenomenon that depends, e.g. on artifacts encountered during the experiment. If the impurities possess an electric charge, the resulting electric field broadens at the Dirac point. However, the Mott-Anderson transition at the Dirac point has never been properly observed[37]. Originally, a metallic transition in a 3D structure of hydrogen atoms with electrons screened the electrostatic binding of the ionic centers and the transitions were influenced by disorder.

In graphene, the conductance varies from logarithmic to exponential and its decrease depends on the sample size L. Disorder excludes the possibility of minimal conductivity, and under all conditions, the conductance $g(L \to \infty) = 0$. Consequently, at temperatures as low as 10 K, and low carrier concentration, conductance equals to zero. For conical band, the situation is the same. Being temperature dependent, the maximum resistivity is 33 kΩ at 10 K[37]. The measurement set-up for double graphene layer parameters is depicted in Fig. 8.24.

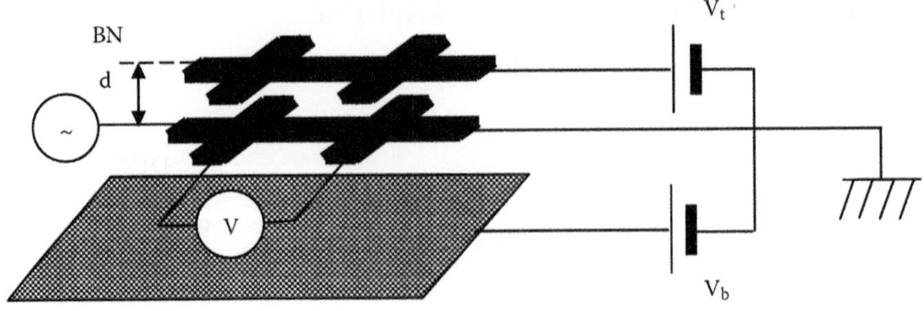

a)

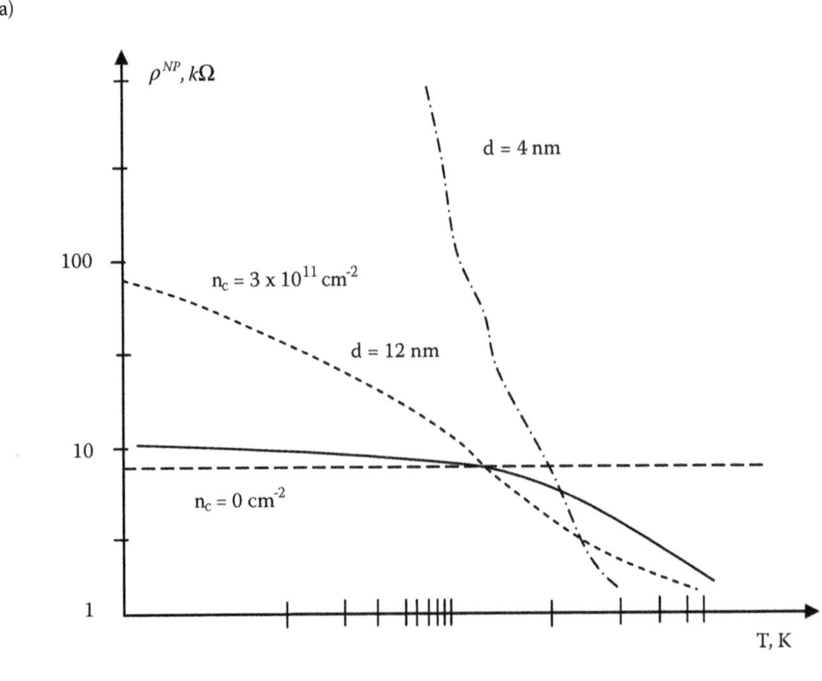

b)

FIGURE 8.25 Measurements of double-layer graphene structure; a) the electric diagram of the Hall voltage measurement; b) the dependence of the neutral point resistivity on temperature[37].

The insulating region in the lower layer has V_t = const, to set up constant carrier concentration n_c in the upper screening layer. In Fig. 8.25 a) two layers rest on h-BN single crystals. The lower layer is placed on oxidized Si substrate. No electron tunneling is possible between the two graphene layers. The measurements of the Hall effect coefficient and layer resistivity are performed as it is shown in Fig. 8.25 a). The applied voltage, V_t, does not induce a current but creates areas of charge with opposite signs. The vertical magnetic field causes the Hall effect that monitors the charges of opposite sings. Fig. 8.25 b) shows the dependence of neutral point resistivity (ρ^{NP}) vs. temperature. The mobility of the upper layer was unstable and was decreasing but the mobility of the lower level was stable[37]. The resistivity is largely affected by the screening concentration n_c. The back-gate voltage V_b varies the carrier concentration in the lower layer. The lower layer neutral point resistivity ρ^{NP} (Fig. 8.25 b)) can reach mega Ohm values having a high carrier concentration. This effect is possible at temperatures below 4.2 K in the screening layer or in the measured layer if the layers are reversed. The screening is applied as in a usual field-effect transistor configuration (Fig. 8.25 a)). The upper film that needs to be measured is fixed above an initial graphene layer and below is a biasing substrate. The tuning is possible by voltage applied to the back gate. Subsequently, the Dirac point should be reached.

Graphene in its intrinsic state has zero conductance at T = 0K[39]. Impurities, artifacts and other irregularities mentioned earlier influence the mobility and conductivity of graphene. The graphene surface is easily penetrated by outside substances. In addition, random electrical charges cause local electrostatical doping. Graphene adheres to a substrate by van der Waals forces. The substrate has an uneven surface with local doping and impurities. The initial popular choice, amorphous SiO_2 does not provide a uniform surface, has a number of local charges and no uniformity at the Dirac point. Since large shifts of Fermi energy are needed to create surface charge, graphene manifests minimum conductance, typical for the material conical density of states. Then, non-polar substrates and electrostatic screening are helpful to minimize electrostatic puddles.

The presence of intrinsic ripples is doubted. Some of them may be caused by thermal fluctuations[40]. Simulations give the height of thermally-caused ripples as $h_{rms} \approx 9$ Å. The height may be calculated approximately:

$$h_{rms} = \sqrt{k_B T / K^*} = L\sqrt{k_B T / 32Yt^3}; \tag{8.52}$$

where L = length of a side of the sample[41].

The root-mean-square of thermal oscillation ω may be calculated:

$$\omega = [Yt^2 / 12\rho(1-v^2)]^{1/2}\pi^2[2/L^2]; \tag{8.53}$$

The frequency for a sample growth sides with length 10 nm is approximately 10 GHz.

8.9 GRAPHENE DISINTEGRATION

At T = 3900 K graphene layers detach from the graphite crystal. Thus, at T = 3900 K and below, we have a local order and disintegration or melting of graphene at T > 3900 K. The disintegration takes place the same way in 2D crystals as graphene disintegrates into carbon strings similar to polymer strands[42]. The fragments of the crystals may be described as "liquid" rather than fragments moving to infinity ballistically. Simulations of the process may be more conveniently done in a fixed crystal volume with imposed boundary conditions. In this case, we have fragments leaving the initial volume that still has some non-zero density. Please note, that leaving a 2D graphene crystal, non-zero fragments go into a 3D direction not confined to the 2D space. One of the features of this disintegration is a local crumpling with appearance of five and seven –link ring defects. Below T = 4900 K, where the local order of elastic membranes holds, the crumpling follows the classical treatment of bending sheets and beams. Fig. 8.26 shows the melting temperature of carbon nanotubes as a function of radius. A smooth curve extrapolates to approximately 5800 K.

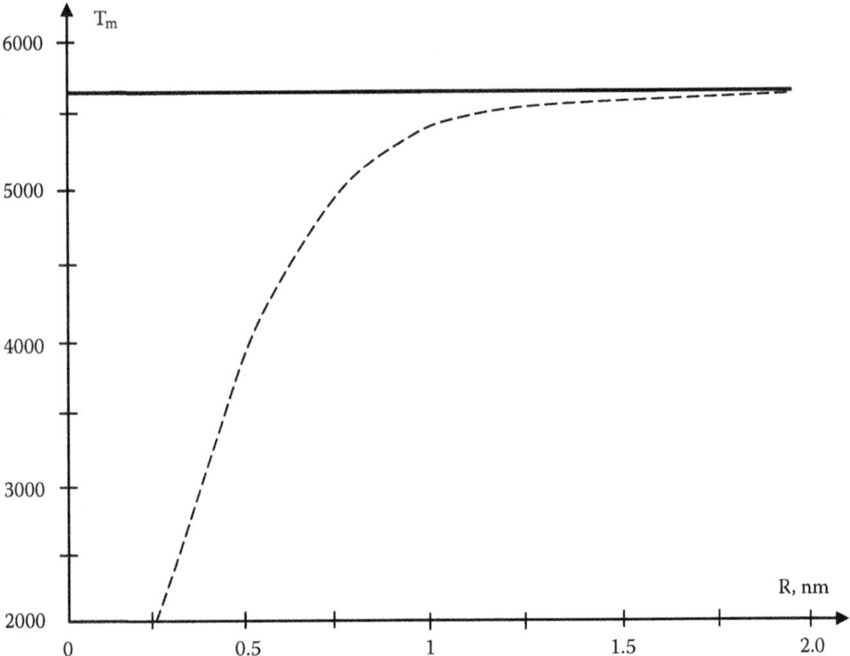

FIGURE 8.26 Dependence of the radius on the melting temperature T_m for a single-wall carbon nanotube, estimated on the basis of the temperature dependence of the radial distribution functions as well as on mean-square deviations and atomic configuration.

Work on liquid carbon has been done for temperatures 5000 K and above on the basis of experimental measurements of radial distribution functions of quench-condensed amorphous carbon. The results from a number of above studies are shown in Fig. 8.27.

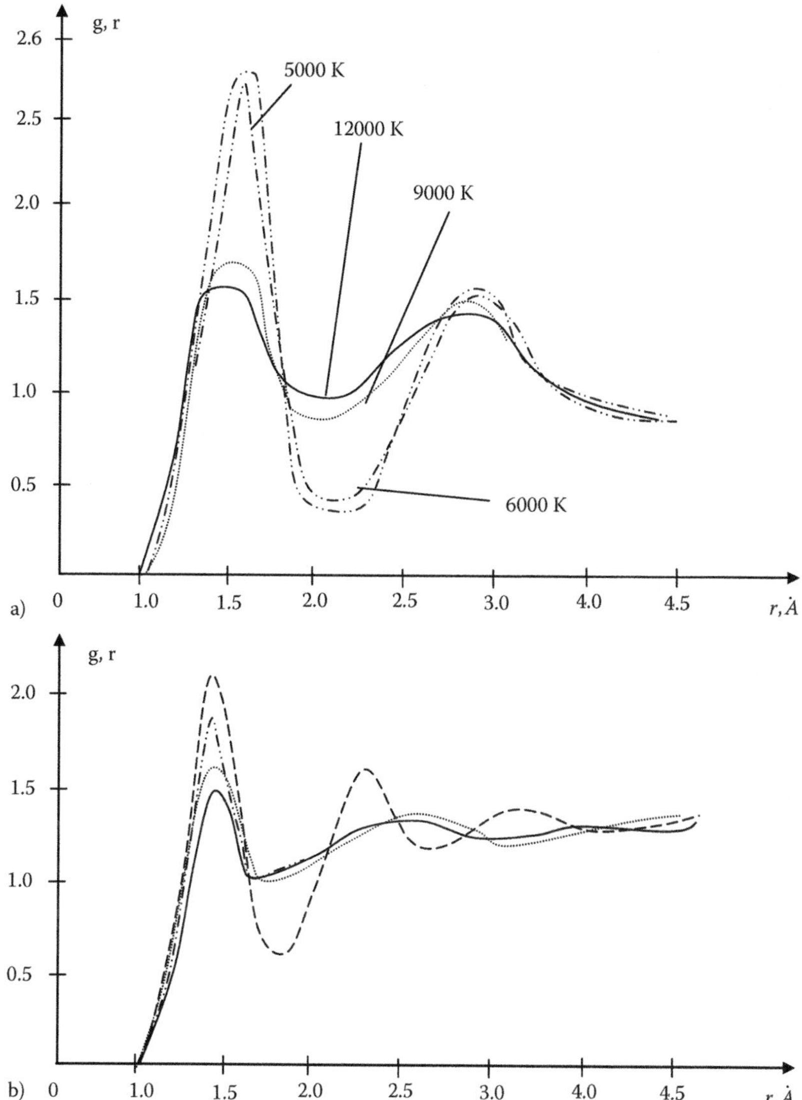

FIGURE 8.27 a) Dependence of distribution function g (r) for liquid carbon at different temperatures; b) A simulation with local density approximation. The density is 2.9 g/cc[43].

Liquid carbon is characterized by the presence of tetrahedral bonding. Graphene disintegrates into chain fragments that have double and triple bonds. The density of amorphous carbon was found experimentally to be 2.9 g/cc. The exact disintegration temperature is not easily determined. Fig. 8.28 gives curves determining transition that reduces the nearest-neighbor distances.

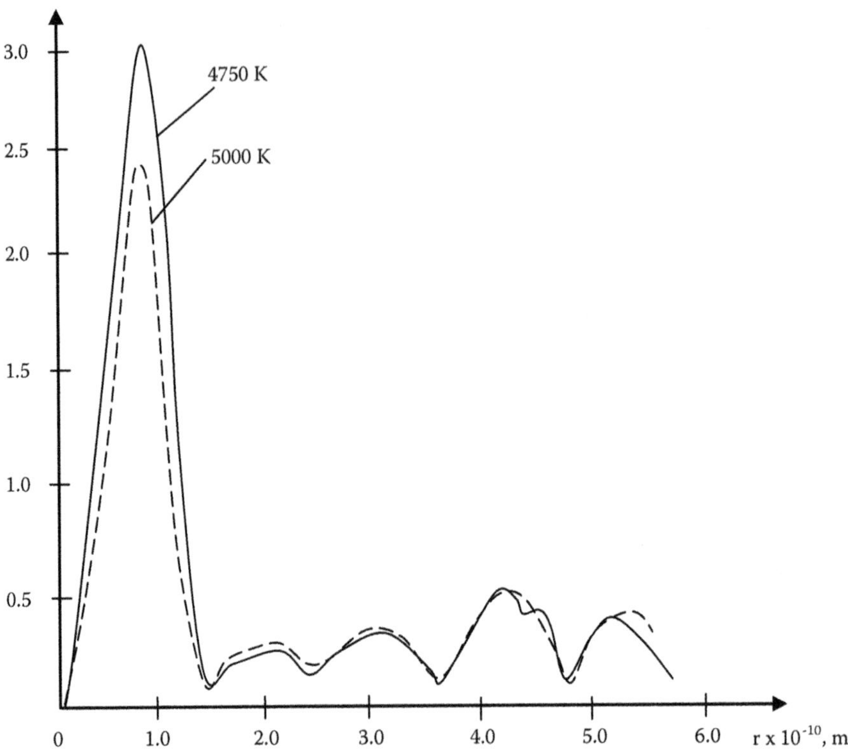

FIGURE 8.28 Radial distribution vs. the distance between the nearest-neighbors. The simulation was done for the fixed number of atoms N if the pressure is zero.

The "liquid" graphene forms 3-dimensional space of "entangled chains" – a phase of disintegration. Large vertical waves form on graphene before disintegration. The disintegration takes place from small local areas to larger strains. The local orders collapse following the appearance of defects (Fig. 8.29). Again, it is not the atoms but chain fragments that constitute the results of disintegration. The density of this "liquid" phase is 2.9 g/cc carbon.

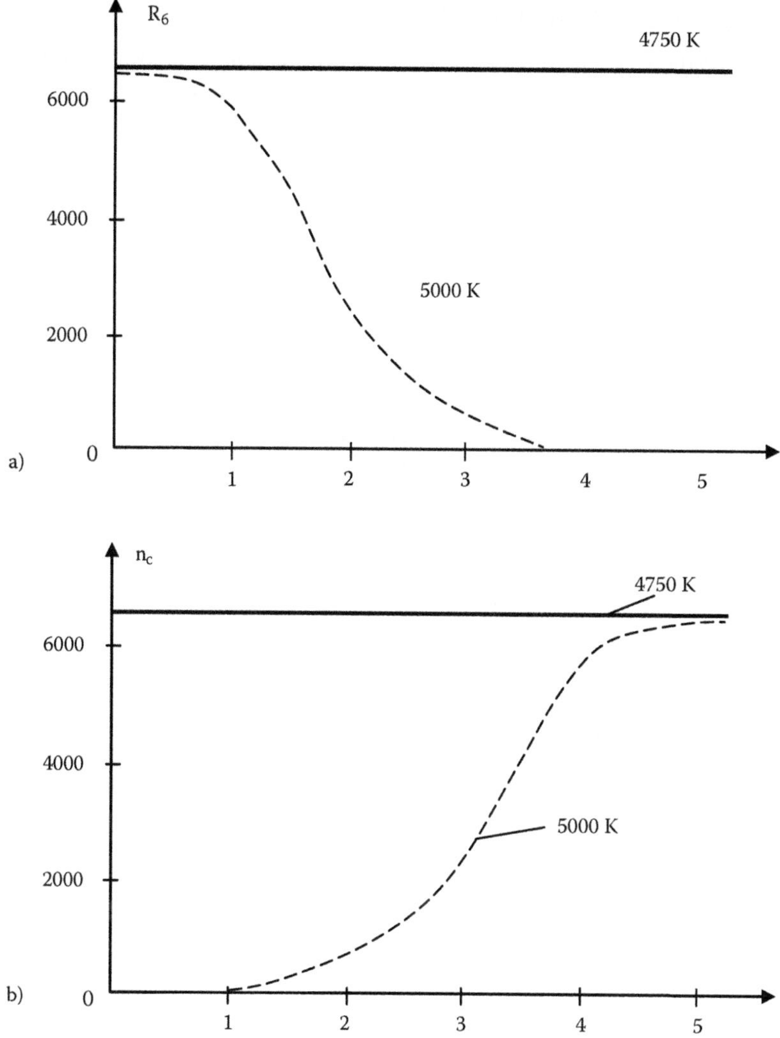

FIGURE 8.29 a) and b) States of graphene: stable state at T = 4750 K, R6 = number of hexagons, n_c = number of chains[42].

8.10 NON-LOCAL IRREGULARITY

Non-local gradient potential appears only at strong magnetic fields. With the Hall effect for the configuration in Fig. 8.30, a longitudinal resistance is:

$$R_{2,3} = V_{2,3}/I_{1,4}; \tag{8.54}$$

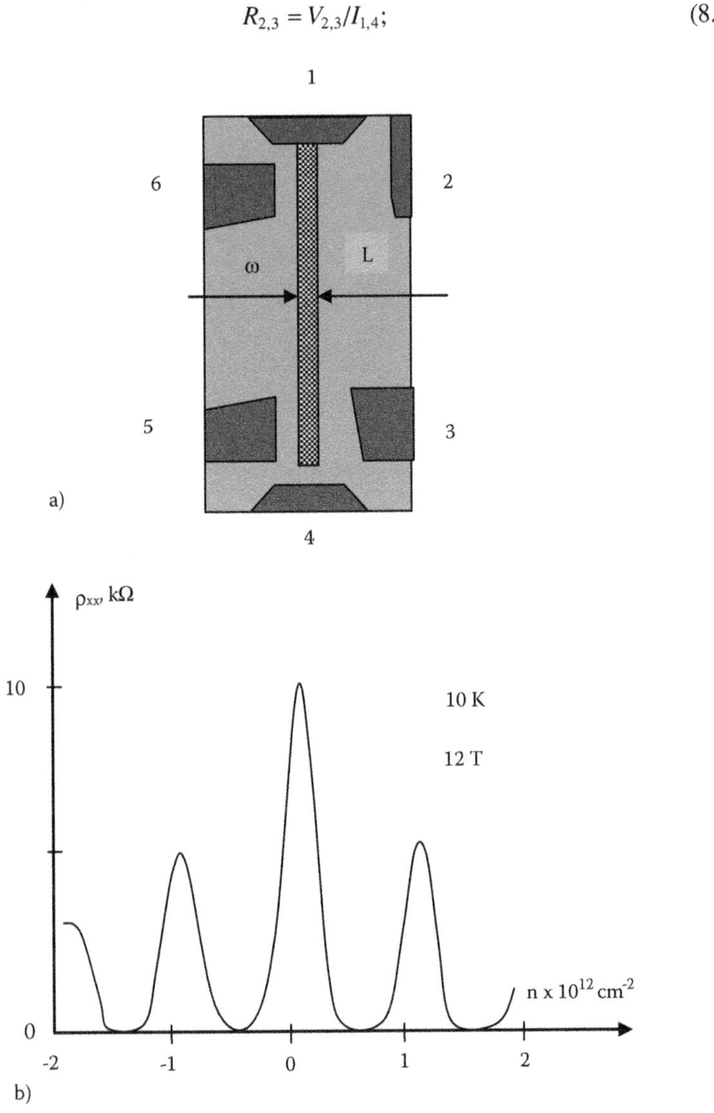

FIGURE 8.30 Device geometry and resistivity at 10 K and 12 T of a device of graphene on SiO_2. a) The structure of the device disclosed by TEM. ω = 1 μm and L = distance between the vertical pins. b) The Figure shows the dependence of the vertical current flow, ρ_{xx} vs. carrier density is when a magnetic field is applied of the intensity of 12 T. The peaks are the Landau level indices ports $\nu = \pm 4$ and $\nu = \pm 8$. *(Continued)*

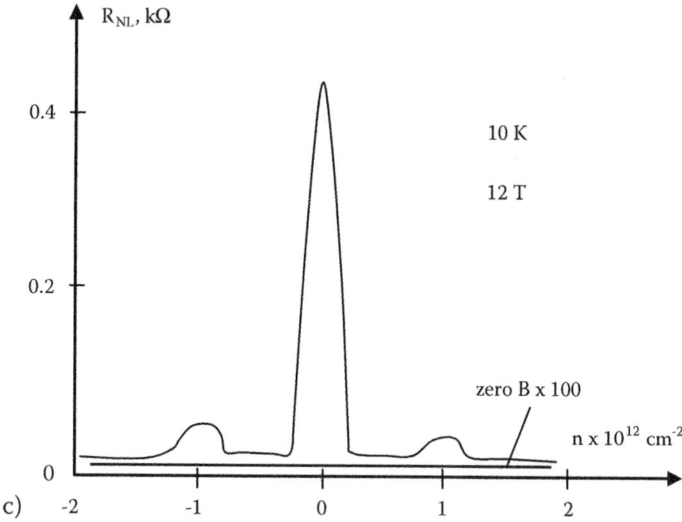

FIGURE 8.30 (Continued) Device geometry and resistivity at 10 K and 12 T of a device of graphene on SiO_2. c) Terminals 2 and 6 are for horizontal current which is measured at 3 and 5 terminals. The bottom curve (almost a straight line) has only local resistance. R_{NL} proportional to ρ_{xx} at 12 T magnetic field[44].

The van der Pauw, method used for these measurements, is a classical approach for measuring resistivity and the Hall coefficient. The advantage of this technique is in the feasibility of measuring a sample of an arbitrary shape. The sample must be two-dimensional and solid. The electrodes are installed as in Fig. 8.30 a) used for these measurements. The current $I_{2,6}$ is measured between 2 and 6. The parallel terminals 3 and 5 are displaced by the distance L. The resistance R_{NL} is calculated from the measured voltage between 3 and 5[45].

$$R_{NL} = V_{3,5} / I_{2,8} \propto \rho_{xx} \exp(-\pi L / \omega); \tag{8.55}$$

When a magnetic field is applied, the electric field between the terminals produces an abnormally large voltage between parallel terminals: a combination of spin-up electrons going in one direction with spin-down electrons going in the opposite direction. Thus, the longitudinal current is diffused and the resulting current has a spin angular momentum type rather than carrying a charge. The described phenomena are illustrated and characterized in Fig. 8.30 and 8.31.

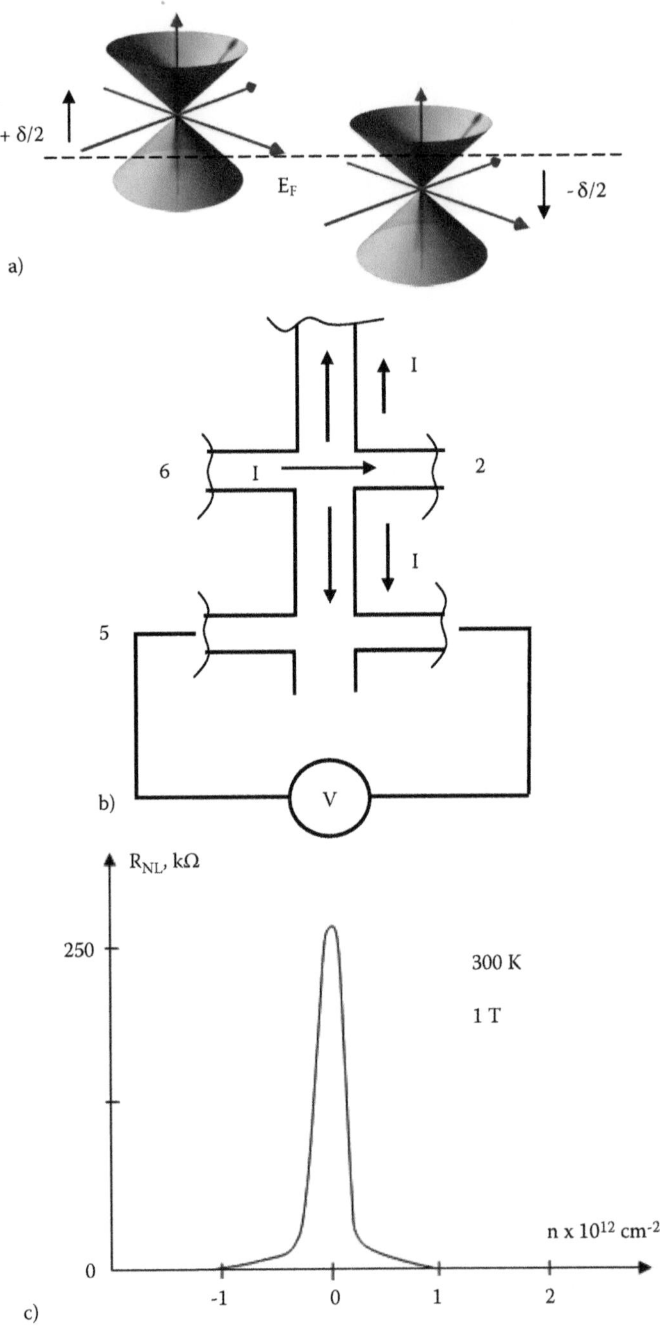

FIGURE 8.31 Spin diffusion for non-local resistivity in the Spin-Hall effect (SHE). Graphene sample is on SiO_2 substrate at 10 K and 12 T. a) Two areas of spin-down and spin-up holes produce Zeeman splitting; b) Horizontal current $I_{2,5}$ carriers no charge in the vertical (longitudinal) direction; c) Another data plot for Fig. 8.30[44].

8.11 KLEIN TUNNELING EFFECT IN GRAPHENE

Oskar Klein applied the Dirac equation to the problem of electron scattering caused by a potential barrier. The result demonstrated that the barrier is almost transparent if the potential is close to the electron mass, V ~ mc^2. It is remarkable that as the potential approaches infinity, the electron is certainly transmitted as the reflection goes to zero. The quantum mechanical meaning of the effect is proton-electron model when a potential well predicts the electron to be inside the nucleus. A potential barrier with height V_0 interacts with a massless particle with energy $E_0 < V_0$ and momentum p (Fig. 8.32).

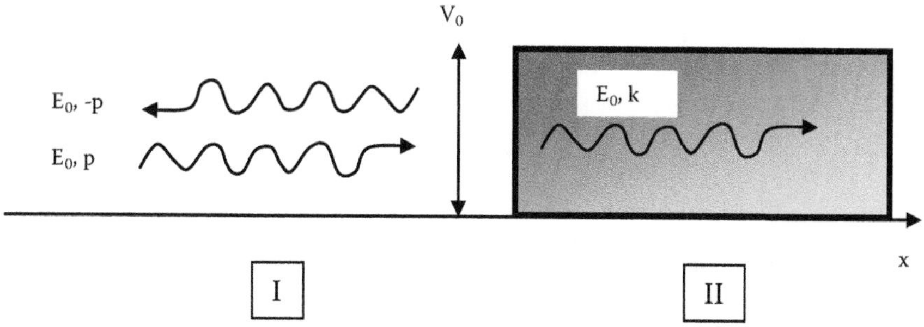

FIGURE 8.32 Massless particle reflection and tunneling.

The time dependent Dirac equation:

$$(\sigma_x p + V)\psi = E_0\psi; \quad V = \left\{ \frac{0,\, x < 0}{V_0,\, x > 0} \right\} \tag{8.56}$$

where σ_x is the Pauli matrix:

$$\sigma_x = \begin{pmatrix} 0 & 1 \\ 1 & 0 \end{pmatrix} \tag{8.57}$$

From Fig. 8.32 (the particle approaches from the left), we have two solutions: one before the step (region 1) and the other one under the potential (region 2):

$$\psi_1 = Ae^{ipx}\begin{pmatrix} 1 \\ 1 \end{pmatrix} + A`e^{-ipx}\begin{pmatrix} -1 \\ 1 \end{pmatrix}, \quad p = E_0; \tag{8.58}$$

$$\psi_2 = Be^{ipx}\begin{pmatrix} 1 \\ 1 \end{pmatrix}, \quad |K| = V_0 - E_0; \tag{8.59}$$

where A, A` and B are complex numbers. The positive velocity corresponds to the incoming and transmitted wave functions (Fig. 8.33).

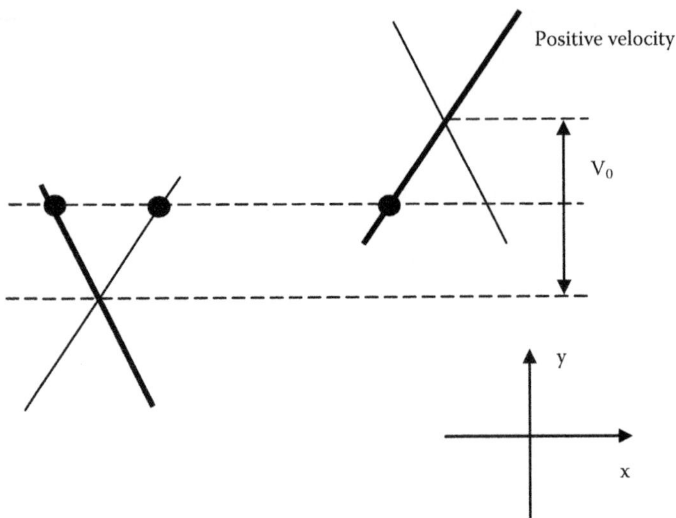

FIGURE 8.33 The dispersion relation for the incoming particles: the momentum increases in x-direction and the energy grows in y-direction.

The transmission T and reflection R coefficients are derived from the probability of amplitude currents. The probability current connected with the Dirac equation:

$$J_1 = \psi_i^t \sigma_x \psi_i, i = 1,2 \tag{8.60}$$

In our case:

$$J_1 = 2[|A|^2 - |A`|^2], \qquad J_2 = 2|B|^2; \tag{8.61}$$

The reflection and transmission coefficient are:

$$R = \frac{|A`|^2}{|A|^2}; \qquad T = \frac{|B\ |^2}{|A|^2}; \tag{8.62}$$

At x = 0, the continuity of the wave function:

$$|A|^2 = |B|^2; \qquad |A|^2 = 0; \tag{8.63}$$

The explanation to the Klein tunneling comes from the fact that a potential step cannot change the direction of the group velocity of a relativistic particle without a mass.

A series of experiments using the Klein tunneling effect has been conducted[46]. The diffusive scattering remains in graphene and the characteristic effects are not possible to distinguish from the bulk resistance that depends on the total transparency of the pn-junction. An applied magnetic field in y-direction causes a pronounced curve that corresponds to a charge motion in x-direction. The structure maintains the device function. The metal gate's width is approximately 20 nm. The thickness and permittivity of the dielectric layer determine the gate voltage. The induced carrier density versus x coordinate and the Klein barrier located at x = 0 is:

$$n(x) = \left\{V_{TG}C_{TG} / [1 + (x/\omega)^{2.5}] + V_{BG}C_{BG}\right\} / e; \tag{8.64}$$

where C_{TG} = 1490 aF/μm^2 and C_{BG} = 116 aF/μm^2, V_{TG} = -10 V and V_{BG} = 50 V, the exponent 2.5 is chosen to match the results of the conducted simulations.

The geometry of the structure for which the above calculations were perform is shown in Fig. 8.34. Transmission through the Klein barrier structure depends on multiple internal reflection similar to those in a Fabry-Perot etalon. This process is shown in Fig. 8.35. The applied magnetic field is equal to zero or larger than zero.

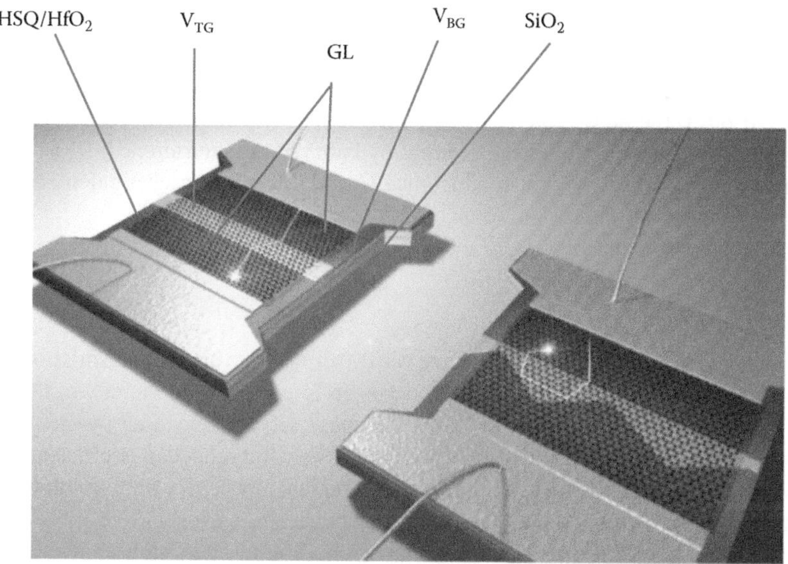

FIGURE 8.34 Geometry of a Klein barrier in a Fabry-Perot etalon device[47]. The top gate is 20 nm wide crosses the graphene layer GL. The gate induces carrier concentration n_2. HfO_2 and poly-hydroxysilane form the dielectric for the top gate. The conductance G = dI/dV (between the source and the drain) is measured.

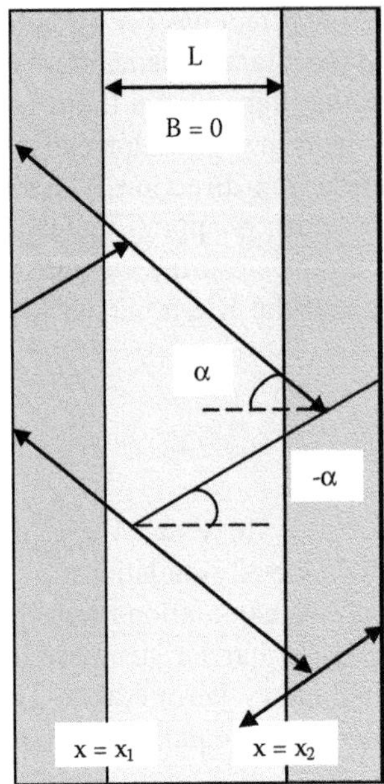

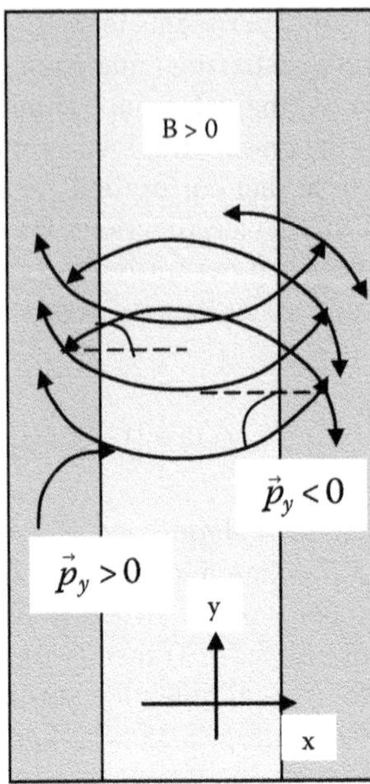

FIGURE 8.35 Electron transmission through semiconductor p-n-p structure at B = 0 (left) and B > 0 (right)[48].

The angle of incidence is α and $L = x_2 - x_1$. The back-reflection coefficient changes at $\alpha = 0$. At $B = 0$ the phase changes when back-reflection amplitudes cancel resulting in a half-period shift in the fringes caused by Fabry-Perot interference.

In a Fabry-Perot etalon, the conductance G of the Klein barrier is calculated:

$$G = 4e^2 / h \sum_{ky} \left| [t_1 t_2 \exp(-L / 2\lambda_{LGR})] / [1 - |r_1||r_2| \exp(i\theta) \exp(-L / \lambda_{LGR})] \right|^2; \quad (8.65)$$

where the summation gives the transverse wavevector, L, t_1, t_2 are the transmission coefficients, L = the width of the barrier, λ_{LGR} = mean free path in the vicinity of the gate, r_1, r_2 are reflection coefficients, $\theta = \Delta\theta$ is the phase difference caused by the particle moving between the two junctions of the transmission and reflection coefficients.

WKB approximation is a technique in mathematical physics for finding solutions to linear differential equation that have coefficients varying in space. The abbreviation stands for the first letters of Wentzel-Kramers-Brillouin. The WKB approximation is used to find an analytical solution for simple tunneling-barrier models. In practice, exact solutions often do not exist and the WKB approximation can give an approximate solution.

In Fig. 8.35, one trajectory is transmitted through both barriers and the second trajectory is reflected at the second barrier. There are two traversals more of the length L that intersect the first trajectory (Fig. 8.35 geometry).

$$\Delta\theta = 2\theta_{WKB} + \Delta\theta_1 + \Delta\theta_2; \tag{8.66}$$

where $\theta_{WKB} = \hbar^{-1}\int p_x(x')dx^1$ and $\Delta\theta_{1(2)}$ are the phase shifts for the interfaces (three shaded regions)[48].

Eq. (8.67) may be somewhat simplified by realizing that r_1 and r_2 are small, giving the oscillatory part of Eq. (8.67):

$$G_{OSC} = 8e^2/h\sum_{ky}|t_1|^2|t_2|^2|r_1||r_2|\cos(\theta)\exp(-2L/\lambda_{LGR}); \tag{8.67}$$

The oscillating part extraction is possible because in the Fabry-Perot model has an oscillating dependence on the phase shift (Eq. 8.67).

$$\theta_{WKB} = \hbar^{-1}\int p_x(x')dx'; \tag{8.68}$$

The shift comes from electron wave accumulation from reflection between the interfaces:

$$\theta_{WKB} = \hbar^{-1}\int p_x(x')dx'; \tag{8.69}$$

where $\Delta\theta_{1(2)}$ are the back-reflection phases for the interfaces 1 and 2. In Fig. 8.34 the net phase contribution $(\Delta\theta_1 + \Delta\theta_2)$ is altered by magnetic field $\boldsymbol{B}$. At zero magnetic field $\boldsymbol{B}$ the phase difference cancels but for the arc-like trajectories, the signs of angles of incidence can be equal. For a given transverse momentum p_y, there exists a magnetic field $\boldsymbol{B}^*$ for the condition $-B^*L/2<p_y<B^*L/2$.[48] And the phase sum $\Delta\theta_1 + \Delta\theta_2 = \pi$. The modeling of the process assuming that the Klein barrier has a parabolic potential $\upsilon(x) = ax^2 - t$ creates pn interfaces at $x = \pm x_\varepsilon(x_\varepsilon = \sqrt{\varepsilon/a})$. Based on these dependences and assuming a magnetic field $\boldsymbol{B}$ with the WKB phase.

$$\theta_{WKB} = \mathrm{Re}\left\{\int_{-L/2}^{L/2}[\pi|n(x)| - (k_y - eBx/\hbar)^2]^{1/2}dx\right\}; \tag{8.70}$$

where $p/\hbar$ is the unit of the integrand expression.

The reflection phases mentioned earlier may be expressed through the Heaviside function. H is zero except for $x > 0$ $H(x) = 1$:

$$\Delta\theta_1 = \pi[H(-k_y + eBL/2\hbar)]; \tag{8.71}$$

$$\Delta\theta_2 = -\pi[H(-k_y - eBL/2\hbar)]; \tag{8.72}$$

The transmission coefficient $t_{1,2}$ and the reflection coefficient $r_{1,2}$ for junction at $\pm L/2$ are:

$$t_{1,2} = \exp[-\pi(\hbar v_F / 2eE)(k_y \pm eBL/2\hbar)^2]; \tag{8.73}$$

$$r_{1,2} = \exp[i\pi H(-k_y \mp eBL/2\hbar)][1-|t_{1,2}|^2]^{1/2} \tag{8.74}$$

where E is the electric field.

The transmission is maximum when $k_y = B = 0$ or $k_y = -eBL/2\hbar$; In Eq. (8.73)

$$eE \approx 2.1\hbar\upsilon_F (dn/dx)^{2/3} \tag{8.75}$$

where dn/dx = the carrier velocity at the junction.

Relying on the approach considered above the observed conductance oscillators may be modelled[47]. The induced carrier density $n_2$at the center of the region with the gate for dG/dn_2 conductance differential is expressed in general as:

$$n(x) = \left\{V_{TG}C_{TG} / [1+(x/\omega)^{2.5}] + V_{BG}C_{BG}\right\} / e; \tag{8.76}$$

where ω is a parameter in the range of 45 – 47 nm.

The prediction shows a fringe shift in the Fabry – Perot oscillation vs magnetic fields data range (Fig. 8.36).

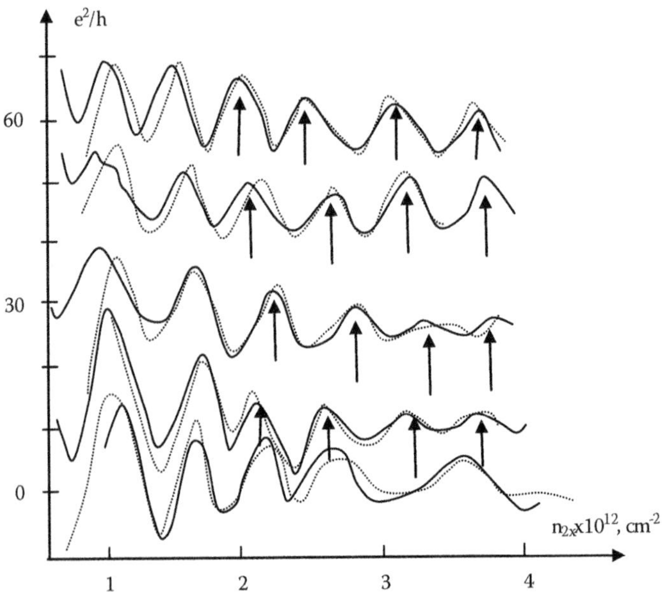

FIGURE 8.36 Conductance oscillating part that corresponds to magnetic field of 0, 200, 400, 600 and 800 mT (from top to bottom). The peaks of the curves correspond to fringe shift on the observed (or modelled) traces of 6o sc[47]. The four arrows for each curve show half-phase shift for the Klein scattering specific half-phase shift.

8.12 SUPERCONDUCTION IN GRAPHENE

Graphene performs well transmitting carriers without scattering. One example is a Josphson junction that acts as a superconductor, a normal-superconductor (SNS) junction. Super conductivity is a quantum mechanical phenomenon that is characterized by expulsion of magnetic fields in some materials if they are cooled below a certain temperature. The superconductivity is accompanied by the Meissner effect that excludes a classical mechanical explanation of the phenomenon. In a usual conductor, an electrical current is a flow of electrons crossing an ionic lattice. The electrons collide with the ions and the collision energy is absorbed by the lattice releasing heat. Thus, the electron energy is partially dissipated – this is electrical resistance and heat dissipation according to the Joule law. In a superconductor, the electronic flow consists of bound pairs (Cooper pairs). This pair creation is caused by an attractive force that exists among electrons from the phonon emission. Depending on the bias (positive and negative) carriers create a charge between electrons and holes. No voltage difference is needed for a supercurrent J:

$$J = J_0 \sin\varphi, \text{ where } \varphi = \theta_1 - \theta_2 \tag{8.77}$$

where θ_1 and θ_2 are coherent phases that correspond to a superconducting current of superconducting pairs on sides 1 and 2 of the superconducting Josephson junction.

Another type of a superconducting junction is a narrow metal layer that serves as a barrier between two superconductors. The superconducting state is stabilized by the coupling energy that exists between the two superconducting electrodes. The N-layer provides ‘the weak link’ in the SNS junction. The superconductor is a single quantum state where the superconducting parts are in the same state with a phase θ. A pair wave function:

$$\psi(\vec{r},t) = \sqrt{n_s}\exp(i\theta); \tag{8.78}$$

Deriving the expression for the pair density, we use flux quantization and the Josephson effects to receive a quantum mechanical-like expression:

$$n_s = \psi^*(\vec{r},t)\psi(\vec{r},t); \tag{8.79}$$

where n_s = pair density.

In the absence of a magnetic field, the phase θ = *const.* The superconductivity pairs move across the N – layer retaining the phase coherence. The pairs consist of two electrons whose binding together distort the crystal lattice. It is a dynamic state as the pair travels in the material[49].

The mechanism of forming pairs involves an exchange of virtual phonons which characterize the lattice distortion (Fig. 8.37). A phonon is an element of a vibrational motion which is produced by a lattice atom oscillating at a single frequency. It is a normal mode in classical physics. The lattice vibration is a superposition of elementary motions. The phonons are described by quantum mechanics, although the normal modes are waves in classical physics. At the same time, the phonon behaves as a particle.

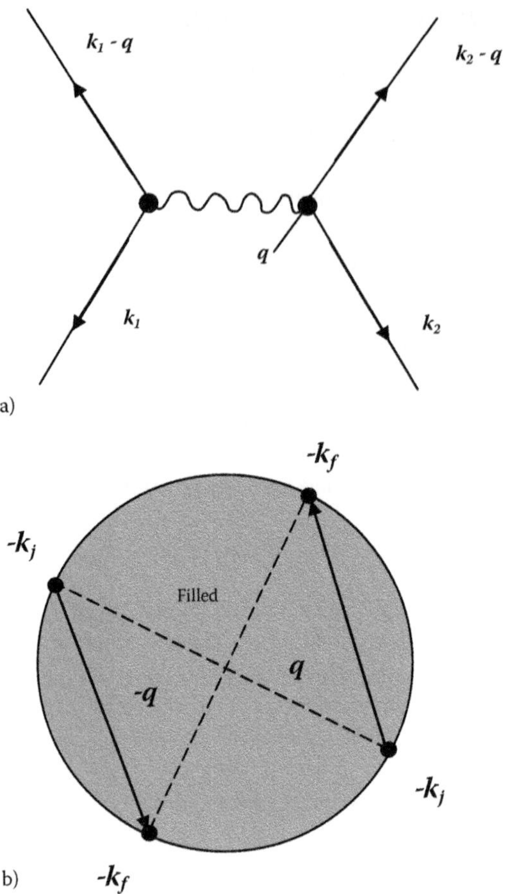

FIGURE 8.37 Interaction of two electrons by giving a virtual phonon (**q**). a) Exchange of a virtual phonon in electron-electron interaction; b) Fermi surface confined in a circle representing a 2D space with a zero momentum at the center ($\hbar k = 0$)[50].

At T = 0 the states are zero but filled inside. The phonon pairs are given by the dashed lines[50]

$$(\vec{k}_i \uparrow, -\vec{k}_i \downarrow) \text{ and } (\vec{k}_f \uparrow, -\vec{k}_f \downarrow).$$

The pairs are Boson with zero net momentum. In Fig. 8.37 a) the $\boldsymbol{k}_1$ electron emits a phonon with wavevector $\boldsymbol{q}$. The momentum changes to $\boldsymbol{k}_1 - \boldsymbol{q}$. The second electron absorbs the phonon $\boldsymbol{q}$ which results in $\boldsymbol{k}_2 + \boldsymbol{q}$ deflection. The energy of the two electrons is lowered if the electrons have the opposite spins and are separated by the superconducting coherence length ξ[49]. The condition of the optimum separation assumes that the first electron that has the speed υ_F gives a momentum impulse to one of the positive ions. When this ion returns to the initial position (and after 0.5 of the time period τ_L of the lattice motion), a second electron will be attracted to a local positive charge. The separation is then:

$$\xi \sim 2\upsilon_F\tau_L = 2h\upsilon_F / h\nu; \tag{8.80}$$

The actual separation is on the order of a micrometer and the typical Fermi velocity is 10^6 m/s. Then the coherence length:

$$\xi = 2\hbar\upsilon_F / \pi\Delta; \tag{8.81}$$

where 2Δ = superconducting gap.

The coherence length can be on the order of a micrometer. Superconducting current takes place by displacing momentum of the total set of pairs, going from $\boldsymbol{k} = 0$ to $\boldsymbol{k} = \boldsymbol{\delta k}$, a very small value. All pairs density n_S are in the same state with $\boldsymbol{k} = 0$.

2Δ is the most common gap width in semiconductors. The pair density (the pair potential) n_S relates to the energy gap parameter as $\Delta = Vn_S$, where V = the pairing interaction. In the SNS junction (Fig. 8.38) the pair density extends into the N region. In order to implement the extension, the interface atoms should be precisely ordered.

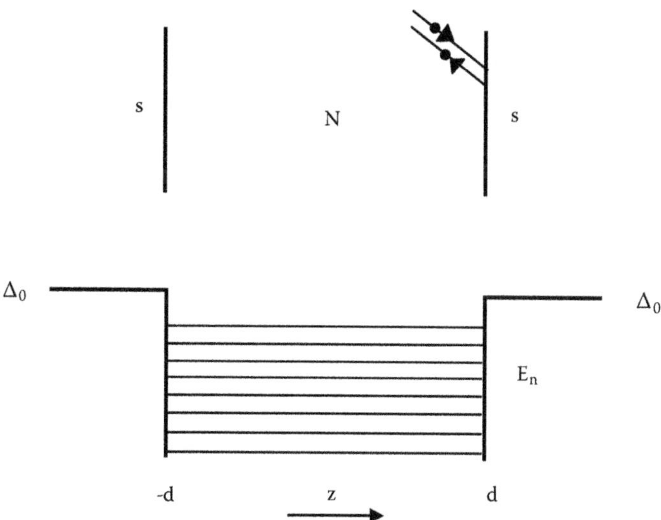

FIGURE 8.38 SNS junction with abrupt pair-potential barriers z = +d and z = -d. A particle in the upper right area is reflected into a hole. The current does not change as the particle reflects at a pair-potential boundary[51].

The pairing interaction V changes abruptly at the SN boundary to a smaller value (down to zero) and the pair potential jumps down at the interface. Thus, the pair wavefunction and pair density n_s change from s to zero deep in the N region. The length of this transition equals approximately to the coherence length and constitutes the superconductivity proximity effect. The pairs that reach the opposite SN boundary insure the interaction between the two pair systems. The interaction between the two pair systems is the Josephson effect which is present even when the pairing interaction $V = 0$ in the N layer. The process of transition through the N region is described by "Andreev reflection" (Fig. 8.39). Andreev reflection describes a kind of particle scattering that takes place at the interfaces between a superconductor (S) and a normal material (N). A charge-transfer process that implies normal current in N is transformed into supercurrent in S. In an Andrew reflection a charge $2e$ moves across the interface, thus, omitting the forbidden transmission of a single particle in the superconductor energy gap.

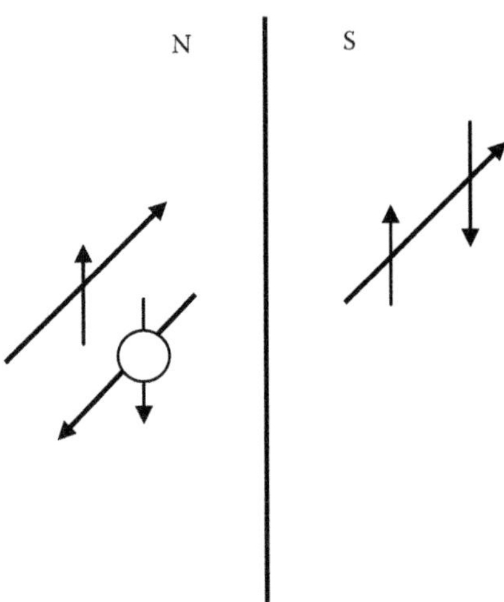

FIGURE 8.39 An electron encounters the interface between a normal conductor (N) and a superconductor (S). A Cooper pair is formed in the superconductor. In the conductor a retro reflected hole is produced. The arrows show the spin band that is taken by each particle.

The incident particle (an electron or a hole) in the superconductor with the retro reflection of a particle (an electron or a hole) of the opposite spin and velocity and equal momentum to the incident particle forms a Cooper pair. The barrier transparency needs to be sufficiently high with no material (such as an oxide or a film) on the superconductor surface that can reduce normal electron – electron or hole – hole scattering that takes place at the interface. The incident pair consists of an electron (or a hole) with up and down spins plus we need a second electron (or a hole) of opposite spin to the incident particle (an electron or a hole). The Andreev reflection greatly depends on the spin. In particular, if the material is completely spin polarized then it is impossible to form a pair in the superconductor or to transmit a single particle.

An electron in the N region encounters the S interface. The pairing interaction V causes this electron to join a second electron from the normal region and a pair is formed, creating a hole in N- at the NS- interface. An array of weakly bound quasiparticle states across the N- region is created. Fig. 8.37 shows the boundary conditions at the interfaces, $z = \pm d$. Since the pair has zero momentum, the momentum of the original electron is opposite to the created hole. Then, the electron and hole motion combined move 2 electron charges through the N – region. The Andreev's model, thus, describes a mechanism for supercurrent transfer in the SNS – structure. The model has recently received a further development[52]. The electron-pair conversion efficiency at the NS – interface is 100%[51]. In Fig. 8.37 Andreev retro reflection transfers 2e into the region S at z = d. The bottom part of the Figure shows the energy levels in the N – region with the width 2d. The quasiparticle states are shown by parallel lines which are combinations of particles and hole states. The matching wavevector conditions are given:

$$q = \left(n + \frac{1}{2}\right)\pi / d; \tag{8.82}$$

The allowed quasiparticle energies are:

$$E_n = \hbar^2 k_{2F}\left(n + \frac{1}{2}\right)\pi / \left(2m_e d^*\right); \tag{8.83}$$

where n = 1,2…, $d^* \sim d$ which means that the particles penetrate the superconductivity region somewhat, and k_{zF} is the z – component of the Fermi vector.

The current density maximum value at T = 0:

$$J_{\max} = \hbar n_e \pi e / 4m_e d^*; \tag{8.84}$$

A practical value of the current density can be on the order of 10^6 A/cm^2. Based on the Andreev reflection theory, a temperature-dependent estimate for the Josephson current density is received:

$$J_1 = \left(6\hbar n_e e / 2m_e d^*\right)\exp[-4\left(d^* / \xi_0\right)\left(k_B T / \Delta_0\right)]; \tag{8.85}$$

where $\xi_0 < 2d^*$ and the estimate was based on the set of quasiparticle states.

The relationship between the current and its phase is periodic with respect to the phase difference φ. The relationship changes from a periodic ramp at T = 0 to the Josephson periodic dependence $J = J_1 \sin\varphi$ at higher temperatures[51]. The quantization of the quasiparticle states in the normal region cause the effect of a much slower decrease with the increase of thickness of the normal layer.

The Josephson effect was observed in one of the experiments yielding Fraunhofer-like pattern with voltage steps (Shapiro steps). The dependence of voltage on current characteristics is shown in Fig. 8.40. The measurements are taken in monolayer graphene SNS without a magnetic field, at T = 30 mK. The horizontal lines in Fig. 8.40 V – I characteristics show the existing supercurrents. The voltage applied to the backgate varies the current concentration in the graphene layer.

The anomalous behavior of the Josephson effect takes place near the Dirac point. In the N – region there should be a density of delocalized states with sufficiently long mean free path. The Fermi level is located in a region of delocalized electron states.

Aluminum deposited in graphene may manifest superconducting properties[53]. A thin Ti intervening layer serves to provide adhesion of Al to the graphene.

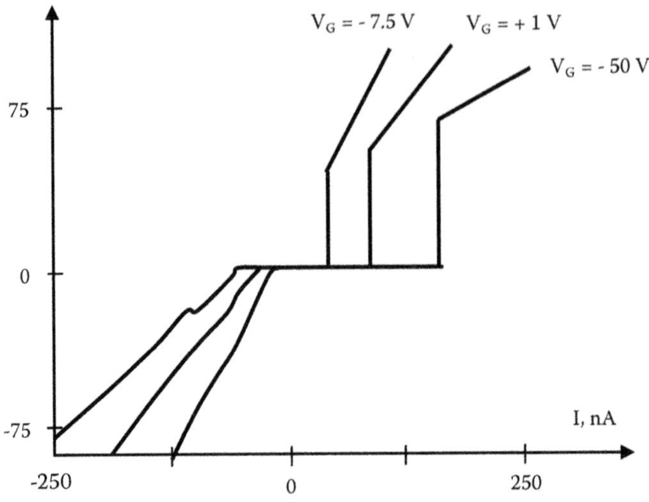

FIGURE 8.40 Josephson effect voltage-current characteristics[53].

At temperature below 1.3 T, the Al/Ti electrodes excite the graphene layer into proximity-caused superconductivity. The intervening graphene N region and the ballistic transmission of electrones and holes between the regions appear to be sufficient for the presence of the supercurrent. The received supercurrent exceed 50 μA. In addition to Al/Ti PbS electrodes on monolayer graphene have been suggested[54]. The backgate voltage tuning has been established as well.

8.13 FURTHER AREAS OF ANOMALOUS BEHAVIOR

The electron behavior in graphene monolayer defies the conventional notion of a donor or acceptor impurity when an ionized impurity becomes a scattering center[55]. For understanding of this behavior the notion of the Rydberg states is introduced. The Rydberg states are the states that are electronically excited with energies that correspond to the Rydberg formula. The Rydberg formula was designed to describe atomic energy levels. It may be used to describe other electronic systems. The threshold ionization energy is the energy necessary to detach the electron from the ionic core of an atom. A Rydberg wavepacket is created by a laser pulse that is aimed at an atom populating a superposition of Rydberg states. The energy of Rydberg states is determined using a correction, the so-called quantum defect. An anticipated behavior of massless particles in 2D-space precludes the establishment of bound states at a charge center and an infinite number of quasibound states are brought about when the Coulomb potential strength surpasses a crystal magnitude $\beta = ½$.

$$\beta = Ze^2 / \kappa\hbar\upsilon_F; \tag{8.86}$$

where the dielectric constant $\kappa = 5$ for $Z \geq 1$.

The impurities with Z > 1 are in "supercritical regime". The examples include Ca and Yb with Z = 2 and La and Gd with Z = 3. The impurities of substantial character include Boron and Nitrogen as well as surface adsorbates[55].

The propagation of a wavepacket in systems with accessible electron/hole states has been characterized as "trembling motion" (Zitterbewegung)[56]. The Gaussian wavepackets have admixtures of positive and negative energy states with the lifetime on the order of picoseconds. Fig. 8.41 shows the Landau levels in bilayer graphene[57]. The level arrangement is the foundation of the Zitterbewegung effect. The laser pulse form may be described by inclusion of a Hamiltonian term (perturbation coefficient):

$$W(t) = -exE_0 \exp(-\alpha t^2 / \tau^2)\cos(\omega_L t); \tag{8.87}$$

where $\alpha = 2\sqrt{2}$; τ = laser pulse duration, E_0 = the electric field strength, and ω_L = laser frequency.

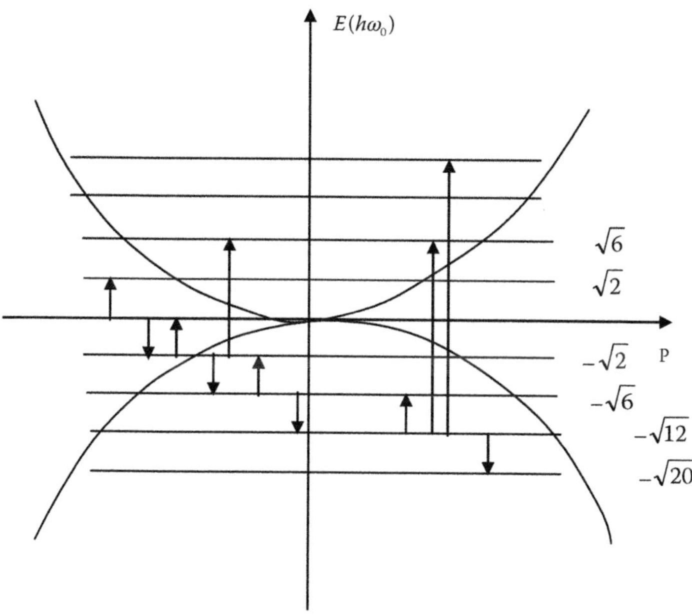

FIGURE 8.41 The Landau levels in bilayer graphene. Magnetic field (intensity 0.2 T – 2 T) causing transitions (shown by vertical arrows) out of states n = - 1, - 2, -3 and – 4 (from left to right)[57]. The source of excitation is a laser pulse.

The laser radiation forces transitions out of the Landau levels with indices from – 1 to – 4 (Fig. 8.41). Further, the simulation initiates interferences between the positive and negative states. The duration of the pulse is taken as 1.6 fs, however, the coherent oscillations last considerably longer. The simulation results are the electric field generated in the graphene bilayer by the carrier motion and the electric dipole (Fig. 8.42).

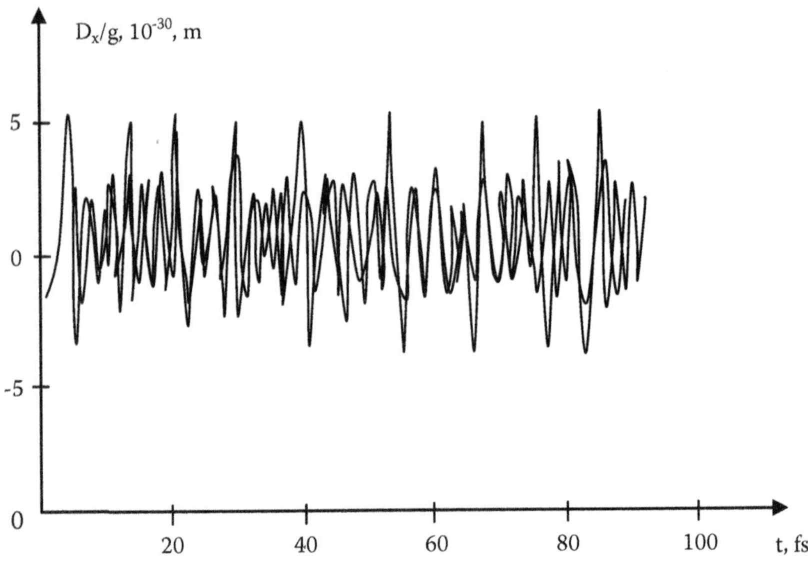

FIGURE 8.42 Oscillations in the dipole moment of bilayer graphene (simulations)[57].

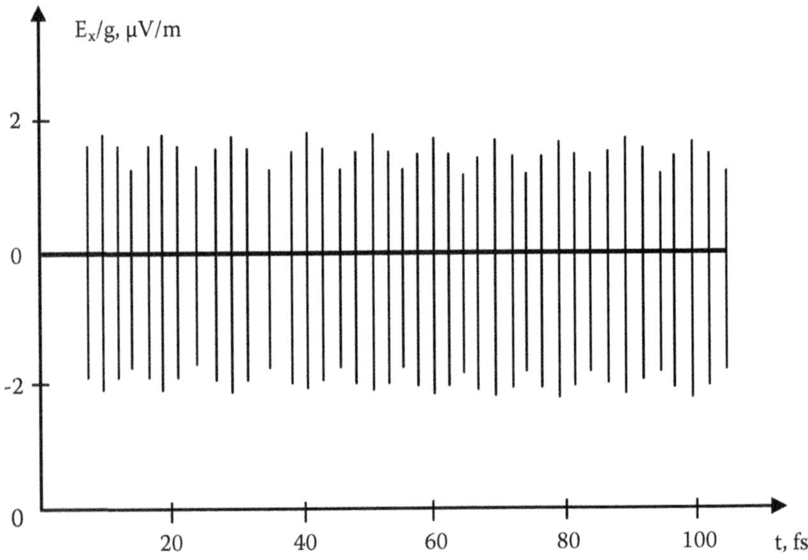

FIGURE 8.43 Oscillations in the electric field of bilayer graphene[57].

In Fig. 8.43 the strength of the induced electric field is scaled up to 1 V/m increments by a degeneracy factor and the intensity of the laser field is 0.1 x 10^7 Vm[57]. The oscillations are subject to periodic delay and regeneration of the delocalized extended states. These electron states constitute the valence and conduction bands making graphene a material with superior conducting characteristics.

REFERENCES

1. Frank, I.W., Tanenbaum, D.M., van der Zande, A.M., and McEuen, P.L., "Mechanical properties of suspended graphene sheet", *JVST* **B113**, Microelectronics and Nanometer Structures (2007)
2. Landau, L.D., Lifshitz, E.M., "Theory of elasticity", Pergamon Press, 1959
3. Nirmalraj, P.N., Lutz, T., Kumar, S., Duesberg, G.S., and Boland, J.J., "Nanoscale mapping of electrical resistivity and connectivity in graphene strips and networks", *Nano Lett.,* **11**(1) pp. 16-22 (2011)
4. Vazquez de Parga, A.L., Calleja, F., Borca, B., Passeggi, M.C.G., Jr., Hinarejos, J.J., Guinea, F., and Miranda, R., "Periodically rippled graphene: growth and spatially resolved electronic structure", *Phys. Rev. Lett.* **100** (2008)
5. Bao, W., Miao, F., Chen, Z., Zhang H., Jang, W., Dames, C., and Lau, C.N., "Controlled ripple texturing of suspended graphene and ultrathin graphite membranes", *Nat. Tech.*, **4**, pp. 562-566 (2009)
6. Castro Neto, A.H., Guinea, F., Peres, N.M.R., Novoselov, K.S., and Geim, A.K., "The electronic properties of graphene", Rev.of Modern Phys., **81**, (2009)
7. Gao, Wei, Huang, Rui, "Thermodynamics of monolayer graphene: ripping, thermal expansion and elasticity", **66**, pp. 42-48 (2014)
8. Gutter, E., David, F., Leibler, S., Peliti, L., "Thermodynamical behavior of polymerized membranes", *Journal de Physique*, **50** (14), pp. 1787-1819 (1989)
9. Tang, Hui, Wang, Bing-Shen, and Su, Zhao-Bin, *"Symmetry and Lattice Dynamics"*, INTECH (2011)
10. Mounet, N., Marzari, N., "First-principles determination of the structural, vibrational and thermodynamic properties of diamond, graphite, and derivatives", *Phys. Rev.* B **71** (2005)
11. Kelly, B.T., "The thermal expansion coefficient of graphite parallel to the basal planes", *Carbon,* **8**, Iss. 4, pp. 429 – 433 (1972)
12. Schelling P.K., Keblinski, P., "Thermal expansion of carbon structures", *Phys. Rev.* B (2003)
13. Marianetti, C.A., Yevick, H.G., "Failure mechanisms of graphene under tension", *Phys. Rev. Lett.,* **105** (2010)
14. Mariani E. and von Oppen, F., "Temperature-dependent resistivity of suspended graphene", *Phys. Rev.,* B **82** (2010)
15. Rasuli, R., Rafii-Tabar, H., Iraji, A., "Strain effect on quantum conductance of graphene nanoribbons from maximally localized Wannier functions", *Phys. Rev.* B **81** (2010)
16. Kirilenko, D.A., Dideykin, A.T., and van Tendeloo, G., "Measuring the corrugation amplitude of suspended and supported graphene", *Phys. Rev.,* B **84** (2011)
17. Meyer, J.C., Geim, A.K., Katsnelson, M.I., Novoselov, K.S., Booth, T.J., and Roth, S., "The structure of suspended graphene", *Nature* **446** (2007)
18. V.K. Sarin (Ed. in chief), D. Mari (Ed.), L. Llanes (Ed.), and C. Nebel (Ed.) *"Comprehensive hard materials"*, Vol. *1 Science* (Elsavier) (2014)
19. Bolotin, K.I., Sikes, K.J., Jiang, Z., Klima, M., Fudenberg, G., Honer, J., "Ultrahigh electron mobility", *Solid State Communication,* **146** (2008)

20. Wang, X., Li, X., Zhang, L., Yoon, Y., Weber, P.K., Wang, H., Guo, J., Dai, H., "N-doping of graphene through electrothermal reaction with ammonia", *Science*, **324** (2009)
21. Lau, C.N., Bao, W., Velasco, J. Jr., "Properties of suspended graphene membranes", *Materials Today*, Vol. 15, Issue 6, pp. 238 – 245 (2012)
22. Gao, W., Huang, R., "Thermomechanics of monolayer graphene: Rippling, thermal expansion and elasticity" *Journal of the Mechanics and Physics of Solids*, pp. 42 – 58 (2014)
23. Katsnelson, M.I., Geim, A.K., "Electron scattering on microscopic corrugations in graphene", *Phylosophical Transactions of the Royal Society, London A: Mathematical, Physical and Engineering Sciences*, **336**, Issue 1863, pp. 195-204 (2008)
24. Fasolino, A., Los, J.H., Katsnelson, M.I., "Intrinsic ripples in graphene", *Nat. Mater.*, **6**, pp. 858 – 861 (2007)
25. Nelson, D.R., Peliti, L., "Fluctuations in membranes with crystalline and hexatic order" *Journal de Physique*, **7**, pp. 1085-1092, (1987)
26. Fedorenko, A.A., Le Doussal P., Wiese, K.J., "Functional renormalization-group approach to decaying turbulence", *J. Stat. Mech.* (2013)
27. Deng, S., Berry, V., "Wrinkled, rippled and crumpled graphene: and overview of formation mechanism, electronic properties, and applications", *Materials Today*, Vol. 19, Issue 4, pp. 197 – 212 (2015)
28. Kobler, U., Hoser, A., "*Renormalization group theory (Impact on experimental magnetism)*, Springer (2010)
29. Malik, U., Kathan, L.S., "Calculation of the frequency distribution function of solids using the 'unfolding technique", *Journal of Physics C. Solid State Physics*, **14**, #5 (1981)
30. Denis, P.A., Iribarne, F., "Computational study of defect reactivity in graphene", *The Journal of Physical Chemistry"*, C (2013).
31. J. Leszczynski, J., *"Handbook of computational chemistry"*, Springer (2007).
32. Wilson, N.R., Pandey, P.A., Beanland, R., Rourke, J.P., Lupo, U., Rowlands, G. and Roemoer, R.A., "On the structure and topography of free-standing chemically modified graphene", *New Journal of Physics*, **12** (2010)
33. Mikhailov, S. (Ed.), *"Physics and Applications of Graphene –Theory"*, Intech (2011)
34. Koenig, S.P., Wang, L., Pellegrino, J., and Bunch, J.S., "Selective molecular sieving through porous graphene", *Science*, **321**, pp. 385 – 388 (2008)
35. Cataldo, F., Milani, P.M. (Series Ed.), Cataldo, F., Graovac, A., Ori, O. (Vol. Ed.) *"The Mathematics and topology of fullerenes"* **04***"Carbon materials: Chemistry and physics"*, Springer (2011)
36. Boukhalov, D.W., Katsnelson, M.I., and Son, Y.W., "Origin of anomalous water permeation through graphene oxide membrane", *Nano Lett.* **13** (8), pp. 3930 – 3935 (2013)
37. Haigh, S.J., Gholinia, A., Jalil, R., Romani, S., Britnell, L., Elias, D. S., Novoselov, K.S., Ponomarenko, L.A., Geim, A.K., and Gorbachev, R., "Cross-sectional imaging of individual layers and buried interfaces of graphene-based heterostructures and superlattices", *Nat. Mat.* **11**, pp. 764 – 767 (2012)
38. Belitz, D., Kirkpatrick, T.R., *"The Anderson-Mott transition"* Reviews of modern physics, (1994)
39. Chen, J.H., Jang, C., Xiao, S., Ishigami, M., and Fuhrer, M.S., "Intrinsic and extrinsic performance limits of graphene devices on SiO_2, *Nat. Nanotechnol.* (2008)
40. Gao, Wei, Huang, Rui, "Thermomechanics of monolayer graphene: rippling, thermal expansion and elasticity", *Journal of the mechanics and physics of solids* **66** pp. 48 – 58 (2014)
41. Liu, Yilon, Xu, Zhiping, Zheng, Quanshui, "The interlayer sheer effect on graphene multilayer resonators", *Journal of mechanics and physics of solids*, **59** (2011)

42. Bae, J., Ouchi, T., and Hayward, R.C., "Measuring the elastic modulus of thin polymer sheets by elastocapillary bending", *ACS Appl. Mater. Interfaces,* **7**(27), pp. 14734 – 14742 (2015)
43. Ileri, N., and Goldman, N., "Graphene and nano-diamond synthesis in expansion of molten liquid carbon", *J. Chem. Phys.* **141** (2014)
44. Abanin, D.A., Morozov, S.V., Ponomarenko, L.A., Gorbachev, R.V., Mayorov, A.S., Katsnelson, M.I., Watanabi, K., Taniguchi, T., Novoselov, K.S., Levitov, L.S., Geim, A.K., "Giant nonlocality near the Dirac point in graphene", *Science* **332**, Issue 6027, pp. 328 – 330 (2011)
45. van der Pauw, L.J., "A method of measuring the resistivity and Hall coefficient on Lamellae of arbitrary shape", *Phylips Technical Review* **26** (1958)
46. Sonin, E.B., "Effect of Klein tunneling on conductance and shot noise in ballistic graphene", *Phys. Rev.* B **79** (2009)
47. Young, A.F, and Kim, P., "Quantum interference and Klein tunneling in graphene", *Nature Physics,* **5**, (2009)
48. Shytov, A.V., Rudner, M.S., and Levitov, S., "Klein backscattering and Fabry-Perot interference in graphene heterojunctions", *Phys. Rev. Lett.* **101** (2008)
49. Srirastava, G.P., "The Physics of phonon", *CRC Press* (1990)
50. Wolfe, E.L., "Graphene: A new paradigm in condensed matter and device physics", *Oxford University Press* (2014)
51. Bardeen, J., Johnson, J.L., "Josephson current flow in pure superconducting – normal superconducting junction", *Physical Review B*, **5**, pp. 72-78 (1972)
52. Beenakker, C.W.J., "Specular Andreev reflection in graphene", *Phys. Rev. Lett.* **97** (2006)
53. Heersche, H.B., Jarillo-Herrero, P., Costinga, J.B., Vandersypen, L.M.K., Morpurgo, A.F., "Induced superconductivity in graphene", *Solid State Communications* **143**, pp. 72–76 (2007)
54. Coskun, U.C., Brenner, M., Hymel, T., Vakaryuk, V., Levchenko, A., and Bezryadin, A., "Distribution of supercurrent switching in graphene under the proximity effect", *Phys. Rev. Lett.* **108** (2012)
55. Shytov, A.V., Kastnelson, M.I., Levitov, L.S., "Atomic collapse and Quasi-Ryberg states in graphene", *Phys. Rev. Lett.* **99** (2007)
56. David, G. and Cserti, J., "General theory of the Zitterbewegung", *Phys. Rev.* B, **81** (2010)
57. Wang, H., Strait, J.H., George, P.A., Shivarannan, S., Shields, V.B., Chandrashekkar, Mvs, Hwang, J., Rana F., Spencer, M.G., Ruiz-Vargas, C.S., Park, J., "Ultrafast relaxation dynamics of hot optical phonons in graphene", Appl. Phys. Lett., **96** (2010)

9 Applications of Graphene

In our technology driven world military applications precede commercial. It happens because of high-rate investments that are typical of any new technology. Defense agencies do not seem to be preoccupied with a high cost of research or of making just a few devices. It is also easier to receive funds from the government for something that has a defense or special application.

In this chapter, we will discuss higher-cost forms of graphene that usually consists of one or several layers. The most promising graphene applications at the present moment and in the near future are in electronics where devices (such as transistor-like devices, interconnects, etc.) may be used with the already existing semiconductor devices. Graphene is currently available in different forms: from single crystals to chemically exfoliated flakes with nanometer-scale thickness. Graphene is widely used as an additive to carbon nanotubes and carbon fibers[1].

The graphene applications aspects discussed in this chapter are more restricted with high-quality few-layer graphene used in or proposed for electronic devices. Graphene superior features include: high current density and high mobility carriers, small thickness not accompanied with the loss of electrical conformity or surface roughness. The mentioned qualities may allow a substantial down-sizing of field-effect transistor logic implemented in computers and some other devices. The material allows adjusting of the work function by an applied electric field (in this case, a gate electrode controls the Fermi level)[2]. For example, photoconductive devices with quantum dots and the Barrister barrier modulation transistor that is meant for fast on-chip logic structures. Another direction of possible implementations is the development of large touch-screens and transparent conductors, e.g. a solar cell[3].

CVD (Chemical vapor deposition) broadly used for experimental graphene devices is, however, unlikely applicable to large-scale manufacturing. The attractive qualities of graphene, such as the usage of only one or several layers of the material, its superior carrier mobility, superior mechanical characteristics and the availability of graphene in large quantities make graphene the material of choice in the field of electronics.

The industry leaders give predictions on when certain applications will become available:

TABLE 9.1
The Current and Prospective Graphene Applications. The Chart Gives a Prediction When a Particular Device May Be Expected[1]

2015	**Graphene transfer** (medium quality) - Touch-screen; - Rollable e-paper; - Foldable OLED;
2020	**Transferred or directly grown** (large-area graphene (high quality)) - High-frequency transistor;
2025	- Logic transistor/thin-film transistor;
2030	- Future devices;

Transparent conductive coaters are commonly used in electronic devices. They comprise e-paper (electronic paper), organic light-emitting diodes (OLEDs) and touch-screen displays, all of which require low sheet resistance and high transmittance (> 90%). Graphene is a suitable material for the above applications (sheet resistance ≈ 30 Ω/square and transmittance > 97.7 %). The electrical parameter requirements, however, differ from application to application. E.g., touch-screen electrodes may have higher sheet resistance, i.e. 50 ÷ 300 Ω/square and lower transmittance, i.e. 90 %. Rolling e-paper requires a bending radius of 5 – 10 nm, the requirement that may be met by graphene. Another quality is uniform absorption in the visible spectrum, necessary for color e-paper. The OLED needs a low (< 30 Ω /square) sheet resistance. The work function and the surface roughness of the electrode determine the OLED performance. Advanced flexible OLEDs are introduced if conformal deposition of graphene on 3D structures and contact resistance satisfy the requirements.

Graphene is a likely candidate for a new generation of high-frequency transistors. However, at the present moment, for graphene there is competition with more mature technologies. Thus, it may take several years before III - V materials lose the predominance. In particular, their cut-off frequency does not exceed f_T = 850 GHz and maximum of oscillation frequency f_{max} = 1.2 THz, the limit of the top frequency for current – the limits that graphene can surpass.

Another opportunity for graphene is to replace Si technology approximately after 2020. Several approaches have been utilized to open a bandgap in graphene: chemically modified graphene, bilayer control and formation of nanoribbons and single-electron transistors. It has been difficult to open the bandgap wider than 360 meV which limits the required 10^6 on/off ratio to only 10^3. This particular problem is overcome in the new transistors, in which the modulation of the work function controls the vertical transport. The integration is still necessary for graphene logic applications, which may be implemented after 2025.

Photonics related applications are summarized in Table 9.2.

TABLE 9.2
Graphene Photonics Applications. The Expected Time Points are Given Based on the Development Prognosis of Industry[1]

2015	**- Transferred or directly grown graphene**
	- Transferred graphene (high quality)
2020	- Polarization controller;
	- Photodetector;
	- Modulator;
	- Solid-state mode-locked laser;
	- Tunable fiber mode-locked laser;
	- THz wave detector;
2025	- Isolator;
	- Mode-locked semiconductor laser;
2030	- THz wave generator;

Below are given explanations for each component in Table 9.2.

Photodetectors have been studied more extensively than a number of other graphene applications. Graphene (at least in principle) has a spectral width from ultraviolet to infrared unlike semiconductor photodetectors with a limited spectral range. Another feature of graphene is high operating bandwidth – the necessary quality for high-speed communication. The high carrier mobility of graphene allows ultrafast generation of photo carriers. The traditional InGaAs and Ge have bandwidths that are 150 GHz and 80 GHz respectively which prevents semiconductor photodetector from high bandwidth operation. Graphene, on the other hand, has a bandwidth of 1.5 THz, calculated at the saturation carrier velocity. In practice, however, the graphene bandwidth is restricted by the time constant caused by the capacitance delay but not by the transit time and is reduced to 640 GHz.

Since graphene does not have a bandgap, there is a different carrier extraction mechanism from that in semiconductor photo process. There is a local potential variation in graphene near the metal-graphene interfaces that extracts the carrier by illumination. The maximum responsivity is still low reaching only a few mAW compared to the required 1 AW. It is explained by the limited absorption caused by the thin graphene sheets and small effective detection areas. The graphene sensitivity may be improved by using plasmonic nanostructures which increase the local optical electric field or using a waveguide for the light-graphene interaction length increase.

Optical modulators encode transmission data by changing the properties of light such as amplitude, phase and polarization by the means of electro-absorption or electro-refraction. Si optical modulators, ring resonators and electro-absorption modulators use interference, resonance and electro-absorption effects for their operation. Their slow switching times restrict their operation bandwidth. In silicon waveguides, a large p –n junction resistance is a problem, causing the bandwidth to be less than 50 GHz.

Graphene can increase the operating speed beyond 50 GHz by utilizing the graphene's ability to absorb incident light in ultra -wide wavelength ranges and produce an ultra-fast response. In order to produce this effect, the interband transitions, the electrons, generated by the incident light in a single layer of graphene, are modulated by a drive voltage in broad spectral ranges. Operating speeds with bandwidth wider than 1 GHz are achieved in the near infrared range. A prospective approach with structural changes (using inter-gated dual graphene layers with resistance reduction in the RC delay time) offers the possibility of achieving bandwidths of hundreds of gigahertz. THz wireless communications where optical losses are substantially smaller than in metals is another application field for graphene.

Mode-locked laser/THz generator. Mode-locked lasers have been widely used for security applications, bio-medicine, spectroscopy and micromachining. The lasers are passively mode-locked and use semiconductor saturable absorbers. Mode-locking is an optical technique producing pulses of very short duration ($10^{-12} - 10^{-15}$ s). The technique induces a fixed phase between the longitudinal modes of the resonant cavity. If this condition is satisfied, the laser is called "phase-locked" (or "mode – locked"). The produced pulses may have an extremely brief duration as short as femtoseconds. Passive mode-locking does not need an external signal (with respect to the laser) in order to produce pulses. The light in the cavity is used to produce a change in an intracavity element which in turn will produce light in an intracavit. A saturable absorber is normally used for such purpose. A saturable absorber is an optical device whose transmission depends on the intensity and which absorbs selectively low-intensity and transmits high-intensity light.

Active mode-locking establishes a standing wave in the laser cavity using an acoustic-optic modulator. A sinusoidal amplitude modulation of the light in the cavity is controlled by an electrical signal.

Unlike its semiconductor counterparts, graphene absorbs an insubstantial amount of photons per thickness and consequently transmits a significant amount of phonons per thickness. The advantages of graphene saturable absorbers are wide spectra range high-thermal conductivity, ultra-fast carrier range relaxation time, controllable modulation depth and high damage threshold. This area may produce commercially available graphene absorbers even during this decade (until 2020) since only a small graphene area is necessary.

Optical interconnection has a wavelength division multiplexion scheme that needs a laser array with different wavelengths. A method to produce many different wavelengths is to use a single-layer graphene with multiple longitudinal modes. A suitable candidate for the above purpose is an actively mode-locked Si hybrid laser. A graphene-saturable absorber can enable a passively mode-locked semiconductor laser. However, a development of this application demands a highly integrated optical interconnection which may be available by 2030.

THz generators are applicable in various devices: security set-ups, medical imaging, and chemical sensors, to name a few. Graphene is used as a gain medium generating stimulated emission by optical pumping. It is difficult to produce a continuous – wave operation because there is no difference in diffusion times of electrons and holes, and, consequently, no formation of a dipole and no rending THz emission. It has been suggested that a pulsed excitation of a single layer or of several layers of graphene by femtosecond pulses generated by a laser may be used to generate a THz wave. A practical THz generator may become available by 2030.

Optical polarization controllers are passive optical components that control the polarization properties. Faraday rotation can control light polarization and Landau quantization in the 2D electron gas in graphene providing a fast response in a broadband range. Multistacked graphene structures can give still larger polarization rotations.

Practical compact optical polarizers have already been in data communication optical fibers with in-line conductive layers of graphene.

Composite materials, coating and paints. Electronic, chemical and mechanical properties of graphene are suitable for composite materials. From the mechanical point of view, graphene still has a serious competitor – carbon fibers. Especially strong is carbon fiber commercial dominance. It may take a long time before graphene is on the market as the main reinforcement component. Also, graphene may need some chemical modification to improve its adhesion properties to the matrix.

Graphene can enlarge the functional qualities of composites. It can serve as a moisture, gas barrier and electric shield providing electrical and thermal conductivity. In addition, graphene can strengthen the surrounding polymer matrix. Graphene additives may give lightning protection, increase the operating temperature levels for composites materials, increase the compressive strength of the composite and induce an antistatic behavior.

Practical applications of graphene composites may be expected in the near future but more substantial progress depends on the availability of graphene flakes exceeding 10 mm in size when the material large Young's module is fully exploited. There are also some applications that can benefit from graphene superior mechanical properties when carbon fiber usage is less advantageous.

Energy generation and storage. The quest for new efficient source of energy and its storage has resulted in incorporating graphene into the existing energy storage and generation solutions. Currently, solar cells are being the principal graphene application. Graphene can serve as the active media and as a transparent distributed electrode material. The transparent material version has the same uniform absorption in a broad spectrum. However, the graphene's low absorption needs plasmonic enhancement or complex interferometry to receive the necessary responsivity which is not easy to implement. The broader applications may be found using graphene transparent electrodes in dye-sensitive solar cells or in quantum dots. The properties of the transparent electrodes are modified by varying the Fermi level in order to make the media electron or hole conductivity. The dye solar cells can also receive a wide use as the cost of thermal exfoliation of graphene is decreasing.

Graphene can enhance some of the already well-established technologies. The widely used in commercial lithium –ion batteries cathodes often lack good electrical conductivity. This drawback may be overcome by addition of carbon and graphite during the electrode formation. The graphene's sandwich-like nanostructure acts, in this case, as a highly conductive component. The increased conductivity removes one of the main lithium-ion battery limitations – its low specific power density. Another advantage of the material is its high thermal conductivity allowing maintaining high currents in the battery. Anodes, in particular, may be made out of graphene nanosheets together with carbon nanotubes and fullerenes to increase the charge capacity of the battery. The so-called "supercapacitors" use electrochemical double-layer capacitors for energy storage. Here, the predominately electrostatic storage of energy is determined by the integration of a high-surface area activated carbon and a nanoscopic charge separation between the electrode and electrolyte (see Fig. 9.1).

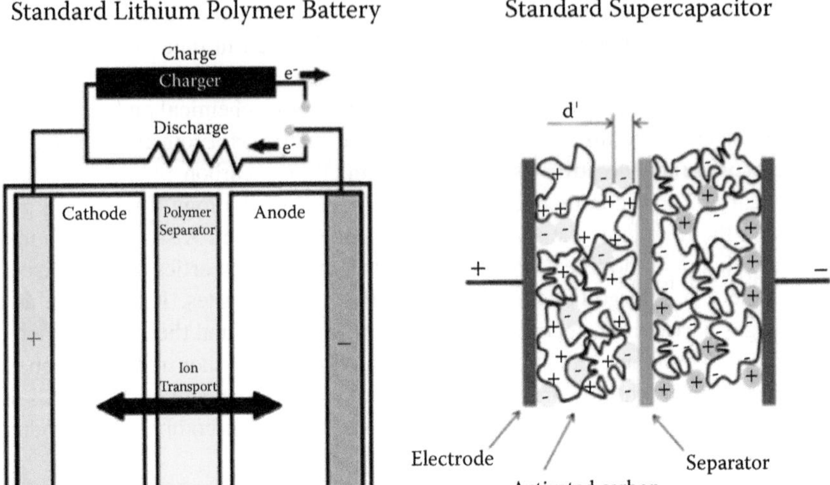

FIGURE 9.1 Comparison of a traditional lithium battery and a supercapacitor.

Graphene has a high intrinsic electrical conductivity, a well-defined pore structure, high - temperature stability and a low- level oxidizing process. In order to make supercapacitors commercial there are several issues that need to be addressed, in particular, the existing exceedingly high irreversible capacitance. The graphene suitable for the above purposes already exists, however, the particular issues mentioned above need to be addressed for graphene to be superior in performance and cost.

Sensors. By the virtue of its superiority the material is excellent for sensors. One of the prospective applications is measuring the velocity of surrounding. The corresponding strain gauges can be optical or electrical. The other strong feature of graphene is its multi-functionality, including the ability to stretch (up to 20%). The simple device can be used in a number of roles. They can measure magnetic fields, gas pressure and mechanical strain simultaneously.

Graphene has a unique bandstructure. The zero-energy and the first Landau level are divided by a large splitting. Such a gap provides the universal resistance standard that is characterized by the quantum Hall effect. Epitaxial graphene is grown on Si of the SiC substrate, the structure that is more effective than the traditional GaAs heterostructure.

Bioapplications. Graphene offers potential bioapplications, chemical purity, a large surface areas and opportunities of functionalization. The material's unique mechanical properties lead to tissue-engineering and regenerative medicine. Because of the mentioned qualities, chemically functionalized graphene may be used for ultra-sensitive measurements of a range of bio-molecules, such as cholesterol, DNA, haemoglobin and glucose. Graphene's large surface area and delocalized p electrons bing drug molecules may be solubilized and be a drug delivery carriers. Membrane barrier penetration can be achieved since graphene is lipophilic (i.e soluble in organic compounds). Graphene-based technologies for drug carriers may become commercially-available in 2030.

Another emerging technology is tissue engineering where graphene can improve the mechanical parameters and selective barrier properties. The biological performance is modulated for cell adhesion, proliferation and differentiation.

9.1 GRAPHENE NEMS (GNEMS)[4]

Small signals used in GNEMS make measurements difficult. There are different methods to actuate GNEMS, among them optical and electrostatic actuation. A laser beam is modulated at the drive frequency and focused on the graphene sheet. Consequently, the temperature is modulated causing periodic contraction/expansion of the graphene layer – a motion of the material. Detection laser beam reflects from the suspended graphene (the substrate), forming an interference pattern. The intensity of the reflected signal depends on the graphene layer position. The graphene motion is determined by measuring intensity modulation of the reflected signal. The detector is a fast photodiode. AFM (Atomic Force Method) may be used to measure the graphene deflection. In AFM, the RF signal is applied to the gate and the modulation frequency f_{mod} is set equal to the resonant frequency of the AFM cantilever. This method can provide a resonance spatial image but, since the measurements are performed in the air, the quality factors are low (~ 5).

The optical readout has its disadvantages, in particular, the set-up takes a considerable space, low temperatures and applied magnetic fields. Thus, the electrical transduction (i.e. action or process of converting energy into another form) techniques are necessary. Fig. 9.2 shows a typical schematics for electrical transduction of GNEMS. Source and drain electrodes are connected to a graphene sheet and are placed over the gate electrode. A DC voltage is applied to the gate electrodes causing static deflection of the graphene sheet toward the gate. A resonant motion takes an additional RF signal applied to the gate. As a result, we have a RF force:

$$\delta F = -\frac{1}{2}\frac{\delta C_g(V_g + \delta V_g)^2}{\delta z} \approx C_g^{'} V_g \delta V_g; \tag{9.1}$$

where δV_g = RF regime at frequency ω, C_g' = first spatial derivative of C_g.

The static deflection toward the gate has a parabolic shape. When resonance is achieved, the graphene sheet oscillates around its static position. The oscillations have a sinusoidal shape in the fundamental mode. It is possible to quantize the graphene sheet motion using z coordinate. An important quality of graphene for NEMS is the possibility of using its charge-dependent conductance G to transduce mechanical motion to a varying current in time domain. The dependence may be expressed as:

$$\frac{dI_d}{dz} = V_d V_g \frac{dG}{dV_g}\frac{C_g^{'}}{C_g}; \tag{9.2}$$

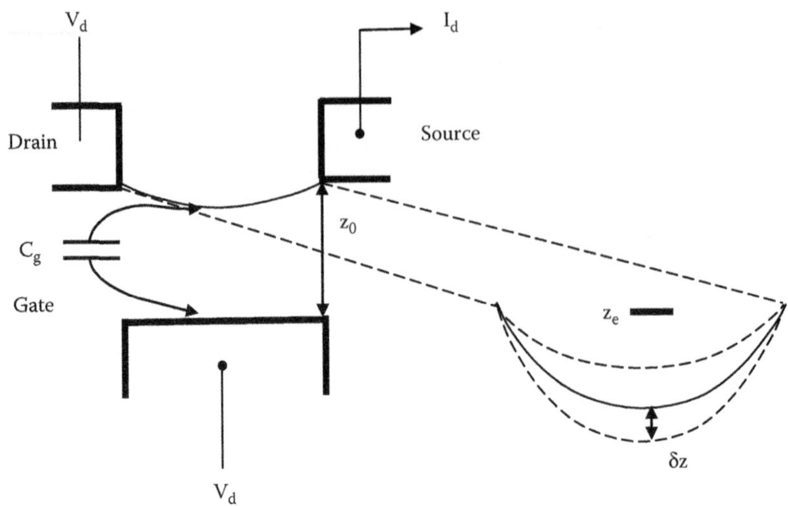

FIGURE 9.2 A typical schematics of electrical transduction for graphene resonators, where I_d = current through graphene, z_0 = distance between graphene and gate; V_g = DC gate voltage, V_d = DC drain voltage, C_g = capacitance between graphene and gate, z_e = maximum static deflection of graphene.

Since graphene is a high mobility material, dG/dV_g can be large which means that current is several magnitudes larger than for a conventional material with the same transconductance. Using the parallel plate capacitor approximation and taking into account diffusive transport limit, we have:

$$G = ne\mu\frac{W}{L}; \tag{9.3}$$

and

$$\frac{dG}{dV_g} = \mu\frac{W}{L}\frac{1}{WL}\frac{d(C_gV_g)}{dV_g} = \mu\frac{C_g}{L^2}; \tag{9.4}$$

Eq. (9.2) is transmitted into:

$$\frac{dI_d}{dz} = \mu V_d V_g \frac{C_g}{L^2}\frac{1}{z_0}; \tag{9.5}$$

Positive feedback transforms mechanical resonators into oscillators. Such self-sustained oscillators do not need an RF-input stimulus and produce compression of more than 200 times the indigenous resonator. A graphene voltage-controlled oscillator (VCO) may be implemented by varying the tension across the graphene membrane by changing V_g.

MEMS and NEMS are used in our everyday lives, starting with airbags with accelerometer triggers and ending with the present gyroscopes for smart phones. Fig. 9.3 shows the properties of the market for MEMS oscillators and RF applications.

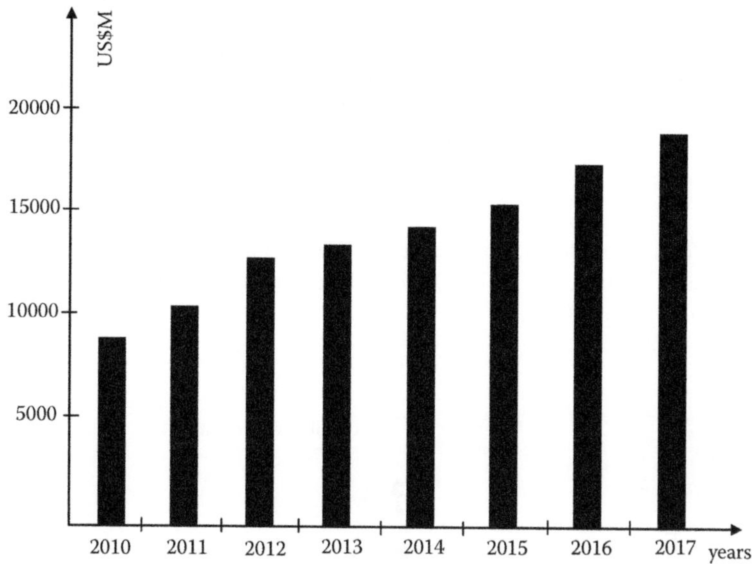

FIGURE 9.3 Market forecast for MEMS[4].

GNEMS, in particular, have two advantages versus Si – based RF MEMS. First is, graphene has bending rigidity which results in resonant frequency tuning in a wide range. Second, copper and aluminum connectors used in electronics would not be able to stand high temperature of RF MEMS fabrication. On the other hand, large graphene devices are CMOS compatible. This feature allows substituting GNEMS with widespread CMOS technologies.

One of the prospective applications is signal processing with RF GNEMS. It is cognitive radio or written communication (e.g. cell phones). A schematics for a typical RF receiver is shown in Fig. 9.4.

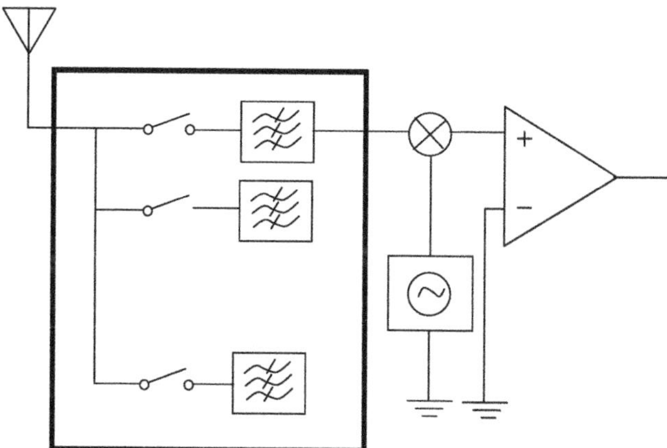

FIGURE 9.4 Schematics of RF MEMS. The electronic components inside the box-frames can be substituted with mechanical components.

Resonators are manufactured placing graphene layers over trenches with thicknesses 0.3 – 45 nm made in oxidized silicon. Their resonance frequencies lie in the range of 1 MHz to 170 MHz with the quality factor of 20 to 850. Having beams double-clamped, the fundamental frequency is:

$$f_0 = \left\{[A(E/\rho)^{1/2} t/L^2]^2 + 0.57A^2\vec{T}/\rho L^2\omega t\right\}^{1/2}; \tag{9.6}$$

where E = Young's modulus, ρ = density, t, L, ω are the dimensions of the beam, $\vec{T}$ = the tension applied., N; A = 1.03 for a beam doubled clamped and 0.162 for cantilever beam; E = 1 TPa[6].

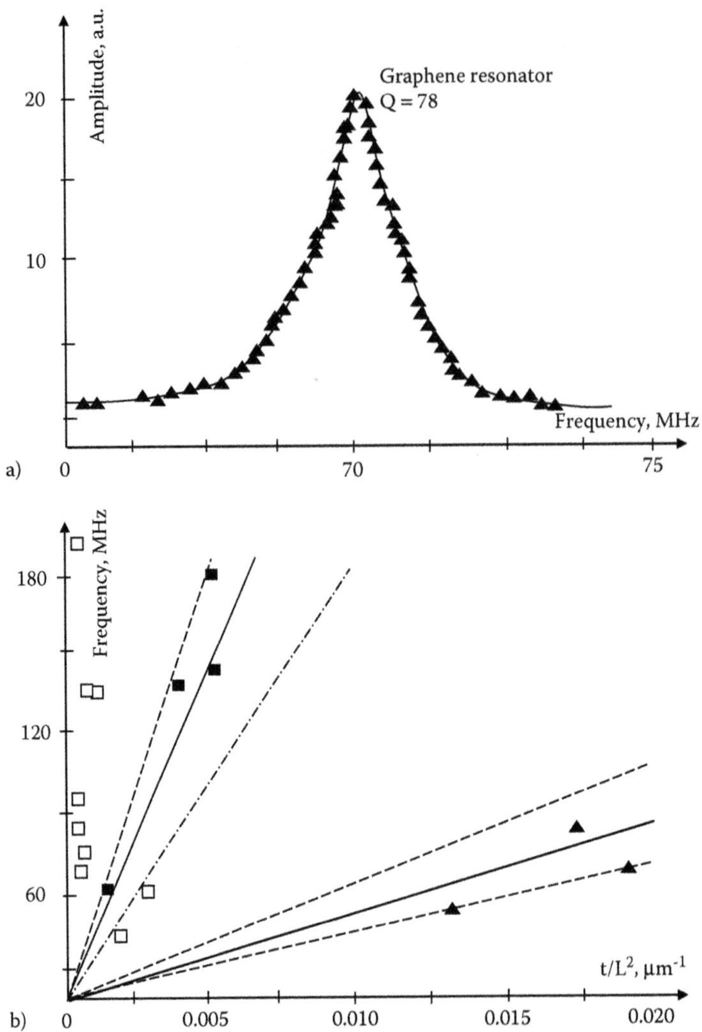

FIGURE 9.5 a) Amplitude dependence on frequency; b) Oscillation frequencies versus the ratio of thickness to the area of a graphene sheet[5].

The graphene sheets, necessary for the graphene electro-mechanical resonator, are exfoliated by mechanical means. Then, the sheets are placed across trenching. The tension $\vec{T}$ appears in this case. Using Eq. (9.6) $\vec{T} = 13n\vec{N}$ will receive the frequency, in particular, the resonant frequency (Fig, 9,5 a). In Fig. 9.5 b) square symbols stand for double-clamped beams, solid squares stand for thicknesses larger than 7 nm. The solid line is prediction calculated from Eq. (9.6) for T = 0 K. The two dashed lines were calculated for Young's module of 2 TPa (upper line) and 0.5 TPa (lower line).

9.2 GRAPHENE FETs

The field-effect-transistor (FET) is the most widespread electronic device which is manufactured in astronomical quantities (more than 10^{18} per year)[7]. A graphene version of the FET is shown in Fig. 9.6. Graphene grown on silicon carbonite (SiC) has inherent electron conductance.

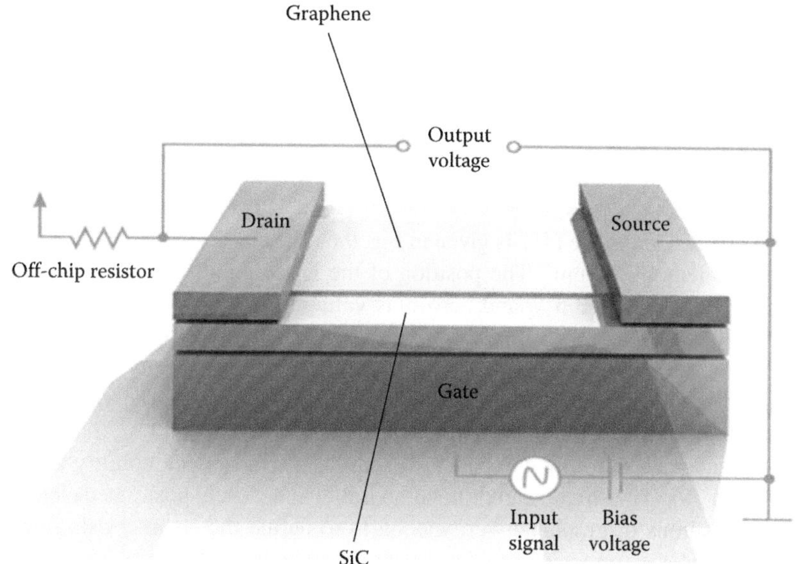

FIGURE 9.6 Graphene FET with a 2 –inch SiC wafer as a substrate[8].

The gate electrode draws carriers to a channel that connects the source and drain electrodes, thus producing an ON-condition of the FET which have two main applications: memory devices, logic that functions as switches and radio – frequency amplifiers. The ratio of current in ON/OFF position should be between 10^4 and 10^7.

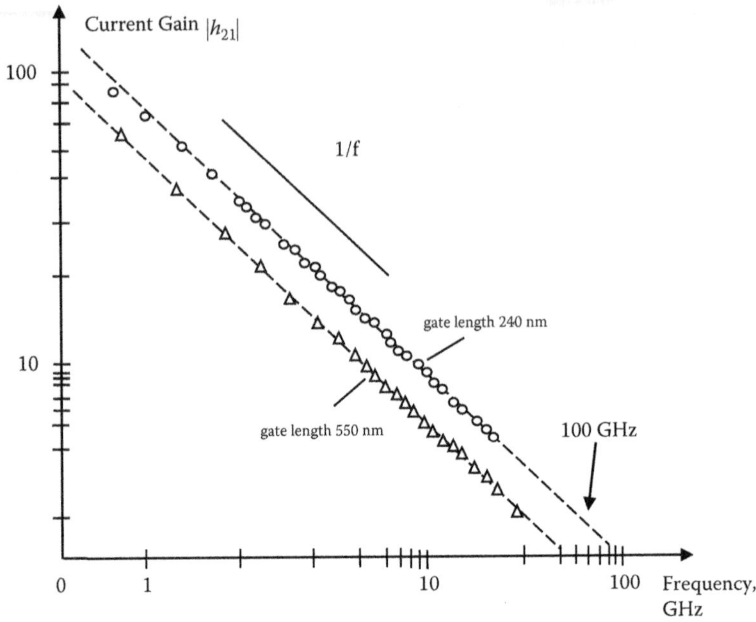

FIGURE 9.7 Current gain frequency dependence[8].

A sketch of a fast graphene FET is given in Fig. 9.6. The cut-off frequency is 100 GHz and the gate length 240 nm[8]. The position of the Dirac point requires a substantial negative voltage $V_G < -3.5$ V and zero bias voltage with n – type material in the channel and $n > 4 \times 10^{12}$ cm^{-2}. These conditions help to obtain a low series resistance of the device. The cut-off frequency is taken as the highest frequency where the current gain of the device equals, at least, unity or higher (Fig. 9.7). The graphene was grown on the Si layer of semi-insulating SiC and annealed at 1450^0 C. The electron carrier density was determined to be 3×10^{12} cm^{-2}. The Hall-effect mobility was from 0.1 to 0.15 m^2/Vs. The source and drain are parallel and closely positioned electrodes (Fig. 9.6). Regions of graphene were etched in to define the channel using oxygen plasma and PMMA (poly methacrylate) as an etch mask.

A number of steps were taken to minimize graphene mobility degradation achieving the values in the range from 0.09 to 0.1520 m^2/Vs. The measured cut-off frequencies depending on the gate length are shown in Fig. 9.8 suggesting that the cut-off frequency 1 THz is achievable at the gate length 20 nm.

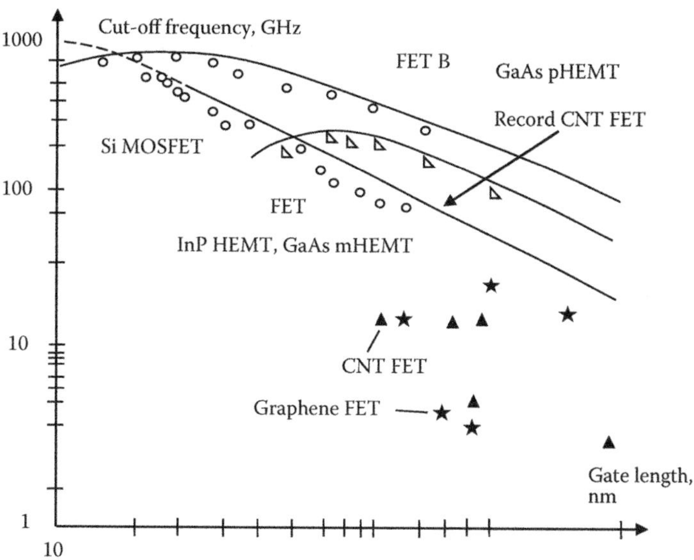

FIGURE 9.8 Dependence of the cut-off frequencies on the gate length[9].

The drawback of the above FET design is high temperature necessary for the growing process (T = 1450^0 C). The higher temperature also leads to the mobility degradation and the problems with the top gate alignment. Self-alignment is already possible[10]. A FET with such alignment and narrow channel is based on a graphene flake which was micromechanically cleaved and transferred to an oxidized silicon wager (Fig. 9.9).

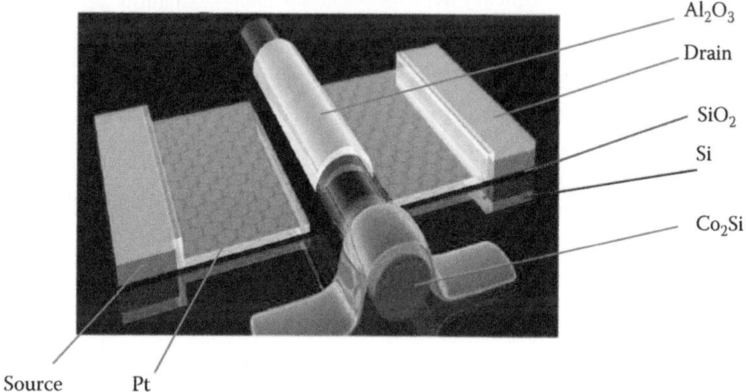

FIGURE 9.9 High-speed graphene FET with self-aligned nanowire gate. Graphene is micro – mechanically cleaved. The top gate is Co_2Si – Al_2O_3 core-shell nanowire.

In Fig. 9.9 platinum thin-film pads optimally localize the gate by extending the source and the chain electrodes. The device in Fig. 9.9 has the cut-off frequency of approximately 300 GHz and the current density 3.32 mA/μm. The scattering mechanism is emission of the substrate surface optimal phonons. The diamond-like-carbon substrate (DLC) has phonons with a higher energy (165 meV), which are less likely to be excited and, therefore, cause less scattering.

At the present moment, growth by CVD on Cu and then transferring cm-scale samples on Cu are available. The cut-off frequency is 155 GHz with the gate length of 40 nm[11]. The dependence of cut-off frequency on the gate length is given in Fig. 9.10.

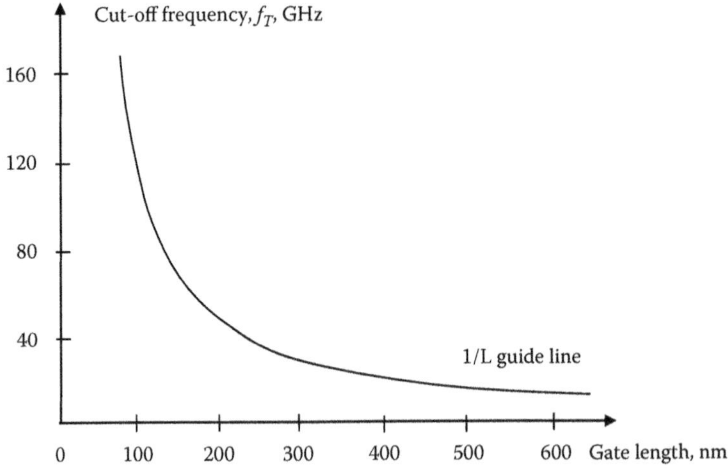

FIGURE 9.10 The cut-off frequency dependence on the gate length[11].

The CVD growth was characterized by Raman spectroscopy after the transfer to a DLC surface. The DLC film was grown using C_6H_{12} (cyclohexane) on an 8-inch Si substrate. The vapor pressure in the CVD chamber was 1.8 psi. The rate of the DLC growth was 32 Å/s at 60^0 C and subsequently the film was at T = 400^0 C for four hours. The source and drain electrodes were made of 20 nm of Pd and covered by 30 nm Au by means of e-beam evaporation.

9.2.1 Graphene FETs Grown on SiC

Epitaxial graphene grown on 4H SiC (0001) that is the silicon polar surface has features that influence the characteristics of the FETs that are grown on it. The adherence of the epitaxial graphene layer is determined by van der Waals attraction and Schottky barriers are formed when graphene (cleaved by mechanical means) is deposited on SiC. The barrier is a strong rectifier. The Schottky barrier is a potential energy barrier which is formed at a metal-semiconductor junction. One of the most important characteristics of the Schottky barrier is its height, Φ_B. The value of Φ_B is determined by the metal and semiconductor work functions. The barrier height may be estimated from the forward-bias exponential curve[12]. The barrier characteristics are different depending on what substrate it is created upon. There are three known cases: epitaxially grown (EG) in situ, on the SiC surface and on the SiC with a micro-mechanically cleaved monolayer graphene. The barrier has been successfully reduced for epitaxially grown graphene (Fig. 9.11).

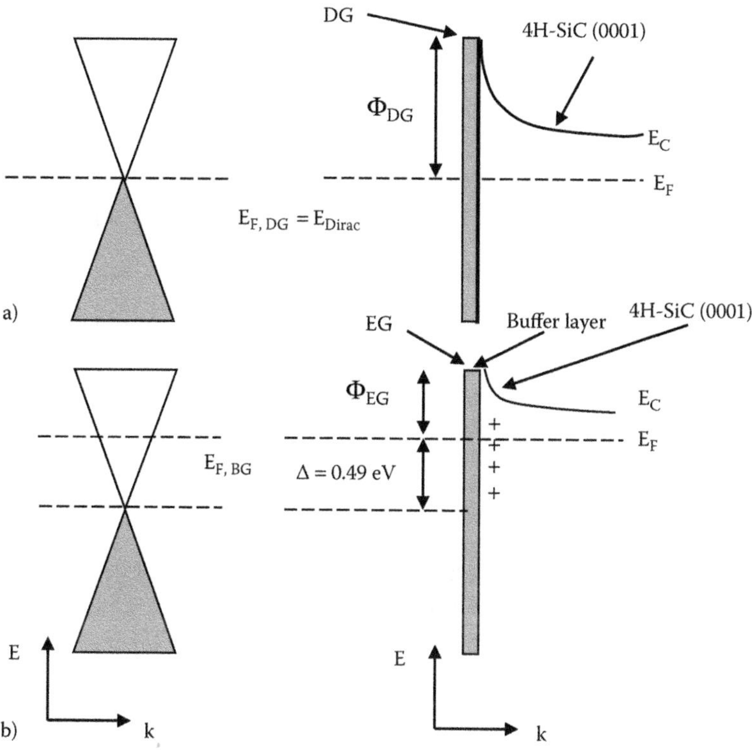

FIGURE 9.11 Graphene DG (deposited graphene) band diagram: a) Intersection of SiC with the (mechanically-cleaved) graphene doping; the barrier height is 0.85 eV. b) The barrier height decreased to 0.49 eV because of the Fermi level pinning to donor states on SiC surface.

The pinning (Fig. 9.11 b) takes place in the carbon-rich buffer layer. The inherent work function for graphene is 4 eV anf the electron affinity of 4H SiC is 3.7 eV[13]. Depending on what metal is used for contacts graphene may be "n" or "p" – type. Al, Ag and Cu give n – type and Pt or Au charge the type to "p". In all the described cases graphene is doped by the adjacent metals. No doping takes place in DG flakes (Fig. 9.11) on SiC. No heat treatment is needed. Then, epitaxial graphene is grown on Si surface of SiC: first, a silicon-rich with $\sqrt{3}x\sqrt{3}\ R30^0$ phase, and the second, the "buffer" C - rich at high sp^2 hybridization[14]. The Fermi level is raised by buffer layer donor centers (Fig. 9.11 b). The graphene epitaxial growth is illustrated in Fig. 9.12[15].

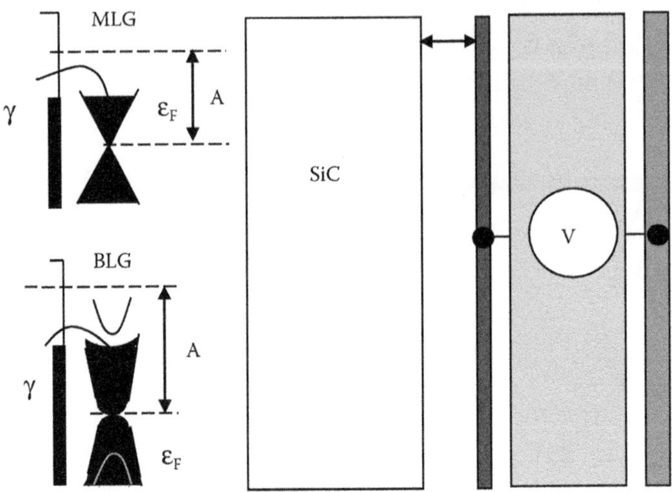

FIGURE 9.12 SiC/G (graphene) – based FET. MLG = monolayer graphene, BLG = bilayer graphene d = spacing between in SiC surface and that of graphene. A = work function and γ = the surface state density[15].

Work function A causes electron transfer from the surface states of density γ and (or) the bulk donor states with density ρ. The surface state densities $\gamma \sim 1 \times 10^{13}$ $cm^{-2}eV^{-1}$. The saturation density of n – type doping of a graphene monolayer is ~ 10^{13} cm^{-2} for d = 0.3 nm.

The following two equations allow receiving numerical solutions for the graphene carrier density:

$$\gamma[A - e^2 d(n + n_G)/\varepsilon_0 - E_F(n)] + \rho l = n + n_G; \tag{9.7}$$

and

$$A' = E_F(n) + U + e^2 d(n + n_G)/\varepsilon_0; \tag{9.8}$$

where $E_F(n)$ = the Fermi level that corresponds to the carrier concentration n; $n_G = cV_G/e$; A and A' are the barrier heights, eV.

$$U = e^2 \rho l^2 / \kappa \varepsilon_0; \tag{9.9}$$

where U is the Schottky barrier height between the material and a metal electrode, SiC permittivity is $\kappa \sim 10$ and l = the depletion layer width.

9.2.2 Electrostatic Doping for Vertical Tunnel FETs

A new approach to the graphene FET configuration has been proposed recently[16]. Graphene's metallic qualities prevent it from forming a high resistance layer. This excludes graphene FET from prolific logic applications. The vertical graphene configuration has a source and a drain placed on an oxidized silicon that has a gate electrode. The forward current of the device flows in the vertical direction. The gate voltage controls the current flow concentration and changes to some extent the tunneling probability across the tunnel barrier. Two monolayer electrodes are produced by micromechanical cleavage and have an almost ideal I(V) characteristics of the tunneling. The tunnel barrier is a few layer film on MoS_2 or BN (Boron Nitride). Such devices are operated by electrostatic doping of the electrode connected to a source and drain. The FET that employs electrostatic doping for Cu/BN graphene structure is depicted in Fig. 9.13.

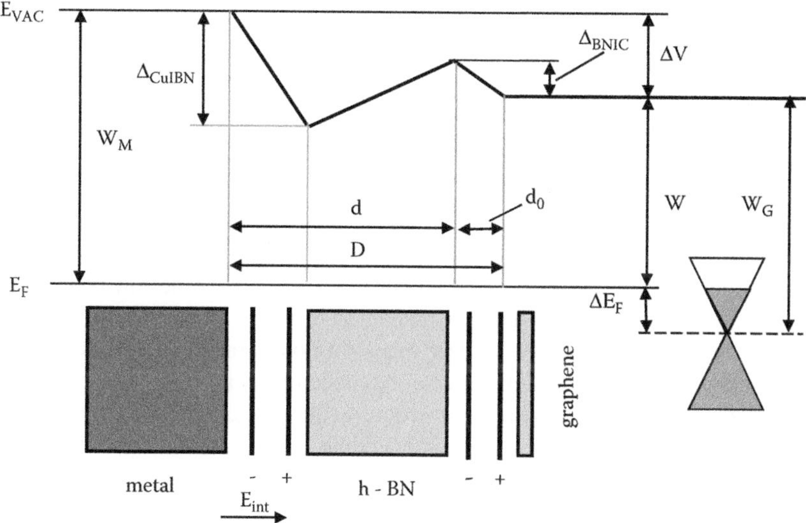

FIGURE 9.13 Geometrical dimensions and electrostatic potentials in Cu/h – BN graphene structure.

In Fig. 9.13 ΔF_F = induces shifts of the Fermi energy; E_{int} is the total electric field across the h – BN piece of thickness d; W_M = the copper work function and W_G is the work function of graphene. h – BN layer is important in forming a sharp change in the electrostatic field which takes place at Cu – hBN interface. The work function of Cu and graphene are different (W_M = 5.25 eV and W_B = 4.48 eV). The drop in the electrostatical field potential is the result of electric dipole layers. It is $\Delta Cu/_{h-BN} = 1.12$ eV in the left of Fig. 9.13 and only $\Delta_{h-BN/graphene} = 0.14$ eV. When Cu is deposited on graphene, their Fermi levels become equal. Electrons from graphene move to Cu resulting in graphene p – type. Sharp change of the electrostatic potential are shown from left to right in Fig. 9.13. The changes take place at Cu – (h – BN) and (h – BN) – graphene interfaces and are caused by electro –dipole layers. An equilibrium balance between graphene and Cu exists. In order to find the change in the Fermi levels ΔE_F (Fig. 9.13), the effective work function W is calculated for Cu – h – BN stack[17]:

$$W = W_M - \Delta_{Cu/h-BN} - \Delta_{h-Bn/graphene} + eE_{\text{int}}d; \tag{9.10}$$

The charge density on the graphene surface, σ is:

$$\sigma = \varepsilon_0 (E_{ext}t - \kappa E_{\text{int}}); \tag{9.11}$$

where E_{ext} is an external electric field caused by gate voltage $V_G = -eE_{ext}d / \kappa$, $\kappa =$ the permittivity of the dielectric BN (Boron Nitride) of thickness d.

The charge density depends on the density of states:

$$\sigma = e\int g(E)dE; \tag{9.12}$$

where $g(E) = g_0 |E| / A'$, $A' = 5.18\ \dot{A}$ - the area of the graphene unit cell, $g_0 = 0.09/eV^2$ unit cell. Then, the shift of the Fermi level is:

$$\Delta E_F = \pm\{[(1 + 2\alpha g_0 d / |V_G - V_0| / \kappa)^{1/2} - 1] / (2\alpha g_0 d / \kappa\}; \tag{9.13}$$

The above expression helps with calculations concerning the design of the vertical tunneling transistor[18]. The device is sketched in Fig. 9.14. Fig. 9.14 a) gives the FET layer composition. In the simplest case, only one graphene electrode, G_{rB}, is important since the outside electrodes can be made of metal. Fig. 9.14 b) shows the effect of the applied voltage to the band structure. Fig. 9.14 c) presents the same band structure for a finite voltage, V_g at zero applied bias, V_b. Fig. 9.14 d) illustrates a graphene's spectrum for the electron tunnel barrier.

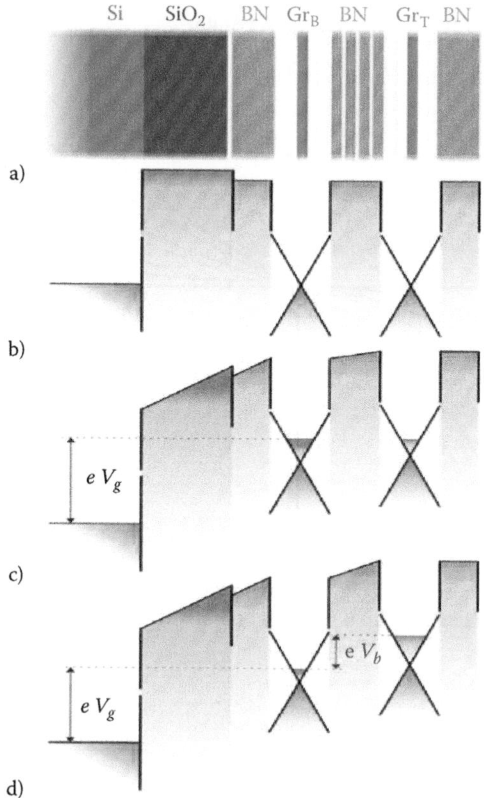

FIGURE 9.14 Vertical FET tunneling transistor. The upper drawing assumes no voltage bias. The middle drawing: positive bias applied. The lower drawing: source – to – drain bias is applied to the sturcture[18].

Fig. 9.15 shows the I/V characteristics of the source/drain graphene layers by a barrier of a few layer h – BN[19]. The electric fields for bottom graphene layer is:

$$F_b - F_g = n_b e / \kappa \varepsilon_0; \tag{9.14}$$

where F_b and F_g are electric fields.

and for the top layer (t)

$$-F_b = n_t e / \kappa \varepsilon_0; \tag{9.15}$$

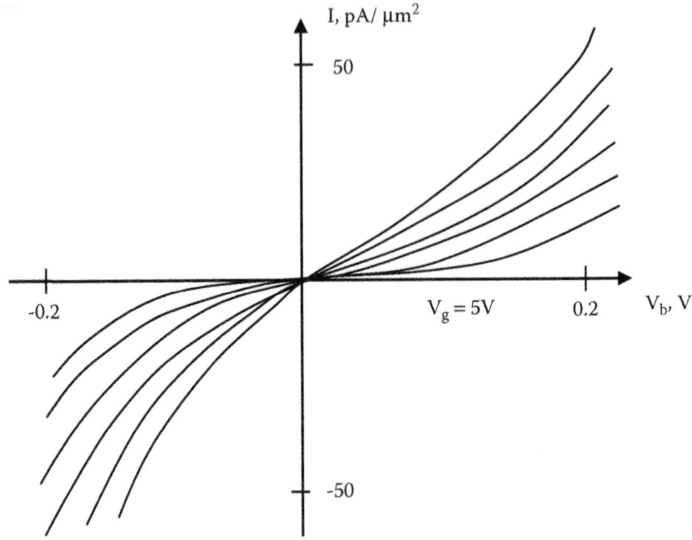

FIGURE 9.15 The dependence of the tunneling current (current density) on the gate voltage at T = 240 K for a device with 6 ± 1 layer with h – BN as the tunnel barrier. V_b = bottom layer voltage.

For the voltage V_b between two graphene layers:

$$eV_b = eF_b d - \mu n_t + \mu n_b; \tag{9.16}$$

The electrostatic doping condition:

$$n_t e^2 d / \kappa\varepsilon_0 + \mu n_t + \mu(n_t + \kappa\varepsilon_0 F_g / e) + eV_b = 0 \tag{9.17}$$

Eq. (9.17) determines the carrier density with the gate voltage V_g applied to the top graphene layer. The Fermi energy is proportional to n (the carrier density) for a 2D electron gas. The spacing d is larger than distances between atoms in Eq. (9.17). Graphene's low density of states and Dirac-like spectrum yield: $\mu(n) \propto \sqrt{n}$[20]. Thus, the forward tunnel current in the vertical device is:

$$\sigma^T \propto g_{bottom}(V_G) g_{top}(V_G) T(V_G); \tag{9.18}$$

The assumption is for the electron tunneling conductance at the voltage V_G, g = the density of states and T = the probability of tunneling between the top and bottom graphene layers at a certain bias. An example of a concrete graphene device, barrister is shown in Fig. 9.16. The current versus gate voltage dependence is given in Fig. 9.17. The switching takes place as the gate voltage changes the sign.

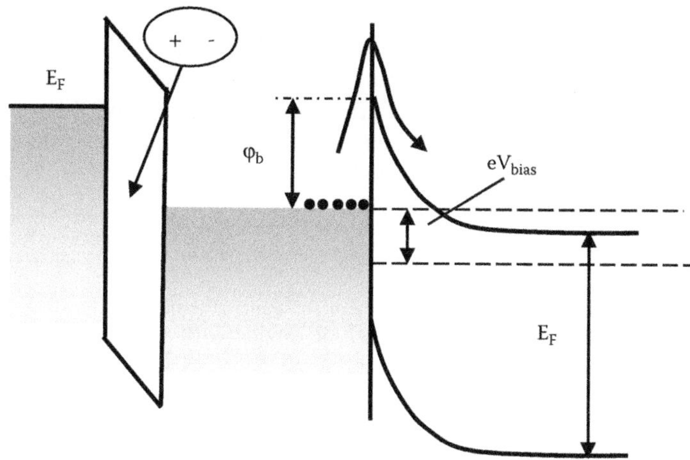

FIGURE 9.16 Band diagram of graphene barrister[21].

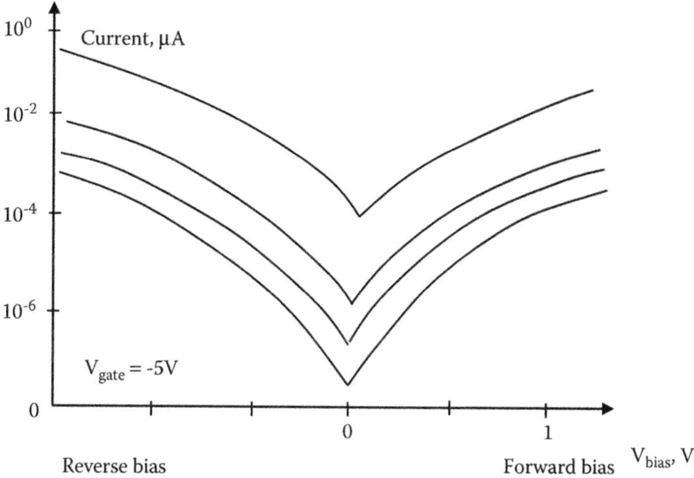

FIGURE 9.17 The dependence of the current of the barrister vs the base gate voltage[21].

The barrister is usually fabricated horizontally on silicon wafers. The gate electrode is formed on top. The negative bias at the gates causes the Fermi level to move into the graphene valence band. As a result, the work function barrier φ_B increases. The current has a strong thermal dependence. The tunneling (Fig. 9.15) takes into account conical density of states in the graphene layers. The predominately hole tunneling takes place with the barrier height $\Delta = 1.5$ eV and mass 0.5. The tunnel barrier width d is shown in Fig. 9.13. The ON/OFF ratio is up to 50. If the h – BN layer is replaced by MoS_2 the ON/OFF ratio may reach 10,000 with a smaller bandgap Δ[22].

Hybrid graphene ferroelectric devices have been described[23]. They are advantageous for low-voltage electronics. Field effect transistors have ON/OFF ratios of 10 to be used in the above devices. Graphene sheets grown by chemical vapor deposition are then attached to the ferroelectric PZT (Pb Zirconate). Such devices may be useful for ultra-fast non-volatile electronics.

9.2.3 The Graphene Barrister

The tunneling FET can be a solid state triode device with a wide operating range. The tunneling FET has difficulties in producing gain[18]. Another way to achieve a wide dynamic range and gain is to create the Graphene Barristor (Fig. (9.16). It is a hybrid silicon-graphene thermionic device. The graphene Fermi levels adjust by a gate electrode whose electric field induces carriers. This way the work function of the barrier height φ_b is controlled. The source that consists of a graphene layer forms a Schottky barrier with the surface of a hydrogenated silicon wafer that is a drain. The Schottky barrier height φ_b is modulated by the thermionic current density J through the top-gate electrode:

$$J = A^{*}T^{2}\exp(-e_{b} / k_{B}T)[\exp(-eV_{bias} / \eta k_{B}T) - 1]; \tag{9.19}$$

where A^* = the effective Richardson constant, k_B = the Boltzmann's constant, η = an ideality factor (close to zero), φ_b = the barrier height and V_{bias} = source-drain voltage applied between the graphene layer and the drain.

In Fig. 9.17 we can see a set of curves for a p – type device. The change of the gate voltage from –5 to +5 V with the forward bias between the source and the drain switches the device. The device current decreases from 1μ to 10^{-4} μA.

Electrostatic control of carrier density has been utilized in the high-gain hybrid quantum dot graphene transistor. A particular example is a replacement of upper platinum coating and Co_2Si nanowires by a quantum dot layer of thickness 80 nm[24]. A negative charge is accumulated through the incident light absorption. The quantum dots then electrostatistically dope the underlying graphene. The illumination leads to photoconductive multiplication effect up to 10^8 electrons per photon. Inherent graphene absorption is insufficient because the graphene layer is thin making graphene a transparent conductor. The fundamental absorption wavelength can be controlled by the nanocrystal length L of one side of a cube. Although, quantum dots can have different shapes, the cube shape makes it easier to perform calculations. The energy of photoconduction necessary to create an electron-hole pair is increased by the containment energy which comes out when an electron-hole pair is formed. Now this photon energy is enough to exceed the bandgap energy E_G. The photon confinement energy that corresponds to the lowest-energy electron half-wavelength in the each cube direction. The confinement energy is:

$$\Delta E(L) = \frac{1}{2}\hbar^{2}3(\pi / L)^{2} / m_{e}^{*}; \tag{9.20}$$

where m^*_e = effective mass, kg.

Thus, we have a size-confined electron in (1, 1, 1) state. The total energy for an excitation creation is:

$$E' = E_G + \Delta E_e(L) + \Delta E_h(L); \quad (9.21)$$

The light's wavelength is shifted toward higher frequencies from $\lambda = hc / E_G$ to $\lambda' = hc / E'$.

The exciton donates the hole and leaves the quantum dot negatively charged whose electric field increases the conductivity of graphene providing large photoconductive gain. The responsivity R, A/W is analogous to photoconductive gain:

$$G_{photo} = [\tau_{Lifetime}(QE)\mu V_{SD}] / A; \quad (9.22)$$

where $\tau_{lifetime}$ = lifetime of the remaining negative charge on the quantum dot (QE); μ = the graphene mobility (~ 0.1 m^2/Vs), V_{SD} = voltage between the source and drain, and A = the area of the graphene flake. G_{photo} can be on the order of 2 x 10^8. The doped Si wafer serves as a gate electrode and the back-gate voltage V_{BG}. The gate voltage controls the phototransistor responsivity and moves the Dirac point in the graphene (Fig. 9.18).

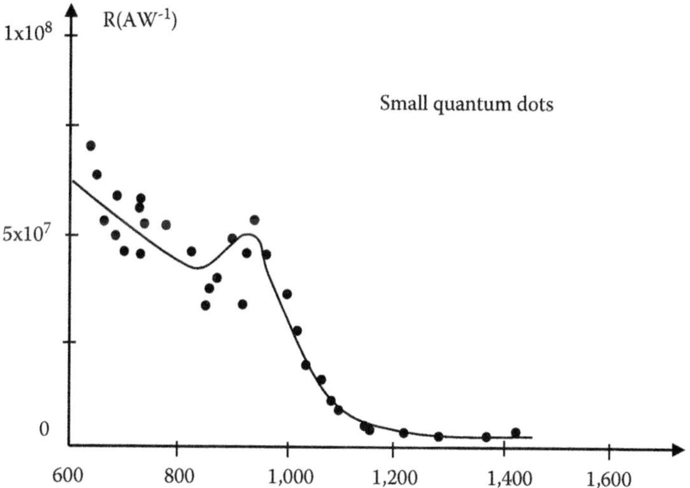

FIGURE 9.18 High photoconductive gain in hybrid quantum dot graphene. The small quantum dots are above the curve. The quantum dots are made of PbS with the height of 80 nm, deposited from a solution[24].

The responsivity dependence on the back-gate voltage is shown in Fig. 9.19. The responsivity's maximum value was close to 10^8 A/W.

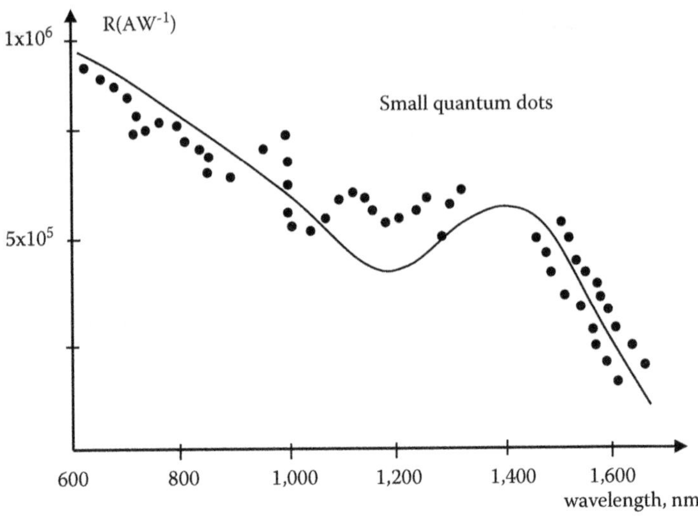

FIGURE 9.19 The responsivity as a function of the back-gate voltage V_{BG} for the graphene phototransistor[24].

9.2.4 Ultrafast Optical Detector[25]

A single – or few-layer graphene sample has the source and drain electrodes on oxidized silicon which is the gate. The above structure is similar to that of the FET. The incident light creates electron-hole pairs in graphene which recombine in picoseconds preventing the process of being useful for photodetection. In the areas adjacent to the contacts where a large electric field exist, the electrons and holes can be separated in the Debye layers or the Schottky barriers between the drain/source and graphene. The quantum efficiency in the photoconducting areas is between 6 and 16%. The photoresponse does not decrease for optical intensity modulations up to 40 GHz with the intrinsic bandwidth exceeding 500 GHz. The device has optical detection with a very large bandwidth, zero dark current, high internal quantum efficiency, and the simplicity of fabrication. Photoconduction in a monolayer-bilayer graphene junction has demonstrated (Fig. 9.20)[26].

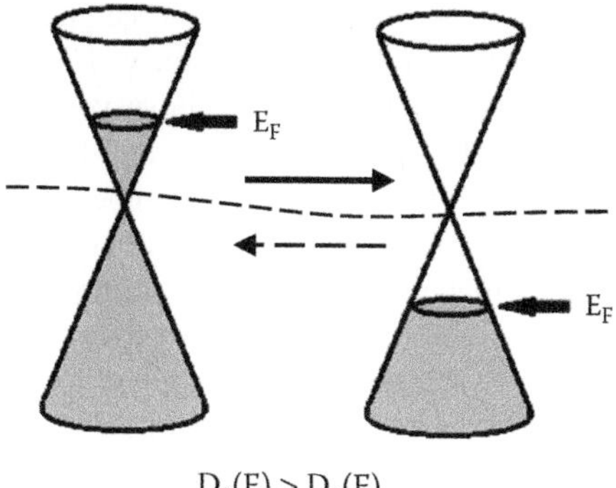

FIGURE 9.20 Bilayer-monolayer junction, the electric field maintains constant carrier density that shifts the Fermi level (the dashed line). The excited electrons are driven by the built-in electric field (the upper arrow). If dominated by a layer entropy, the excited electrons are moved by the layer density of states $D_2(E) > D_1(E)$[26].

The gate electrode controls the position of the Fermi level, the optical absorption and reflectivity. This dependence is illustrated in Fig. 9.21.

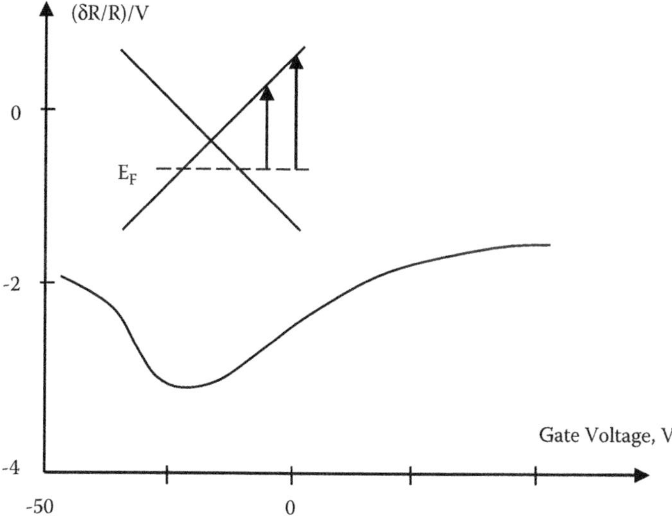

FIGURE 9.21 The reflectivity vs the gate voltage for a single-layer graphene. The Fermi level is changed by the gate voltage.

Fig. 9.21 shows the dependence of $d(\delta R / R) / dV$ on the changing gate voltage and the probing photon energy between 330 and 400 meV. The IR photons were generated by an optical parametric amplifier in the above voltage range pumped by a femtosecond T: sapphire laser. The pulse's energy was ~1 nJ. The spectrum of the pulses was purposefully narrowed to ~ 10 cm^{-1} by a nanochromator and was focused to a square graphene area with a side of 10 μm on a square sample with a 20 μm side. An optical detector was made of mercury – cadmium -telluride. The reflected radiation was sensed by the detector with the maximum absorption of a monolayer of about 2% and a bilayer of 6% for the normally incident IR.

The most important band-to-band absorption near the Fermi level in the graphene is given in Fig. 9.21 in the upper left corner. The carrier concentration n near the field electrode is:

$$n = \alpha(V + V_0); \tag{9.23}$$

where n is positive for hole doping. The capacitor model determines α. Here $\alpha = 7 \times 10^{-2} V^{-1}$.

The Fermi level:

$$E_F = \text{sgn}(n)\hbar v_F [\pi / n)]^{1/2}; \tag{9.24}$$

The transition energy (see Fig. 9.21, the upper left corner):

$$\hbar\omega = 2|E_F| = 2\hbar v_F [\pi\alpha |V + V_0|]^{1/2}; \tag{9.25}$$

The uncertainty (δV_0 = +/- 16 V) is introduced in order to take into account the homogeneity of carrier density in the exfoliated flakes leaving amorphous silicon. The electron-hole puddles distort the intrinsic nature of the conductivity in the vicinity of the Dirac point. The maximum absorption in the IR is 2% only. However, a waveguide 100 μm –long placed on undoped graphene allows complete IR absorption, thus creating the basis for IR detectors, emitters and modulators.

A *graphene optical modulator* has been reported[27]. The electro-absorption device is waveguide-integrated, single-mode waveguide situated on a SiO_2 substrate. The waveguide is doped silicon. The waveguide is tuned to the range of 1.35 – 1.6 μm and has the modulation frequency over 1 GHz. The modulator consists of a graphene monolayer on which a 250 nm – thick waveguide is placed with $Al_2 0_3$ insulator. The graphene monolayer is connected to the waveguide by a golden pat on one side and on the other side, by a 50 nm-thick layer of the doped Si. The Fermi lever in the graphene is adjusted by a bias voltage between the graphene and silicon (Fig. 9.22).

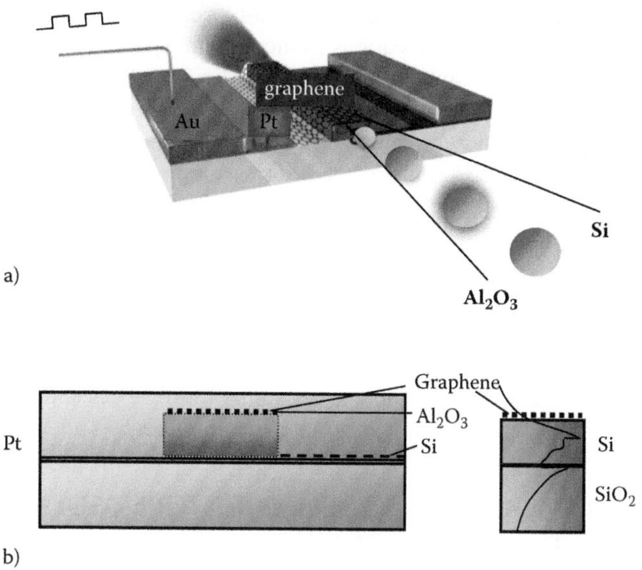

FIGURE 9.22 Broadband optical modulator, a) Si waveguide; b) Right pattern of a single-mode light pattern[27].

Voltage pulses in Fig. 9.22 a) induce charge and shift the Fermi level of the monolayer graphene and the absorption of the single mode photons, whose energy is about 55 eV. The pattern of the optical mode fields with different voltages applied to the electrodes are given in Fig. 9.22 b). The operation of the device is shown in Fig. 9.23.

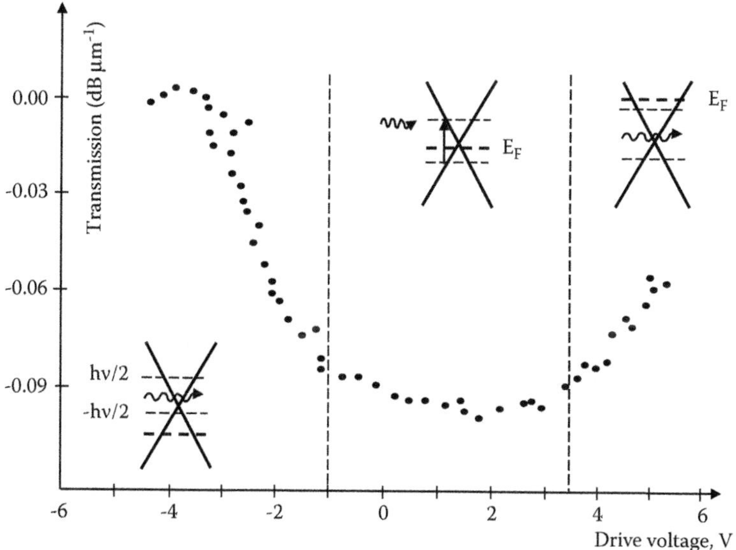

FIGURE 9.23 Two optical-modulator static characteristics at different drive voltages[27].

In the central region of Fig. 9.23, E_F (the Fermi level) is close to the Dirac point. The Dirac points represent size locations in momentum space. The points are situated on the edge of the Brillouin zone and are being divided into two non-equilibrium sets of three points each. Momentum space, in its turn, is the set of all momentum vectors **p** that a physical system can possess. The momentum vector of a particle, included in the above system, is characterized by its motion, $[mass][length][time]^{-1}$. The interband transition causes abruption. If high voltages are applied the Fermi level is at such position that interband transitions are not possible.

A few-layer graphene samples may be used for making of blue-light graphene devices and the subsequent creation of white light sources[28]. The waveguide of the emission is 440 nm. A second emission peak around 550 nm can be produced by adding ZnO nanoparticles. Several peaks result in almost white light. The sources of emission may be viewed as quantum dots that come from particles in solutions. The bandgap plus confinement energy determine the photon energy[29].

9.3 COMPACT SOURCES OF ENERGY: BATTERIES, SOLAR CELLS, AND CONNECTIONS

Large-area graphene samples are necessary for graphene-based device making. 300 Ω/square four one-layer stacks of graphene monolayer sheets have been reported[30]. Graphene is grown by CVD, chemically doped and transferred sequentially. Graphene high conductance, low resistance and transparence are suitable for conductive touch screens in cell phones and other touch screen devices. Presently available 30-inch sheets are large enough for tablets and computer screens but are still small for solar cells (Fig. 9.24) and some other applications where low-resistance graphene is applicable. Some related chemical qualities of graphene are still investigated. For example, it is not clear whether liquid –phase or chemical exfoliation give sufficient conductance for solar cells. Also, annealing and recrystallization are problematic for graphene in the light of its refractory qualities.

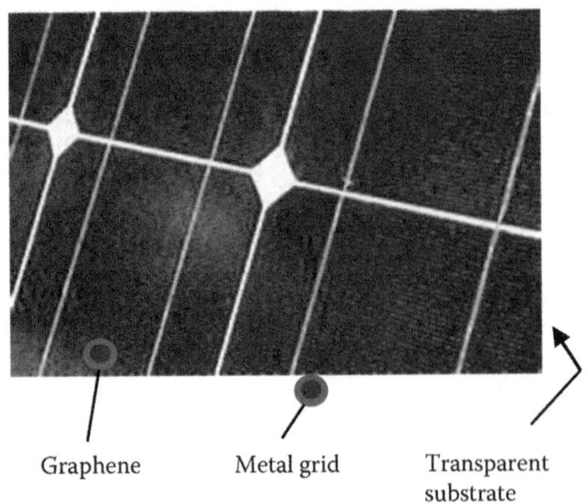

FIGURE 9.24 Graphene solar cell.

Nano-ink-deposited CuInGaSe solar cell manufacturing process uses an annealing/recrystallizing step to substantially enhance the conductivity of the thin film. Using graphene, the annealing/recrystallizing step is not possible. Solar cells require large working areas since the incident radiation averages only 200 W/m^2. The typical rate of graphene deposition is 6.25 mm/min. The market for thin film solar cells is rapidly growing, however, it is not clear whether low-resistance graphene can compete with large-size conventional solar cells.

Improvement of lithium-ion batteries is another graphene application (Fig. 9.25). The anode is usually made of $LiCoO_2$ and graphite Li ions are located at intercalated sites. Intercalation assumes reversible insertion of ions inside layered structures. Charge capacity depends on the available surface area. The recharging rate depends on the rate of ion flow in and out of the storage.

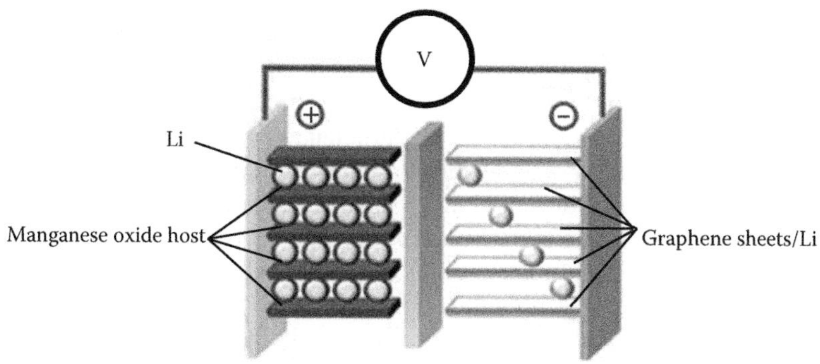

FIGURE 9.25 Graphene-based lithium battery.

State-of-the art compound C_6Li has the charge capacity that corresponds to 372 mAh/g. Graphene nano-sheets (GNS) are used in rechargeable lithium ion batteries[31]. The GNS are 6 to 15 layers thick, having the lithium content of a Li battery 540 mAh/g. The GNS + carbon nanotubes have 730 mAh/g. GNS + C_{60} implies adding C_{60} molecules, addition of which provides wider interplanar spacing for the Li ions whose radius is approximately 0.06 nm. LiC_2 stoichiometry may allow potentially up to 1116 mAh/g.

It is found that an increase in the interlayer spacing increase the storage capacity of chemically exfoliated nanosheets. The average interlayer spacing of 0.34 nm in graphite increased to almost 0.4 nm in the GNS + C_{60}.

Chemical techniques for producing low-sheet resistance still give low quality of the sheet and the available dimensions are insufficient. A 7 nm – thick graphene sample has a sheet resistance $R_{sh} = 800\ \Omega / square$ and optical transmission of 82 % at the wavelength of the incident light of 550 nm. The sheet resistance is much larger than that of the ideal conductance ($R_{sh} = 62.4\,\Omega / N$, N = the number of layers). The substantial increase in sheet resistance is explained by grain boundaries, and traps associated with oxidation and lattice defects are impeding carrier mobility. Spin-coating, the process used for the described method, is only useful for surfaces with dimensions less than several centimeters. An alternative approach, doctor-blade technique tends to produce thicker but less-even deposits with a large substrate.

The necessity to have large areas has led to an integrated graphene-nanotube structure[32]. Single layers are separated by arrays of single-wall nanotubes positioned perpendicular to the layers. An ohmic contact exists between the carbon tubes and the graphene layers. The surface area extends beyond 2000 m^2/g which is close to the theoretically possible limit (2630 m^2/g). The structure capable of holding hydrogen but is not likely to be practically available. Basic organic photovoltaic cells having graphene deposited on quartz have been reported[33]. The achieved resistance per square was about 800 Ω/square. The cells performance was rather poor, probably, due to high sheet resistance. Spin coating is used for graphite oxide particles. The deposit is subsequently annealed by heating to 1100^0 C.

Solar cell performance has been improved substantially in Schottky barrier graphene structures on n-type Si[34]. Graphene grown by CVD on copper foil was transferred to the surface of n – Si and then doped by TFSA, which transferred charge to a graphene layer. Electrons are transferred to the graphene by the work function difference between the graphene and Si. The forming built-in potential separates the electrons and holes. An efficiency of 8.6% was reported. The described technique may be applicable to Schottky barrier solar cells on CdS, CdSe and some other semiconductor. At the moment, solar cells used in calculators, wrist watches and similar devices have the size on a millimeter scale. Solar cells on roofs or side-of-the road units require the single cell to have a dimension of 15 cm on a side. Solar energy provides about 100 W/m^2 (energy averaged in time). On this scale, cell areas of square kilometers may be necessary for a noticeable economic impact[35].

Interconnects are essential to Si chip functionality with Al and Cu have been classical interconnects. A computer based on these chips usually has six layers of metallization. Graphene's low sheet resistance of 30 Ω/square of a four-layer p-type graphene sheet is advantageous for making on-chip interconnects. The International Technology Roadmap for Semiconductors (ITRS) predicts the basic width of wiring of 22 nm by 2020 which requires the current density of 5.8×10^9 A/cm^2 for interconnects. The current density exceeds the copper wiring possibilities. It was reported that the thickness increases with the higher methane concentration used to dope the graphene[36]. Graphene films 10 – 20 nm – thick measured 500 – 1000 Ω/square with the breakdown current densities up to 4×10^7 A/cm^2 which exceeds approximately 10 times the breakdown current density of Cu.

Graphene was grown directly on Si or on Cu foil by CVD yielding $\mu = 1.5$ m^2/Vs at carrier density of 1×10^{12} cm^{-2}. Fig. 9.26 compares current density of graphene grown Cu by CVD on SiO_2 to graphene mechanically exfoliated on SiO_2.

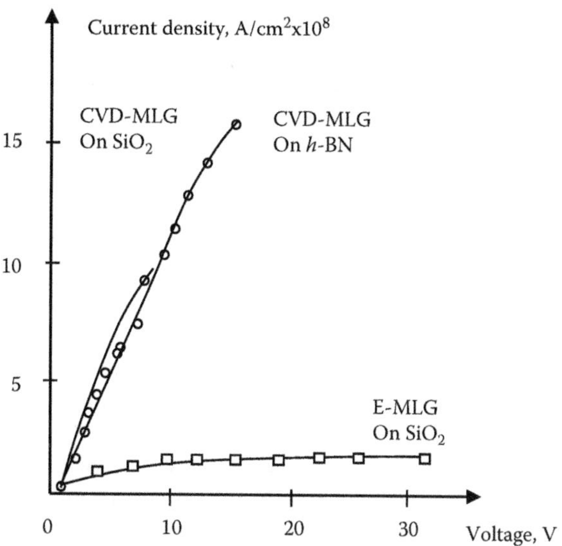

FIGURE 9.26 Comparison of current density vs. voltage of monolayer graphene h – BN, CVD of graphene on SiO_2 and of mechanically exfoliated graphene on SiO_2.

9.4 GRAPHENE STORING CAPABILITIES

Hydrogenated graphene can contain ~ 5 wt % hydrogen, is stable and as such may remain in a storage condition over a long period of time. sp^3 C – H bonds are present in the hydrogenated graphene. At the temperature range between 200 and 500^0 C the hydrogenated graphene releases all of its hydrogen. The above qualities of graphene suitable for chemical storage of hydrogen make it possible to create a chemical tank for an automobile with hydrogen replacing petrol. The fuel cell proves more efficient than a combustion engine, driving electric motor of the vehicle. Another advantage is emission reduction and brake energy utilization. These principles are already implemented in numerous hybrid automobiles. The storage of solar cell energy may be put to practice using the above energy for water decomposition releasing hydrogen.

The Birch reduction process was used for adding of hydrogen to the pristine graphene. The Birch reduction is an organic reaction in which two hydrogen atom are attached to the opposite ends of the molecule. However, the process is not practical involving lithium in liquid ammonia. Once added, the hydrogen could be released by heating of the already hydrogenated graphene. Another drawback of the technique was oxygen blocking the absorption sites. Since oxygen is almost always present the oxygen penetration would be prevented by a polymeric coating or by a nanoporous agent that can bar oxygen from the penetration[37]. There are, e.g. some alloys as Palladium-silver that are passable for hydrogen but resistant to other gases including oxygen.

Supercapacitors require large surface areas with absorption of hydrogen for the capacitance enhancement. At the present moment, graphene's monolayer can provide the largest surface area per gram of any other substrate[38]. The supercapacitors based on graphene possess high energy density (85.6 Wh/kg at 20⁰C and 136 Wh/kg at 80⁰C) with a current density of 1 A/g (Ampere/gram). Although, the same parameters can be achieved by Ni-metal-hydride batteries, the charging time is substantially lower, on the order of seconds to minutes[39].

Graphene's large spin mean free path may be used in spintronic devices. Spintronics is a physical phenomenon involving the intrinsic spin of the electron, the magnetic moment and charge. The spins are manipulated by magnetic and electric fields. Spintronics emerged as a study of electron transport phenomena including spin-polarized electron injection coming from a ferromagnetic to a usual metal. The long mean free path for the spin is augmented by the spin-orbit interaction in the carbon atom. An example of a popular spintronic device is MTJ (Magnetic Tunnel Junction) magnetic disk reader in modern computers. In the early spintronic devices. Co or Ni ferromagnetic electrodes are the sources of spin-polarized electrons. In the Tunnel Valve (or Tunnel Magnetoresistor, TMR) device, a MgO tunnel barrier keeps the spin-alignment for tunneling electrons. The device's materials should have a long spin lifetime which corresponds to long spin diffusion lengths which have been found in graphene[40]. The geometry of a spintronic device is shown in Fig. 9.27.

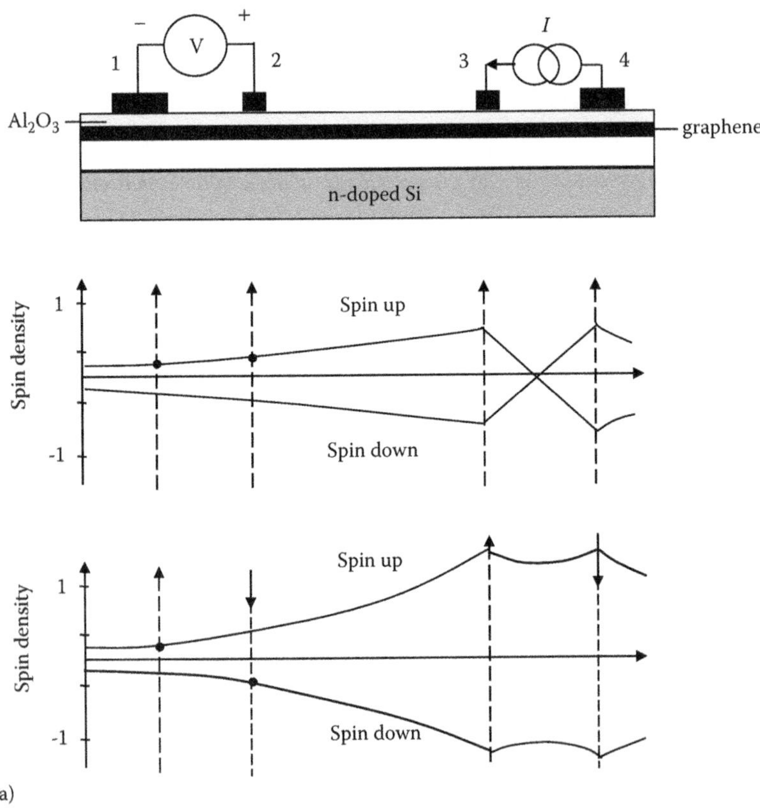

FIGURE 9.27 a) Graphene spin-valve device. *(Continued)*

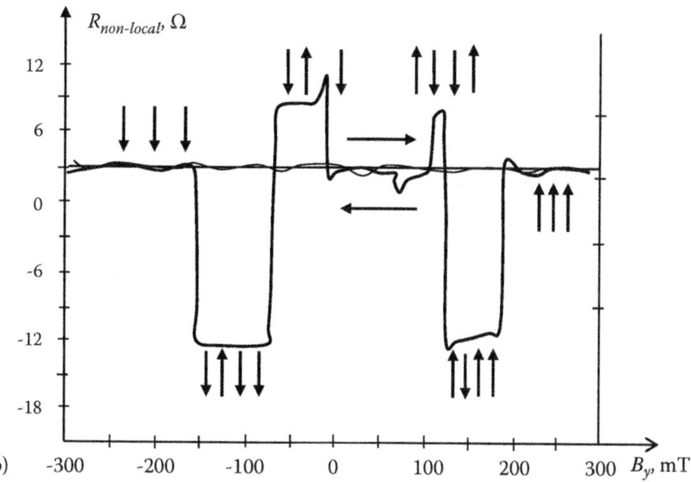

FIGURE 9.27 (Continued) b) Measurements of graphene spin-valve device. Magnetic field B_y are directed into the page upward in the drawing[41].

Ferromagnetic electrodes denoted 1 – 4 are separated from graphene by aluminum oxide barriers that allow spin injection into graphene. The voltage is applied to electrodes 3 and 4 and is measured at electrodes 1 and 2. The electrodes are made long and contain only a single magnetic domain. Electrode 3 is the narrowest and demands the largest reverse magnetic field to direct its magnetization out of the page. The spin density curves are symmetrical around the spin density zero line in Fig. 9.27 a). The Co electrode magnetization directions are denoted by up and down- pointed arrows which correspond to into- and out-of page physical directions. The current flows only between the electrodes 3 and 4. The left and right section of the graphs in Fig. 9.27 a) have the value drop because of electron diffusion. The middle of the graph, the falling spin-up density is caused by the imposed spin-up electrons extraction. The middle bottom increase is caused by the injection electrodes having the opposite magnetization. The large increase is also due to the spin –up electron injection: electrode A3 injects spin-up electrons, and electrode 4 extracts spin-down electrons. In Fig. 9.27 b), the measurements were made at T = 4.2 K of the V_{21}/I_{43} verses polarizing magnetic field B_y. Thus, non-equilibrium magnetization in graphene (paramagnetic metal) can be detected applying an open-circuit voltage between the graphene and Co (ferromagnetic). The sign of the measured voltage depends on the spin sign. Electrodes 3 and 4 (Fig. 9.27 a)) inject electrons into graphene whose polarization depends on the magnetization direction of each Co electrode. The polarization is controlled by the applied magnetic field B. The measurements of the no –local-valve device shown in Fig. 9.27 b) were made at T = 4.2 K detecting 4 resistance levels with the switching of the electrode magnetization (horizontal arrows).

The largest change in Johnson-Silsbee voltage takes place when the magnetization of the current electrodes 3 and 4 is switched from parallel to anti-parallel. The spin-up electrons are injected and spin-down electrons are extracted, thus upsetting the equilibrium between the electrodes 3 and 4 (Fig. 9.27 b)). The magnetization

$M = N_{\mu B}$ caused by individual magnetic spins with an applied magnetic field B, precesses a static field (the Hanle effect) at the Larmor frequency:

$$\omega_L = g\mu_B B / \hbar; \tag{9.26}$$

where $g = 2$ for the electron and μ_B is the Bohr magneton.

Precession is a change in the direction of the axis of rotation of a rotational body. There are two kinds of precession: torque-induced and torque-free. The Hanle effect implies reduction in the polarization of light: the atoms emitting the light are influenced by a magnetic field in a certain direction. The Larmor frequency refers to the rate of precession of the proton's magnetic moment around the external magnetic field. The frequency of precession is connected with the strength of the magnetic field B_0. The precession effect takes place when a magnetic field is perpendicular to the graphene device plane (Fig. 9.27 a)) and is applied so that the electron spin precesses at the Larmor frequency (Eq. 9.26). In Fig. 9.27 b) B = 100 mT. When the spin-up electrons are injected through electrode 3, they diffuse to electrode 2 and precess around the vertical magnetic field. The spin diffusion length was found between 1.5 and 2.0 μm[41]. A non-local magnetoresistance $\Delta R_{NL} = 130\ \Omega$ was observed by tunneling spin injection from Co through MgO/TiO_2 barrier[42].

9.5 GRAPHENE'S PROSPECTIVE SPECIAL APPLICATIONS

The electric field effect in graphene is the basis for the material's exceptional sensitivity to the pressure of single molecules on the graphene surface. The local Fermi energy level is shifted by an adsorbed molecule. This effect is confirmed by an electric parameter charge when graphene is cleaned by heating in order to remove adsorbed substance[43]. The FET and analogous devices, based on the effect have been used to measure the solution quantity pH, pH is a numeric scale used in chemistry to quantize the activity of alkalinity of an aqueous solution. Approximately, pH is the negative of the $\log_{10}$ of the molar concentration with units of moles/liter of the hydrogen atom activity.

Specifically, SGFET (Solution Gated Field-Effect Transistor) was employed to yield 99 meV/pH sensitivity[44]. Gas-phase sensors, called "electronic noses" were described as being effective to detect changing of gas phase[45].

The quantum Hall effect in graphene mentioned earlier can provide a resistance standard based on graphene[46]. The von Keitzing constant $R_K = h/e^2 \sim 25.813\ \text{k}\Omega$ with a graphene sample. Raman effect is also used for nano-measurements of graphene[47].

9.6 MEMORY DEVICES

A possible flash memory on graphene is based on capacitor-like storage which is a capacitor of minimal lateral dimensions (about 45 nm at the present moment). An example is a polysilicon floating-gate device with a p-type Si wafer as a substrate[48]. The above device is a state-of-the-art memory which may substitute the current laptop computer flash memory and USB flash-memory drives.

Another example is a graphene flash memory device with a capacitor-like structure CVD-grown on an oxidized *p*-type Si substrate[49]. A single capacitor consists of graphene/(5 nm SiO_2 tunnel oxide)/*p* – type Si). Its discharge rate is shown in Fig. 9.28 a). Fig. 9.28 b) shows advantages of graphene flash memory comparing simulations of usual floating gate (FG) polysilicon flash memory (the upper curve) with graphene flash memory (the lower two curves).

The height of the barrier between graphene and quartz in the graphene storage capacitor is approximately 3 eV. The capacitor is charged from an upper gate electrode on which a gate is grown being separated by an insulator. The gate's top is covered by sapphire (35 nm of Al_2O_3). The read/write gate-electrode has ± 6 V for storage and ± 7 V for erasing. The critical parameters are the desired retention time (across the tunneling barriers) of charge and reproducibility (across the gate barrier) in the voltages to read and erase.

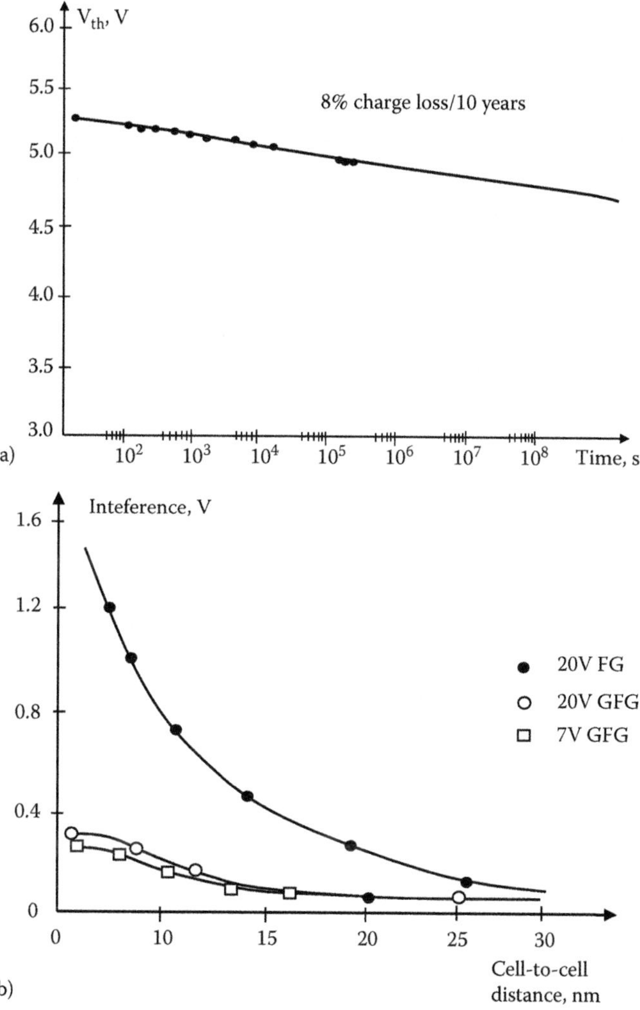

FIGURE 9.28 a) Dependence of decrease of voltage on the memory device on time[49]. b) Dependence of cell-to-cell distance interference voltage (shift in threshold voltage of an unprogrammed cell by its two nearest-neighbor cell) on cell-to-cell distance[49].

The gate oxide was grown on 1.1 nm of Al. The gate electrode consists of Ti/Al/Au with thickness 10/500/10 nm deposited by photolithography and electron-beam evaporation with areas in the range of 2.5 x 10^{-5} – 7.4 x 10^{-4}. The density of flash memory cells is reduced by interference voltage between the adjacent cells (crosstalk). The conductance layer thickness is 0.34 nm – 1 nm (Fig. 9.27 b). The reduction of thickness is imported to the crosstalk minimization. The graphene flash memory is advantageous with a lower voltage interference, lower operating voltage and lowering the energy per bit by 25 %. The graphene's advantage as a material over polysilicon is its high work function, high density of states and lower dimensions.

9.7 GRAPHENE FOR FET SWITCHES

It is obvious at the moment that graphene-based devices will play a significant role in electronics. However, an integration into the existing silicon may be more practical since a large proliferation of existing silicon technology with investments has taken place. The CMOS FET is the major component of digital logic. A high level microchip contains up to 2 x 10^9 FETs while annually some 10^{18} FETs are made. The manufacturing facility's cost often exceeds 10^9 dollars[7]. The existing tendencies of the semiconductor industry imply still further development of Si-based devices for which graphene-based components will be complementary. The basic structure and the operation principle of an FET is given in Fig. 9.29. The voltage applied between the gate and source controls the channel conductivity of the FET determining the latter's operation.

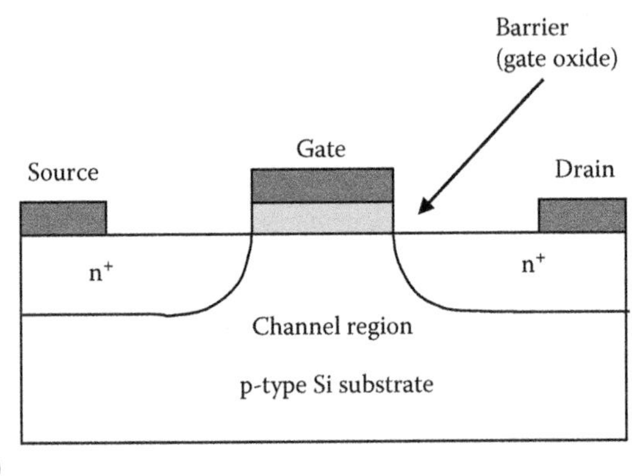

FIGURE 9.29 a) an N – FET device on *p* – type Si crystal. *(Continued)*

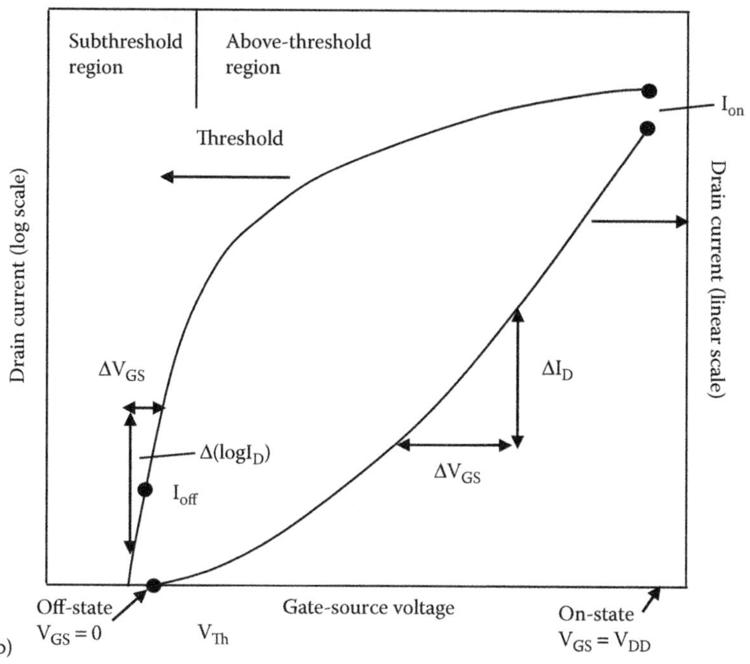

FIGURE 9.29 (Continued) b) graphs of switching of the FET.

S = inverse subthreshold slope. The steepness of the curve logI/V_g depends on the device switching capability.

$$S = \frac{dV_{GS}}{d(\log I_D)} = \frac{\Delta V_{GS}}{\Delta(\log I_D)}; \tag{9.27}$$

The schematics of the drain current on the left is logarithmic and the linear – on the right[50].

At room temperature the ideal value S for a MOSFET, $S = k_B T \ln 10 = 59$ mV/decade (decade is a ratio of 10 between two numbers when the ratio is measured on a logarithmic scale). For high speed applications a fast response to a voltage change V_g demanding fast carriers and short gates. Down-scaling brings short-channel effect. The gates can alleviate the problem until very short gate lengths are achieved (approximately 12 nm)[51]. The probability of using graphene as only one-atom-thick channel is the most attractive feature of the material with regard to FETs while in the silicon technology thinner channels have rougher surfaces that decrease the carrier mobility. CMOS logic consumes power even in the OFF state, thus the goal is to increase to ON/OFF ratio to 10^4 and more. The latter condition is met by graphene having a symmetrical p-n structure and a big difference in electron and hole effective masses. The energy consumption is total energy for switching operations. For acceptable small OFF current I_{OFF}, switching energy scales:

$$p \propto I_{OFF} V_{dd}^3; \tag{9.28}$$

where V_{dd} = the supply voltage.

So far silicon technology has been successful immeasurably coming close to the limit of miniaturization and integration. The mass chip market is likely to be in a static situation. Only, perhaps, cloud computing on a large scale may move towards superconductivity logic to reduce power consumption. Notwithstanding the vastness of the CMOS logic market, there are portion of it where the graphene tunnel FET may take over[52].

A family of MOSFET devices with channels of no more than 5 nm – long were made[53]. The dependence of drain current on gate voltage for n – and p – type devices are given in Fig. 9.30. The drain currents reached 0.1 mA/μm. Drain voltage was restricted to 0.4 V, the saturation was negligible due, probably, to parasitic resistance. The cut-off characteristics are good at 0.4 V. The gate electrodes were made of polysilicon.

For logic applications, it is important for a FET to have a large ON/OFF resistance ratio which requires a large bandgap. Graphene is metallic and has almost no bandgap. The solution is to use nanoribbon devices for which the ribbon has to be very narrow (less than 10 nm) in order to receive a bandgap of at least 0.4 eV. Such devices are able to produce ON/OFF ratios of approximately 10^6 [Ref. 54]. Manufacturing difficulties, however, are likely to exclude the graphene nanoribbons from following the Moore's Law. Further options remaining for graphene logic structures are in wide-area graphene devices that do not need a narrow width in order to produce a suitable bandgap.

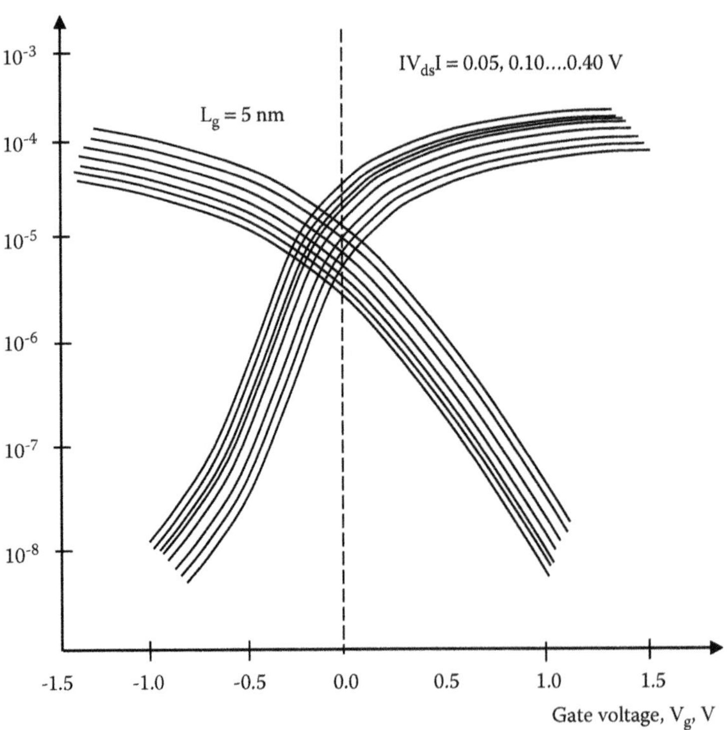

FIGURE 9.30 I_{DS} vs V_G dependences for n-FET and p-FETs with gate length of 5 nm[53].

So far, three types of the described graphene developments have been suggested:

1. Traditional FET structure with an electrically-induced gap in bilayer graphene;
2. Bilayer graphene FET structure;
3. Two-dimensional tunnel transistor based on graphene.

One of the first graphene FET (Tunable-gap bilayer graphene FET) is depicted in Fig. 9.31[55]. The height of the FET is about 5 nm with an applied vertical electric field to induce a gap. The vertical height excludes the width of top and bottom gate-electrodes (made of metal or graphene). The overall length, L = 15 nm is small in comparison to the channel length. Two single layers of graphene are separated by 0.35 nm. The source and drain are doped with n^+. The overall height is much less than the length of the channel – an advantageous feature that helps to avoid short-channel effects.

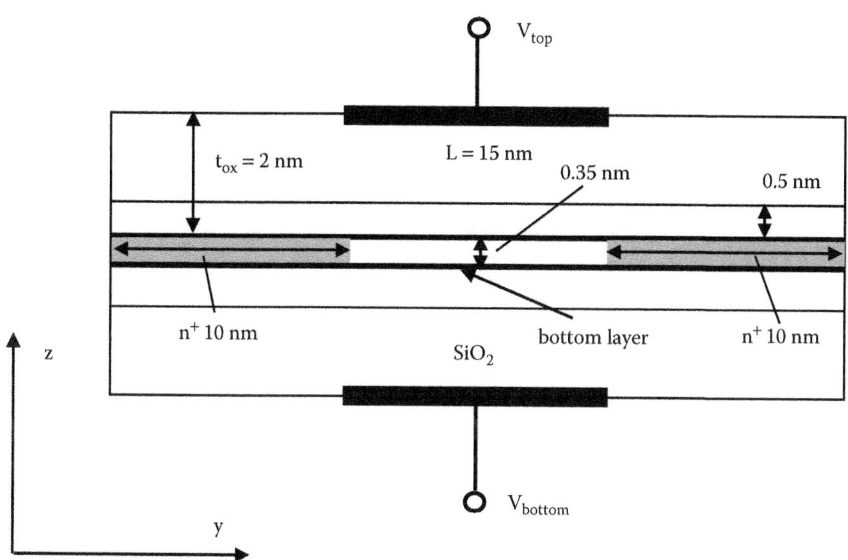

FIGURE 9.31 Tunable-gap bilayer graphene FET[55].

9.8 GRAPHENE TUNNELING FETs

Graphene may be used for tunneling transistors, tunneling FETs (TFET) that operate by interband tunneling. A commercially available tunneling device already exists, it is the Zener diode that has a reverse biased p – n junction. The threshold voltage applied to the junction becomes highly conductive when the volume band carriers penetrate across the depletion zone to the conduction band with empty states. The current is independent of temperature and given for n:

$$n = [F^2 m_\tau^{1/2} / (18\pi \hbar E_G^{1/2})] \exp[-\pi m_\tau E_G^{3/2} / 2\hbar F]; \quad (9.29)$$

where n is the rate of electron transfer with an applied electric field E, with $F = eE$, with a bandgap E_G, m_τ = the reduced effective mass is equal $m_\tau = m_e m_h/(m_e + m_h)$.

The mass of the carrier charges as the carrier crosses the depletion region. Another example is the Esaki which us also a semiconductor junction with highly degenerate n – and p – layers and a narrow depletion region.

TFET device based on interband tunneling has a $p - i - n$ structure[56]. P and n regions are heavily doped with i – region between them (weakly doped p – silicon). The bands are shifted in i – region allowing Zener tunneling (Fig. 9.32).

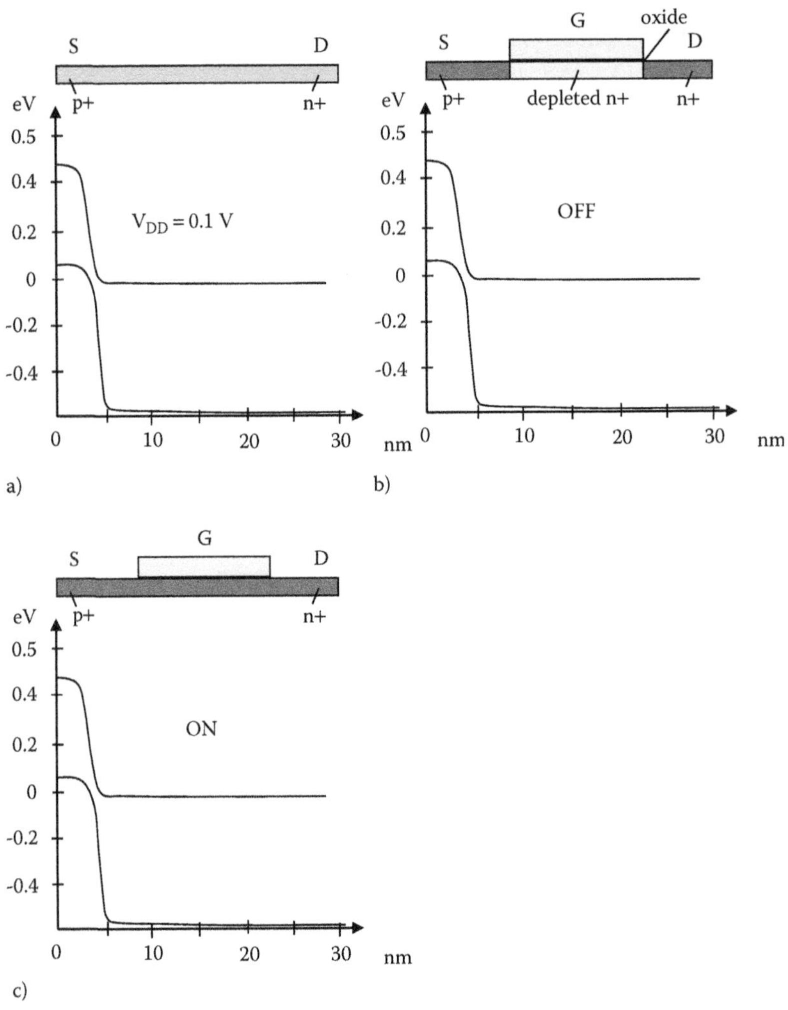

FIGURE 9.32 T – FET (Tunneling Field – Effect Transistor) energy band diagram, a) no gate effects, the channel with sufficient source-drain bias $V_{DD=0.1}$ for the Zener tunneling; b) the OFF condition created in the full depletion of the channel; c) T-FETs have a lower power dissipation and faster switching with a smaller s- factor (which is not limited by k_BT/e) than conventional FETs[57]. The s – factor of graphene is proportional to the graphene's material index.

The p –i – n FETs were grown on a Si surface by diffusion of dopants. The gate is located over the intrinsic I region. It controls the n –i – p FET that operates as a MOS-gated reverse biased $p – i – n$ diode. $P – i – n$ structure also minimizes the leakage current. A TFET with a different mode of operation with respect to the TFET in Fig. 9.31 is presented in Fig. 9.33. Both TFETs geometries are similar.

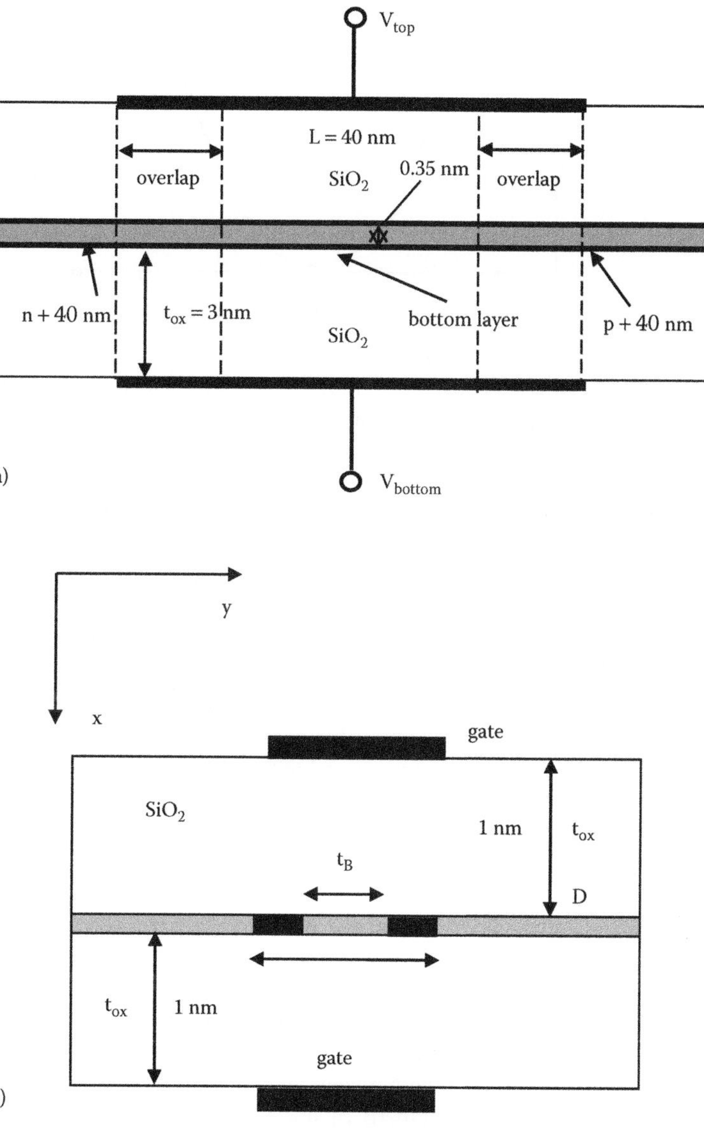

FIGURE 9.33 a) T – FET device with a bilayer graphene tunnel. Two graphene layers are separated by 0.35 nm. The bilayer is undoped in the central region, L = 40 nm wide. The band energies are shifted by fractional dopings. b) transistor with high I_{ON}/I_{OFF} ratio, monolayer graphene with two gates. p– insulator- p (PFET). *(Continued)*

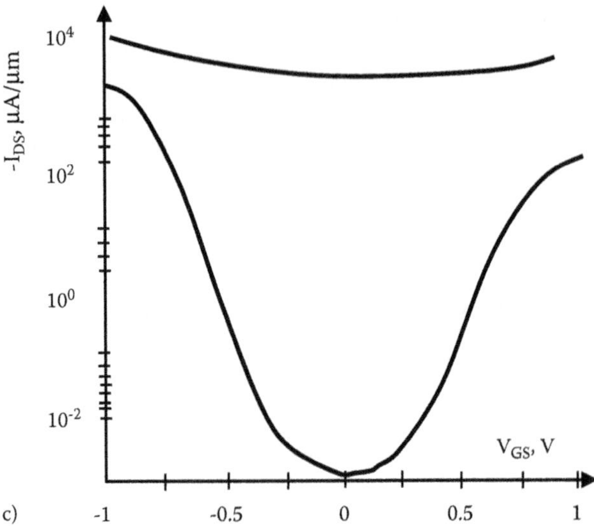

FIGURE 9.33 (Continued) c) simulation of drain current I_D vs gate voltage. The lower data curve is for graphene – h – BC_2N tunnel FET. $I_{ON}/I_{OFF} > 10^4$ with source-drain voltage 0.6 V at room temperature. The upper curve represents a simple graphene FET transistor.[55]

With a drain to source voltage of 0.1 V, an ON/OFF > 1000 is possible[55]. The switching for the described tunneling FET is superior with *s*- factor 20 meV/decade vs 60 meV/decade for the best thermionic *s* – factor. In Fig. 9.33 a) the gates form a large vertical field. In heavily doped regions (beyond the gates) the bilayer graphene bandgap is zero. The 6 nm wide SiO_2 layer's breakdown voltage reaches 14 MV/cm. The Klein tunneling does not take the place because bilayer graphene massive electrons and holes are in parabolic bands. The Klein tunneling implies overcoming of the barrier if the potential is on the order of the mass of an electron ($V \sim mc^2$). This result was achieved by applying the Dirac equation to the effect of electron scattering caused by a potential barrier. The transistor's manufacturing process may involve a graphene bilayer growth on Cu foil, then masking for *n* – and *p*- type deposition to form the source and drain region, all under the usual CVD conditions.

A monolayer-graphene tunneling transistor with two gates is presented in Fig. 9.33 b). A single graphene sheet (a wide ribbon) contains the source and drain. The ends of the sheet are heavily doped *p* – type (the narrow barrier is hexagonal BCN)[58]. An insertion of the BCN barrier is adaptable: the BCN barrier region can be adjusted changing the barrier's composition: the barrier height range is 1 – 5 eV[59]. The composition adjusting can be used to block one or the other type of carrier. The structure in Fig. 9.33 b) is a hybrid h-BCN – graphene transistor that has a chemically – formed barrier inserted between the two gates (the source and drain). The barrier length is epitaxial length tB of h – BCN. Hexagonal BCN can have energy gaps from 1 eV to 5 eV, adjustable in this range. The h – BCN can be grown in two or three layers on copper foil by conventional CVD. Methane and ammonia-borane ($NH_3 - BN_3$) are used in the proposed device that has the PFET form, i.e. the drain and source (*p* – type heavily doped graphene). The barrier region is likely to have parabolic but not conical shaped bands. The simulated performance is given in Fig. 9.33 c) where the barrier composition is BC_2N. The dashed line is a value of the steepness of the (logI)/V_g plot concerning the digital logic switching capability of the proposed device. The s-value was measured to be 80 mV/decade which is superior to the conventional graphene FET (the curve in the upper part of Fig. 9.33 c)). The authors have suggested also resonant tunneling devices with negative resistance phenomenon. The structure of such a device is similar to that in Fig. 9.33 b) but the barrier region is split into two BN barriers (t_B) along with an intervening narrow graphene region. The localized states align in energy with the Fermi level of the source resulting in a peak in the $I_D - V_{GS}$ characteristic. It is argued that the achievable characteristics are superior to those in III – V -based devices since graphene offers more effective operation control. TFET graphene is superior to the silicon FET by the virtue of having a conducting channel with the height much larger than the thickness since the depletion width of any silicon transistor is much larger than 0.34 nm of a graphene device. The height of the proposed FET is also an advantage: 2.5 nm in Fig. 9.33 b). With the gates, the total height of the FET would be smaller than 4 nm. The forward current is large and OFF current is low – superior characteristics as well. The gates are grown by CVD (electrical vapor deposition) with polycrystalline silicon as the deposition material. The insulating qualities of BN help to replace silicon.

9.9 GRAPHENE-BASED SENSORS[60]

Graphene composites can be employed as a high sensitivity sensor suitable for temperatures as high as 1000^0 C. The graphene films are grown on 6H-SiC (0001) surfaces with halogen-based plasma etching followed by rapid thermal annealing. Lithography methods are employed to produce interdigitated finger-like sensor structures. The nucleation of either Pt, Au, Ag or Ir nanoparticles is followed on the graphene surface by solution-based methods. After the growing process, the graphene films were patterned by a two-step lithography process. The produced pattern is shown in Fig. 9.34.

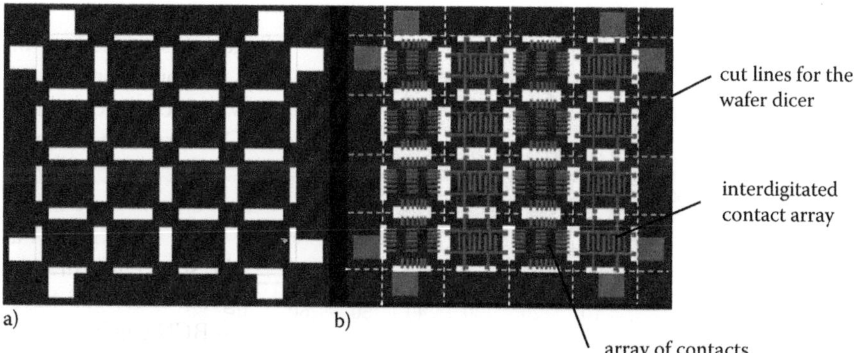

FIGURE 9.34 a) the mask with the selected areas, b) electrical contact deposition pattern (courtesy of C.D. Steinspring).

Nanoparticles of Pt, Au, Ag, Ir were attached to the dice in the graphene areas, with conventional solution. The chemical reaction for Ag, for example, is written as:

$$AgNO_3 + NaBH_4 \Rightarrow Ag + 1/2H_2 + 1/2B_2H_6 + NaNO_3 \quad (9.30)$$

After the process, the particles remained attached to the substrate. The overall RMS (root mean square) roughness of the particle-formed surface way was 6 nm. Fig. 9.35 shows the particle height and diameter distribution.

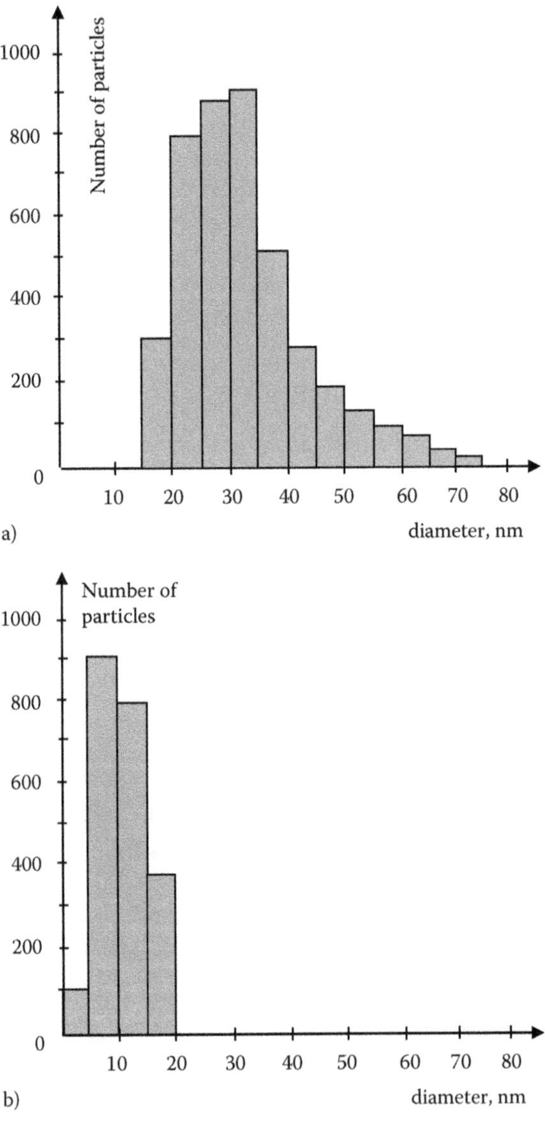

FIGURE 9.35 a) The distribution of the particle diameter after deposition for Ag nanoparticle composite. b) The distribution of particle heights for Ag nanoparticle composite.

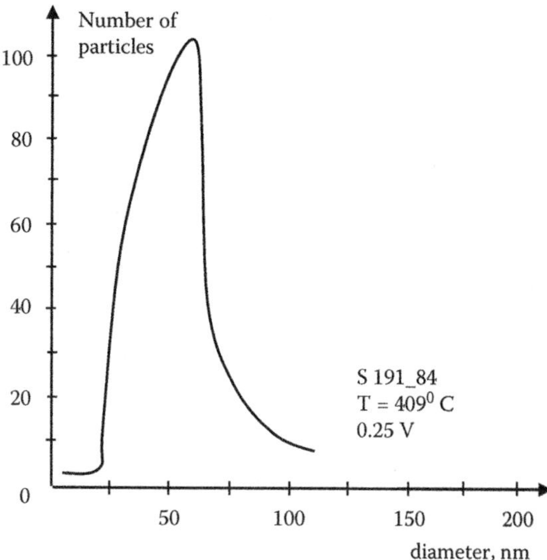

FIGURE 9.36 Response to Au particle graphene to long H_2 pulses.

Figure 9.36 gives the sensor response as a function of time. The sensor's reaction, the appearing pulse is the reaction to alternating pulses of Ag and H_2 at 409^0 C. The initial rise of H_2 pulse can result from H_2 contact with the graphene active regions. The subsequent decrease with time is the H_2 restriction during Ar exposure.

Gas sensors are important in different industrial and special applications. High temperature sensors may be efficient control components for missiles and jets. Gas sensors can be in the form of an array of separate graphene chemoresistive sensors, as it was described above.

9.10 OPTOELECTRONIC MODULATOR[61]

Optical modulators are needed to encode information at THz frequencies (~300 GHz to 10 THz). THz range has the potential of communicating data rates orders of magnitude higher than the existing ones. For a given value of the carrier frequency, bit rates may be increased substantially using advanced modulators, such as QAM (quadrature amplitude modulator). The main QAM's benefit is the ability to independently modulate the amplitude of an in-phase (I) and quadrature (Q) set of signals. The achieved rate > 100 Gbits are for the distances of 20 m and more[62]. However, for further rate improvements, the processing electronics have to be an order of magnitude more efficient.

Graphene can serve as a tuning medium in a frequency- dependent optoelectronic amplitude modulator set to work at 2 Hz, at room temperature[61]. The modulator is an array of plasmonic dipole antennas enveloped by graphene. The modulator reflection characteristics are adjusted by electrostatically doping the graphene through only a back gate electrode.

An optical THz amplifier has been reported with a potential for implementing in an optical THz QAM. Waveguides are available for low-loss transferring of THz waves between emitters and receivers[63]. Optoelectronic THz amplitude modulators, however, are restricted to the frequencies of 10s of kHz with some modulators achieving very high modulation depths.

A modulator with a cut-off frequency in the range of 5.5 ± 1.5 MHz and ~ 30% total modulation depth for 100 V potential difference has been reported[61]. The single-mode QCL (Quantum Cascade Laser) was then externally modulated. Multiple electronic elements were grown on one chip. They were characterized only by the parameter *d* (20, 22, 24, 26), μm which corresponds to the length of a dipole. The device was simulated by the finite element using the commercially available software COMSOL. The basis element is given in Fig. 9.37. The structure from the lower to upper layer consists of: p – Si substrate (6 μm), SiO_2 (300 nm), graphene (15 nm), Au (70 nm) and air (6 μm). The THz waves are incident from the top z – surfaces with Port (S_{11}) electromagnetic boundary conditions. The upper angle is incident and the lower angle is transmitting. Graphene is modeled as an optically thin layer.

Using the coordinated system in Fig. 9.37, the top z surface was defined as an electromagnetic port and the bottom z surface was defined as a listening port allowing control of transmission and reflection waves by taking $|S_{21}|$ and $|S_{11}|$ respectively. The x and y directions were assumed to be periodic, called Floquet periodicity. Floquet theory is based on the theory of ordinary differential equations connected to the solutions of periodic differential equations:

$$\dot{x} = A(t)x; \tag{9.31}$$

where $A(t)$ = piecewise continuous function that has a period T.

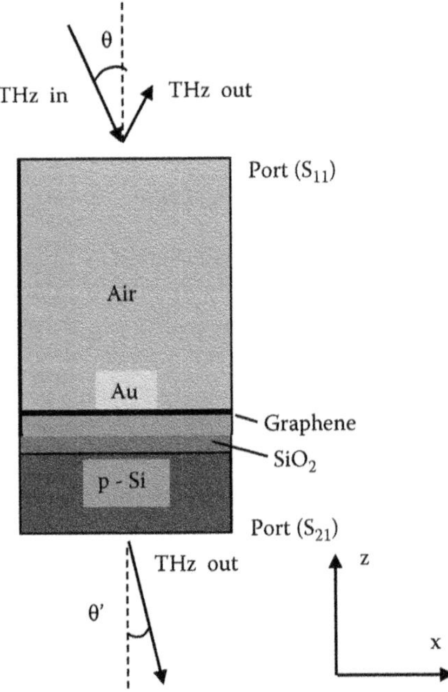

FIGURE 9.37 Basic unit cell of the optoelectronic modulator.

In Fig. 9.38, the stimulated spectral response of reflectivity with respect to the frequency of the incident radiation is shown. The non-resonant area is below 2.2 THz frequency. Below 2 THz, a standard quasi Drude is observed. Graphene sheet conductivity is designated as σ_G^S. The reflectivity increases with an increase of the graphene conductivity. The increase of σ_G^S at the resonance region around 2.8 THz leads to a change in reflectivity.

In order to change the conductivity of the modulator material, a potential V_G is applied to the gate dielectric SiO_2 which changes σ_G^S because of the electrostatic doping of the graphene. The σ_G^S change causes a change in reflectivity. The dependence of reflectivity on conductivity is shown in Fig. 9.39.

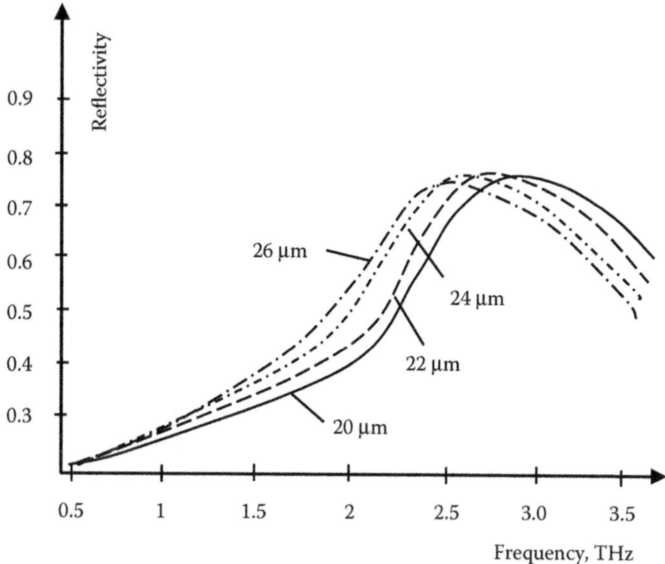

FIGURE 9.38 Simulated reflectivity response to the frequency of the incident radiation.

Electrically, the modulator may be modeled with lumped elements as it is given in Fig. 9.40. V_{AC} = the AC signal value with an DC offset. The lumped element model is a simplified representation since, for example, the capacitive and inductive components of the module cannot be physically separated.

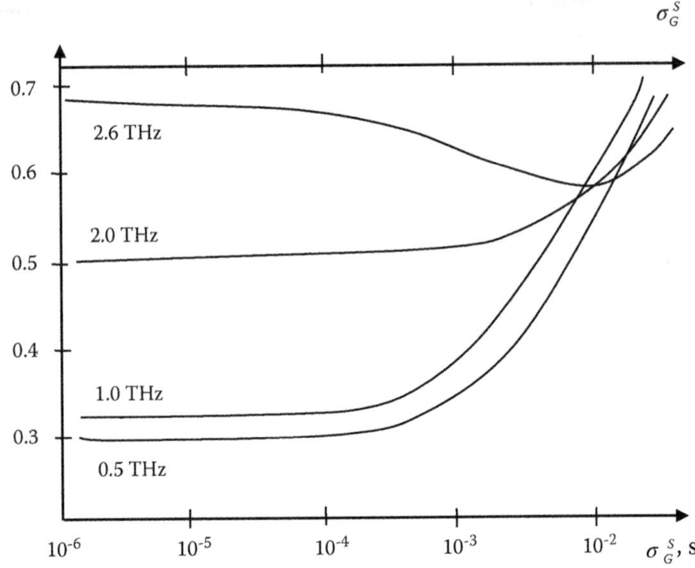

FIGURE 9.39 Simulated dependence of reflectivity on the conductivity of a graphene sheet.

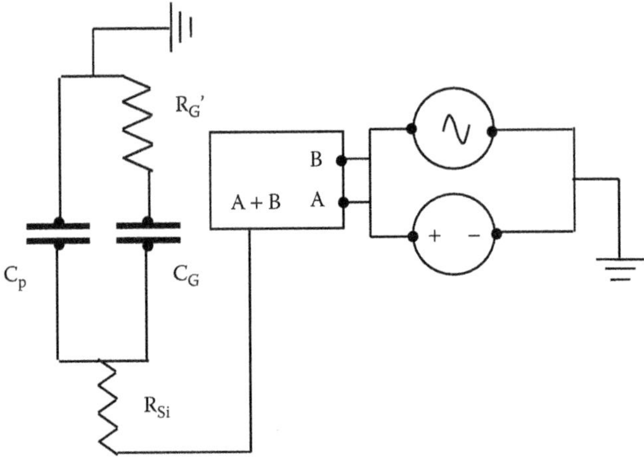

FIGURE 9.40 Lumped element circuit diagram of the proposed modulator. C_p – the source and drain capacitance, R_G – effective graphene resistance, C_G – graphene capacitance, R_{Si}- p-Si substrate resistance.

The current flows from the graphene to the source and drain thus allowing parallel paths for the current decreasing the effective resistance $R_G^{'}$. The capacitance between the source and drain is found from $C = \varepsilon_r \varepsilon_0 A / d$, where $\varepsilon_r = 3.9$, $A_G = 1.2$ mm^2, $A_p = 0.7$ mm^2 and $d = 300$ nm.

The QCL operated in continuous pulse mode with a 30% duty cycle and a repetition rate of 100 kHz. The peak power was estimated to reach 50 μW. The gate was biased by a DC offset V_{DC} + a square wave 10 V peak-to-peak V_{AC}. The source and drain were grounded. The difference in optical power was extracted by a lock-in amplifier (9 Hz the reference frequency). Fig. 9.41 a) shows the dependence of the modulation depth on V_{DC} (a = 26 μm). The maximum modulation depth (which is equal to 7%) corresponds to, approximately, 100 V peak-to-peak modulation of the signal. Fig. 9.41 b) presents the results for the high-speed modulation.

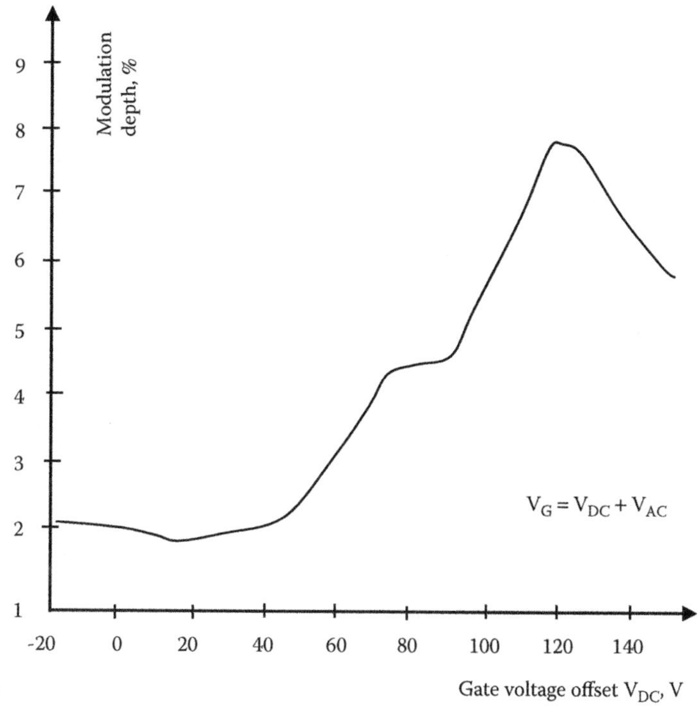

FIGURE 9.41 a) Dependence of the modulation depth on the peak-to-peak signal. *(Continued)*

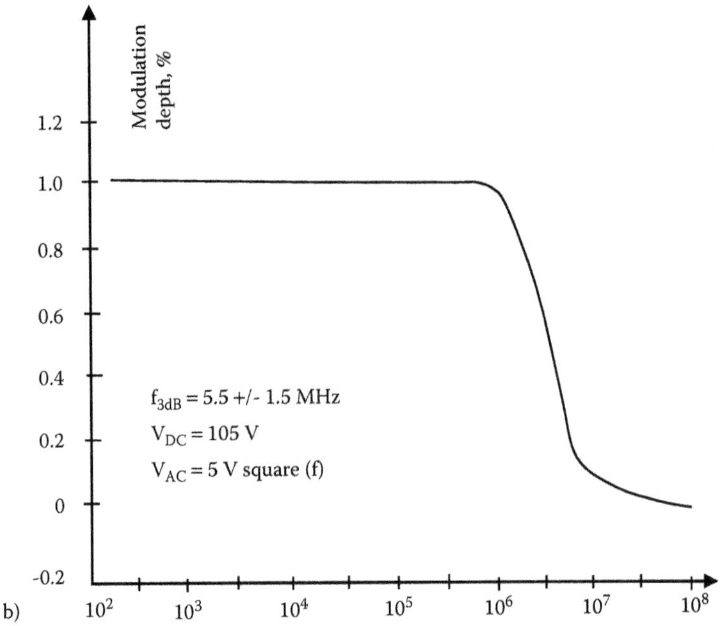

FIGURE 9.41 (Continued) b) Dependence of the measured power vs. frequency.

Lowering the resistivity of the substrate results in the cut-off frequency increase since the resistance R_{Si} of the p – Si substrate decreases. The cut-off frequency effect depends on the ratio of the graphene resistance R_G and R_S, and is greater with the lower values of R_G. To insure the R_S decrease, p – Si substrate resistivity is optimized for reflectivity and transitivity. However, the cut-off frequency increases from the substrate is limited as R_G. Another factor is the capacities of the optical modulator. The large area of the device is needed for large capacitance values. Improved optics focusing may help with the problem. Another approach is to increase the thickness of the SiO_2. The Dirac point may be set at a much lower bias.

With the above adjustments, the modulation frequency can achieve hundreds of MHz at room temperature.

9.11 PASSIVELY Q-SWITCHED Nd:GdTaO$_3$ LASER BY GRAPHENE OXIDE SATURABLE ABSORBER[64]

Graphene may be used as a laser component, in particular as a multilayer graphene oxide as the saturable absorber (GOSA). When the GOSA was placed in the plano-plano laser cavity, a Q- switched laser stable operation was achieved. A maximum average output power 0.382 W, and the repetition rate 362 kHz was reached. The shortest pulse was 194 ns with the single pulse energy of 1.05 μJ. Diode-pumped nanosecond solid state lasers have a wide range of applications for nonlinear frequency conversion, range finding, remote sensing, and material processing. Very promising are carbon nanotubes and SESAMs (semiconductor saturable absorber mirrors) for fiber and solid state lasers. More recently, single-wall carbon nanotubes (SWCNTs) as saturable absorbers (SA) for passive mode-locking for fiber lasers have drawn attention. Such SWCNTs may be easily fabricated while fabrication of SESAMs is quite complicated. On the other hand, carbon nanotube's morphology causes strong scattering and consequently high nonsaturable losses. In addition, ultrashort pulse radiation causes the long-term stability degradation of the carbon nanotube absorbers if the radiation energy is high.

Graphene has been attracting attention because of its outstanding wide broad band saturable absorption, the conventional fabrication, and comparatively low cost. The graphene oxide (GO) possesses the same saturable absorption qualities as graphene in the range of 800 to 2000 nm, suitable for passive Q-switching and laser mode-locking[65]. GO saturable absorbers (GOSA) are easier to produce compared to the graphene SAs. They also offer a wide-broadband or multi-wavelength laser radiation.

High photoluminescence output makes Nd-doped $GdTaO_4$ single crystals suitable solid state laser materials. Its wider absorption at 808 nm FWHM (full width at high maximum) makes the above crystal a preferred choice to such crystals as Nd:YAG and Nd:LYSO. The decreased sensibility to the pumping wavelength results in a higher-power output. Although, the crystal has a shorter FWHM in comparison to Nd:YVO_4, the shorter fluorescence lifetime of the excited state than that of Nd:YAG and ND:GYSGG, it is advantageous for making mode-locked laser pulses. The fluorescence time of the described crystal was 178.4 μs and the absorption coefficient was 11.73 cm^{-1}. The Nd:$GdTaO_4$ laser set-up is given in Fig. 9.42.

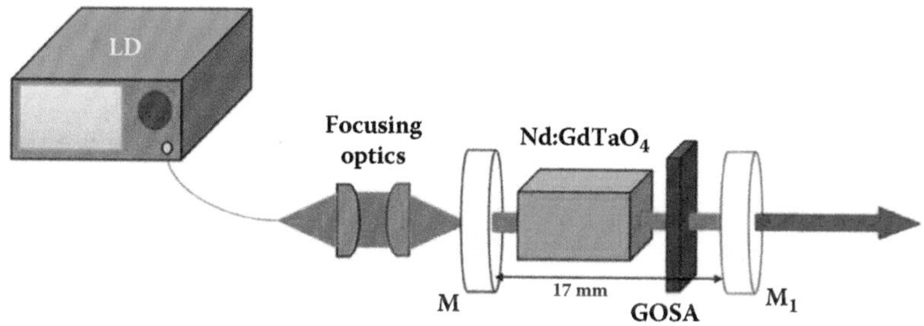

FIGURE 9.42 The Nd:$GdTaO_4$.schematics.

The laser's central line was 808 nm. The fluorescence spectrum of Nd:$GdTaO_4$ is shown in Fig. 9.43.

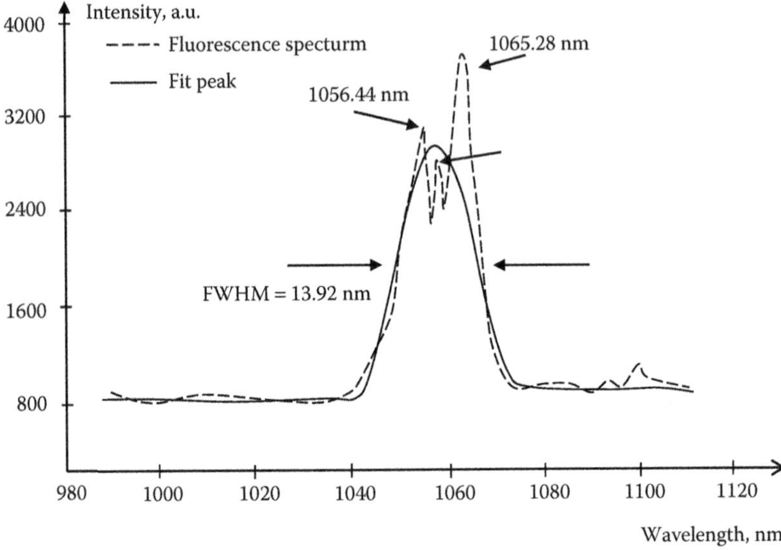

FIGURE 9.43 The fluorescence spectrum of Nd:$GdTaO_4$.

The fluorescence spectrum in Fig. 9.43 ranges from 990 to 1100 nm. Fig. 9.44 compares power efficiency of the CW (continuous wave) and passively Q-switched Nd:$GdTaO_4$ laser with GOSA.

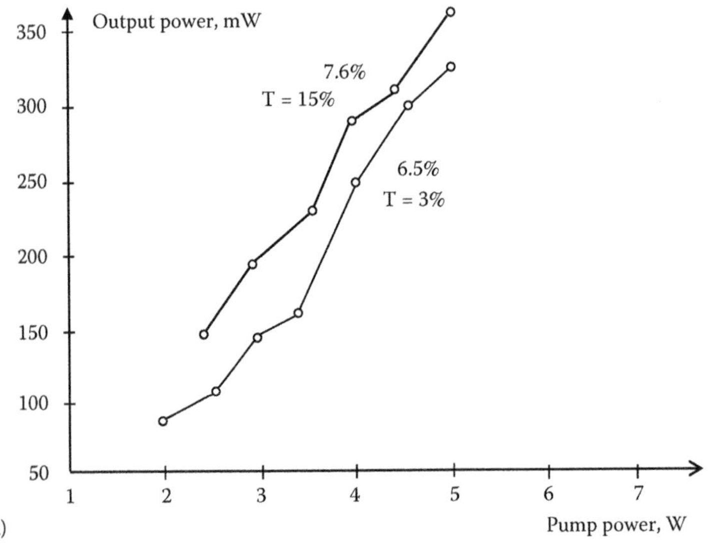

FIGURE 9.44 a) The output power vs the incident power for the CW. *(Continued)*

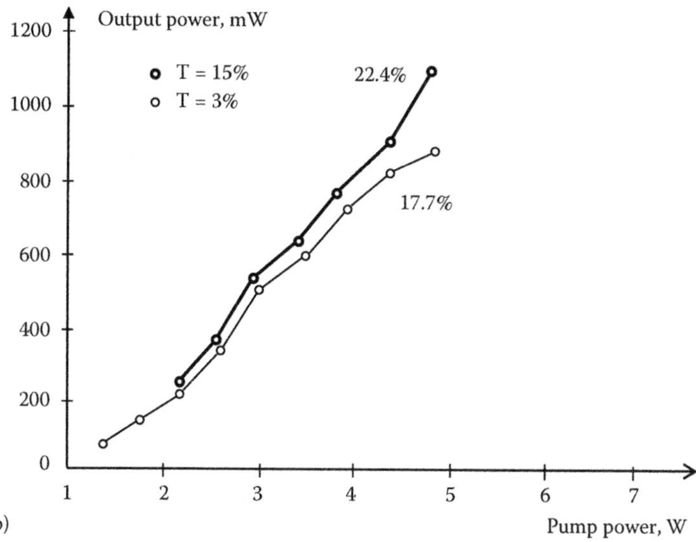

FIGURE 9.44 (Continued) b) The dependence of the output power on the incident pump power of Q – switched operation.

$T_{1/4}$ 3% and 15% couplers were used to achieve a free-running CW mode with the maximum power of 0.89 and 1.2 W when the maximum pump power was 5W. The threshold pump power was approximately 1.5 and 2.3W, respectively, with optical conversion efficiency of 17.7% and 22.4% plus the CW laser slope efficiencies were 28.5% and 37%.

The inserted into the laser cavity a prepared GOSA obtained a stable Q-switched operation with the threshold pump power increasing from 3% to 15% and the output rate increasing from 3% to 15%. The conversion efficiency increased from 6.5% to 7.6%.

The output spectra are presented in Fig. 9.45. Both CW and Q-switched spectra have peaks at 1066 nm with approximately the same FWHM. The GOSA insertion in the laser decreased the intensity of the Q-switched operation.

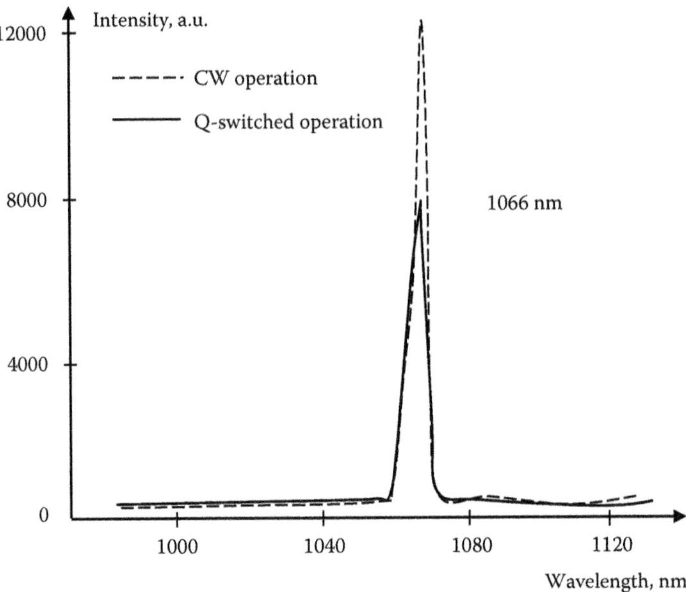

FIGURE 9.45 CW and Q – switched Nd:GdTaO_4 spectrum.

9.12 GRAPHENE MODE-LOCKED AND Q-SWITCHED 2 - μm Tm/Ho FIBER LASERS[66]

Another example of a Q-switched laser with an SA is presented here. Graphene due to its excellent properties with an ultraband operation wave range (from visible light to midinfrared), ultrafast recovery time, controllable modulation and low saturation intensity has been used in a number of laser-based applications, in particular for passive Q-switching or mode-locking. Tm or Ho-doped fiber pulsed laser (T-HLs) emitting at 2 μm have been used in medicine, laser spectroscopy, defense and midinfrared laser generation. As it was mentioned before, an SA is usually effective technique. Thus, such an absorber is advantageous for high performance. Passively Q-switched/ mode-locked T-HFLs graphene-based have been used to exploit Q-switching or mode-locking in the range from 0.8 to 2.93 μm. The Q-switched T-HFLs have yielded an average output power up to 5.2W[67] and the single-pulse energy up to 6.7 μJ[68]. Usually, T-HFL are pumped by laser sources in the range of ~790 nm to 1560 nm. However, it is difficult to develop a high-power laser diode with a single-mode output at ~790 nm. Although a high power laser can be made from Er/Yb at the wavelength ~ 1560 nm, a Tm/Ho-doped fiber laser has a low absorption at the above wavelength. The diagram showing a comparison between 1212 and 1565 nm is given in Fig. 9.46.

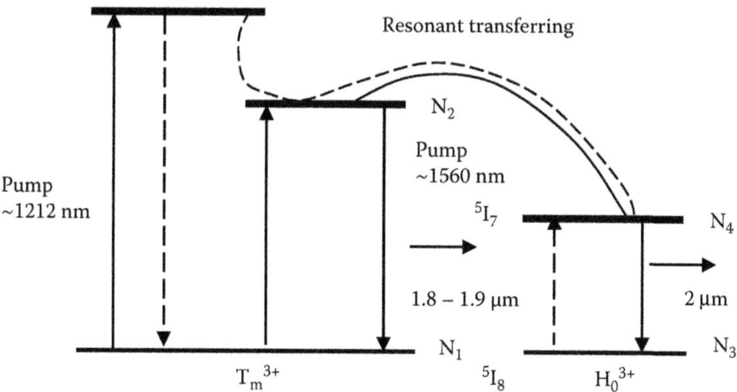

FIGURE 9.46 Energy levels of Tm/H_0 silica fiber pumped by a laser at 1212 or 1565 nm wavelength.

If the pumping occurs at 1565 nm, the energy level shifts from the $^3H_6$ to $^3F_4$ level of Tm^{3+}. The resonant energy exchange takes place from the 3F_4 level of Tm^{3+} to the $^5I_7$ level of H_0^{3+}. By choosing an oscillation wavelength from the spontaneous radiation a lasing at ~2 μm can be achieved through the stimulated radiation of H_0^{3+} ions. If the pumping takes place at 1212 nm, the energy transfers from the $^3H_6$ to $^3H_5$ level of Tm^{3+}. Then, either originating from the $^3F_4$ and to $^3H_6$ level of Tm^{3+} the ion wavelength is between 1.8 and 1.9 μm or originating from $^5I_7$ and giving to $^5I_8$ level of Ho^{3+} the ion wavelength corresponds to 2 μm laser wavelength. The experimental set-up is given in Fig. 9.47.

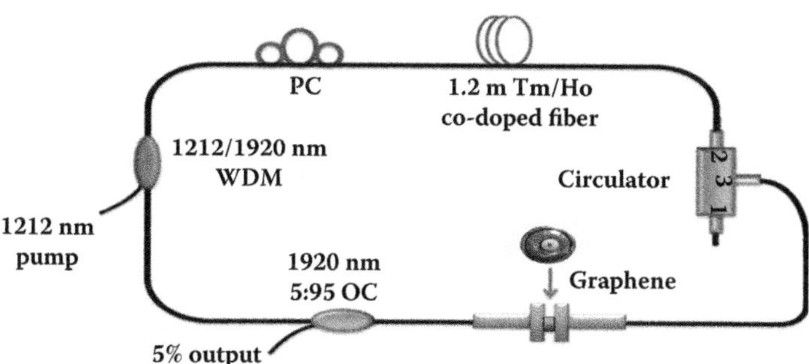

FIGURE 9.47 The set-up of the graphene passively mode-locked Tm/Ho fiber laser (1212 nm pumping).

First, the graphene film fiber is removed from the cavity in order to exclude the possibility of the laser self-Q-switching. Then, the graphene film is spliced and inserted into the cavity causing a stable Q-switching at the pump power of 109.8 mW. Typical characteristics of the Q-switching are depicted in Fig. 9.48.

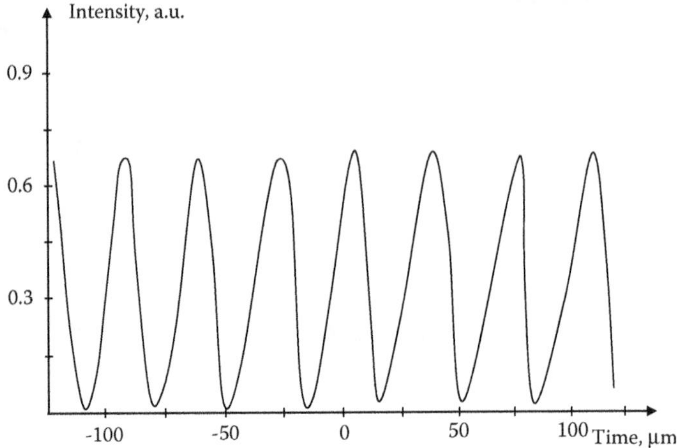

FIGURE 9.48 Typical pulse trains for the Q-switched mode.

The pulse duration can be reduced by reducing the cavity length and improving the modulation depth of the graphene SA. The output power increased almost linearly with the pump power increase reaching ~ 1 mW. This dependencies shown in Fig. 9.49. If pumped by a 1565 nm laser, the efficiency is lower resulting in a lower output power (Fig. 9.50).

With the pumping power increasing from 120 to 200 mW, the (average) output power was within the range of 0.1 to 0.4 mW. The energy of a single pulse reached approximately 20 pJ. The measurements showed that 1212 nm pumping was more efficient than 1565 nm. Graphene-polymer film SA provided Q-switching operation at 1976.0 nm and the mode-locking mode at 1913.7 nm. The Q-switched laser's output power reached approximately 1 mW with the repetition rate 32.2 to 43.0 kHz with the pulse duration of 1.41 μs.

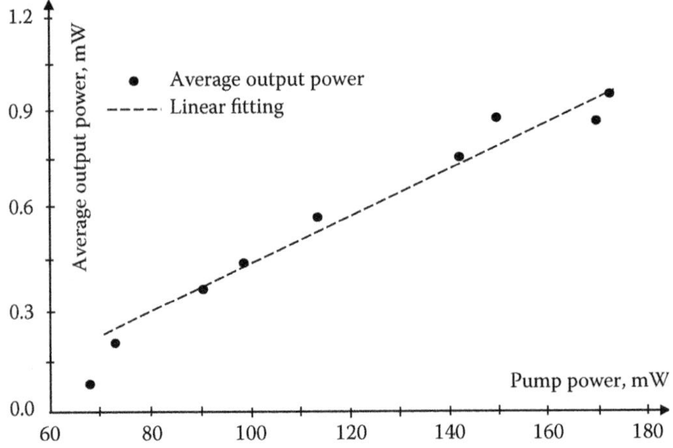

FIGURE 9.49 The average output power as a function of the input power.

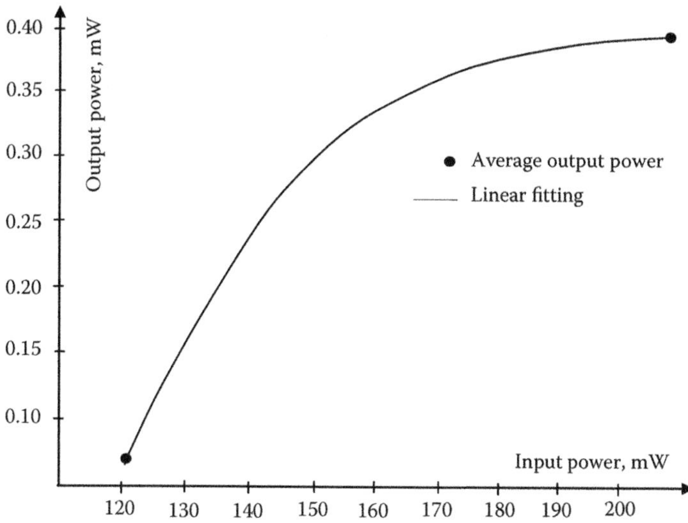

FIGURE 9.50 The average output power of the mode-locked Tm/Ho 1565 nm fiber laser.

9.13 HARMONIC MODE-LOCKING AND WAVELENGTH-TUNABLE Q-SWITCHING GENERATION[69]

Graphene-Bi_2Te_3 heterostructure can be used as a practical high-nonlinear photonic device. Here again, graphene is used in saturable absorbers (SA), this time graphene-Bi_2Te_3 heterostructure with excellent nonlinear optical properties as an effective broadband SA in fiber lasers for short pulse generation. The described 2D structure has unique qualities of being planar, versatile energy band, broadband absorption spectrum and controllable light-matter interaction structure. Graphene as a Dirac-material helps with making SAs effective for short optical pulse generation. The Dirac materials (including graphene and Ti), however, have some drawbacks: the low light absorption ($\pi\alpha = 2.3$ % for monolayer graphene) limits high-matter interaction and light modulation. The Ti on insulating gap on the bulk state has heavily populated intrinsic defects that make their transport properties controlled by the bulk state instead of being dominated by the massless Dirac surface states. Thus, the nonlinear optical absorption coefficients (such as the modulation depth and saturation intensity) decrease. Recently, some non-Dirac materials and black phosphorous (BP) have exhibited a stronger nonlinear optical response and have been incorporated as SA into mode-locked fiber lasers with the goal to obtain ultrashort pulse radiation. Heterostructures allow generating of short optical pulses using nonlinear optical response of graphene – Bi_2Te_3 films (heterostructures) and are important for superior laser parameters. A practical device was developed with Bi_2Te_3 heterostructure. The laser source of incident radiation was a mode-locked YDF laser that at 1055 nm wavelength gave 8.4 ps pulse duration with the maximum output power of 11.12 mW. The received parameters satisfy Eq. 9.32 and given in Fig. 9.51:

$$T(I) = 1 - \Delta T \exp(-I / I_s) - T_{ns}; \tag{9.32}$$

where $T(I)$ = the transmission rate, ΔT = the modulation depth, T_{ns} = the nonsaturable absorbance.

The sample was covered with Bi_2Te_2 with a variable extent that influenced the modulation depth. From Fig. 9.51, the modulation depth was 23.28% and the saturable intensity 3.32 MW/cm^2. The output power reached up to 20 mW. The linear optical response of the heterostructure was measured by ultraviolet-visible spectrometer. The result (Fig. 9.52) was a broadband operation. The curve is almost flat in the range of 900 to 2000 nm.

Another advantage of the broadband operation is the possibility of wavelength optical pulse tuning.

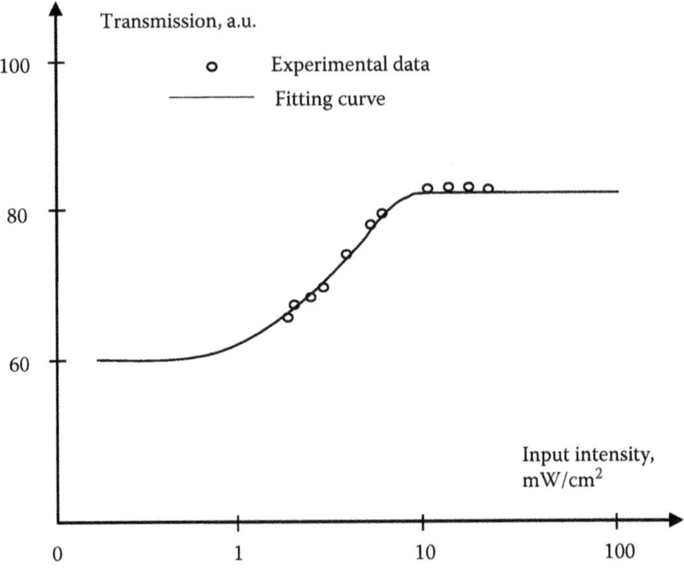

FIGURE 9.51 Nonlinear transmission curve for graphene-Bi_2Te_2 sample with 15% coverage of Bi_2Te_2.

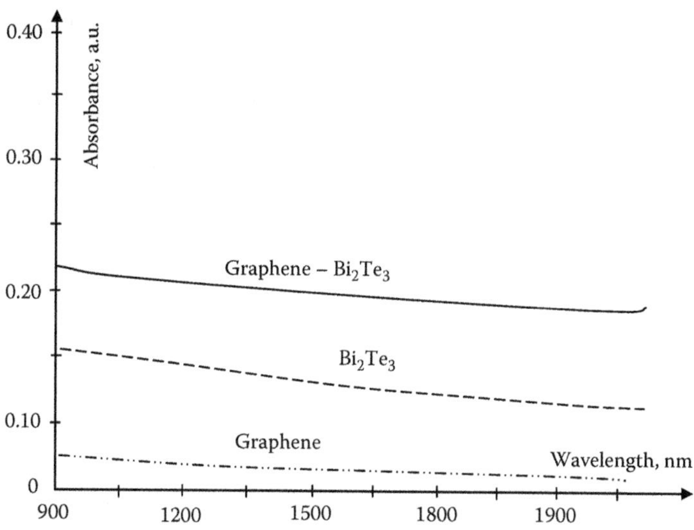

FIGURE 9.52 Graphene-Bi_2Te_3 linear absorption spectrum (85% coverage of Bi_2Te_3).

REFERENCES

1. Novoselov, K.S., Falko, V.I., Colombo, L., Gellert, P.R., Schwab, M.G., and Kim, K., "A road map for graphene", *Nature,* **490** (2012)
2. Yu, Y.J., Zhao, Y., Ryu, S., Brus, L.E., Kim, K.S., Kim, P., "Tuning the graphene work function by classic field effect", *Nano Lett.* **10** (2009)
3. Nair, R.R, Blake, P., Grigorenko, A.N., Novoselov, K.S., Booth, T.J., Strauber, T., Perese, N.M.R., Geim, A.K., "Fine structure constant defines visual transparency of graphene", *Science*, **320** (2008)
4. Chen, C., "Graphene Nano Electro Mechanical Resonators and Oscillators", *Ph.D. Dissertation,* Columbia Univ. (2013)
5. Bunch, J. S., van der Zande, A.M., Verbridge, S.S., Frank, I.W., Tanenbaum, D.M., Parpia, J.M., Graighead, H.G., McEuen, P.L., "Electromechanical resonators from graphene sheets", *Science,* **315** (2007)
6. Prada, M., Platero, G., Pfannkuche, D., "Unidirectional direct current in coupled nanomechanical resonators by tunable symmetry breaking", *Phys. Rev.,* B, **89** (2014)
7. Pinto, H., Markevich, A., "Electronic and electrochemical doping of graphene by surface adsorbates", *Beilstein J. Nanotechnol.* **5** (2014)
8. Lin, Y.M., Dimtrakopoulos, C., Jenkins, K.A., Farmer, D.B., Chiu, H.Y., Grill, A., Avourius, P., "100 – GHz transistors from wafer-scale epitaxial graphene", *Science,* **327** (2010)
9. Schwierz, F., "Graphene transistors", *Nature Nanotechnology,* **5**, pp. 487 – 496 (2010)
10. Liao, L., Lin, Y.-C., Bao, Y.Q., Wang, K.L., Huang, Y. and Duan, X., "High-speed graphene transistors with self-aligned nanowire gate", *Nature*, **467**, pp. 305 – 308 (2010)
11. Wu, G.Y. and Lue, N.Y., "Graphene-based qubits in quantum communications", *Phys. Rev.* B **86** (2012)

12. Vecchio, C., Sonde, S., Bongiorno, C., Rambach, M., Yakimova, R., Raineri, V., Giannazzo, F., "Nanoscale structural characterization of epitaxial graphene grown on off-axis 4 H-SiC (0001)", *Nanoscale Research Letters* **6**(1) 269 (2011)
13. Wehling, T.O., Novoselov, K.S., Morozov, C.V., Volovin, E.E., Katsnelson, M.I., Geim, A.K., and Lichtenstein, A.I., "Molecular doping of graphene", *Nano Lett.* **8**, pp. 173 – 177 (2008)
14. Emtsev, K.V., Speck, F., Seyller, Th., Ley, L., and Riley, J.D., "Interaction, growth, and ordering of epitaxial graphene on SiC $\langle 0001 \rangle$ surfaces. A comparative photoelectron spectroscopy study", *Phys. Rev.,* B **77** (2008)
15. Kopylov, S., Tzalenchuk, A., Kubatkin, S., and Fal'ko, V.I., "Charge transfer between epitaxial graphene and silicon carbide", *Appl. Phys. Lett.* **97** (2010)
16. Britnell, L., Gorbachev, R.V., Geim, A.K., Ponomarenko, L.A., Mishchenko, A., Greenaway, M.T., Fromhold, T.M., Novoselov, K.S., and Eaves, L., "Resonant tunneling and negative differential conductance in graphene transistors", *Nat. Comm.,* **4**, #1794 (2012)
17. Bokdam, M., Khomyakov, P.A. Brocks, G., Zhong, Z., and Kelly, P.J., "Electrostatic doping of graphene through ultrathin hexagonal boron nitride films", *Nano Lett.* **11**, pp. 4631 – 4635 (2011)
18. Britnell, L., Gorbachev, R.V., Jalil, R., Belle, B.D., Schedin, F., Mishchenko, A., Georgiou, T., Katsnelson, M.I., Eaves, L., Morozov, S.V., Peres, N.M.R., Leist, J., Geim, A.K., Novoselov, K.S., Ponomarenko, L.A., "Field-effect tunneling transistor based on vertical graphene heterostructures", *Science*, **335** (6071) 947-950 (2012)
19. Dean, C.R., Young, A.F., Meric, I., Lee, C., Wang, L., Sorgenfrei, S., Watanabe, K., Taniguchi, T., Kim, P., Shepard, K.L., and Hone, J., "Boron nitride substrates for high-quality graphene electronics", *Nat. Nanotech.,* **5**, pp. 722 – 726 (2010)
20. Ponomarenko, L.A., Geim, A.K., Zhukov, A.A., Jalil, R., Morozov, S.V., Novoselov, K.S., Grigorieva, I.V., Hill, E.H., Cheianov, V.V., Fal'ko, V.I., Watanabe, K., Taniguchi, and Gorbachev, R.V., "Tunable metal-insulator transition in double-layer graphene heterostructures", *Nat. Phys.* **7**, 958 – 961 (2011)
21. Yang, H., Heo, J., Park, S., Song, H.J., Seo, D.H., Byun, Kim, P., Yoo, I., Chung, H.-J., Kim, K., "Graphene barrister, a triode device with a gate-controlled Schottky barrier", *Science*, **336**, 1140 (2012)
22. Sciambi, A., Pelliccione, M., Lilly, M.P., Bank, S.R., West, K.W., and Goldhaber, D., "Vertical field-effect transistor based on wave-function extension", *Phys. Rev.,* B, **84** (2011)
23. Zheng, Y., Ni, G.X., Bae, S., Cong, C.-X., Kahya, O., Toh, C.-T., Kim, H.R., Im, D., Yu, T., Ahn, J.H., Hong, B.H., Ozyilmaz, B., "Wafer-scale graphene/ferroelectric hybrid devices for low voltage electronics", *Europhys. Lett.,* **93** (2011)
24. Konstantatos, G., Badioli, M., Gandrean, L., Osmond, J., Bernechea, M., Pelayo, F., de Arquer, G., Gatti, F., and Koppens, F. H. L., "Hybrid graphene-quantum dot phototransistors with ultra-high gain.", "*Nat. Nanotech.,* **7**, pp. 363 – 368 (2012)
25. Xia, F., Mueller, T., Lin, Y.-m., Valdes-Garcia, A., and Avouris, P., "Ultrafast graphene photodetector", *Nat. Nanotech.,* **4**, pp. 839 – 843 (2009)
26. Xu, K., Cao, P., Heath, J.R., "Graphene visualizes the first water Ad layers on mica at ambient conditions", *Science,* **329** (2010)
27. Lin, Y.-M., Valdes-Garcia, A., Han, S.-J., Farmer, D.B., Meric, I., Sun, Y., Wu, Y., Dimitrakopoulos, C., Grill, A., Avouris, P., Jenkin, K.A., "Wafer-scale graphene integrated circuit", *Science*, **332**, Iss. 6035, pp. 1294 – 1297 (2011)
28. Chowdhury, F.A., Morisaki, T., Otsuki, J., Alam, M.S., "Annealing effect on the optoelectronic properties of graphene oxide thin films", *Appl. Nanosci.*, Vol. 3, Issue 6 (2012)
29. Yao, X., and Belyanin, A., "Giant optical nonlinearity of graphene in a strong magnetic field", *Phys. Rev. Lett.*, **108** (2012)

30. Bae, S., Kim, H., Lee, Y., Xu, X., Park, J.-S., Zheng, Y., Balakrishnan, J., Lei, T., Song, Y., Kim, Y.-J., Kim, K.S., Ozyilmaz, B., Ahn, J.-H., Hong, B.H., and Lijima, S., "Roll-to-roll production of 30-inch graphene films to transparent electrodes", *Nat. Nanotech.,* **5**, pp. 574 – 578 (2010)
31. Yang, S.-L., Sobota, J.A., Howard, C.A., Pickard, C.J., Hashimoto, M., Lu, D.H., Mo, S.-K., Kirchmann, P.S., Shen, Z.-X., "Superconducting graphene sheets in CaC_6 enabled by phonon-mediated interband interaction", *Nat. Comm.* # 3493 (2014)
32. Zhu, J, Duan, R., Zhang, S., H., Jiang, Y. Zhang, and Zhu, J., "The application of graphene in lithium ion battery electrode materials", *NCBI,* **3**, 585 (2014)
33. Wu, J., Becerril, H.A., Bao, Z., Liu, Z., Chen, Y., and Penmans, P., "Organic solar cells with solution-processed graphene transparent electrodes", *Appl. Phys. Lett.,* **92** (2008)
34. Miao, X., Tongay, S., Petterson, K., Berke, K., Rinzler, A.G., Appleton, B.R., and Hebard, A.F., "High efficiency graphene solar cells by chemical doping", *Nano Lett.* Pp. 2745 – 2750, 12(6), **2012** (2012)
35. Crane, D., and Kennedy, R., "Solar panels for every home", *The New York Times* (2012)
36. Bae, S., Kim, H., Lee, Y., Xu, X., Park, J.-S., Zheng, Y., Balakrishnan, J., Lei, T., Kim, H.R., Song, Y.I., Kim, Y.-J., Kim, K.S., Ozyilmaz, B., Ahn, J.-H., Hong, B.H., and Lijima, S., "Roll-to-roll production of 30-inch graphene films for transparent electrodes", *Nat. Nanotech./Letter,* **5**, pp. 574 – 578 (2010)
37. Sigal, A., Rojas, M.I., Leiva, E.P.M., "Interferents for hydrogen storage on a graphene sheet decorated with nickel: A DFT study", *International Journal of Hydrogen Energy,* Vol. 36, Issue 5, pp. 3537 – 3546 (2011)
38. Kumar, P., Subrahmanyam, K.S., Rao, C.N.R., "Graphene pattering and lithography employing laser/electron-beam reduced graphene oxide and hydrogenated graphene", *Materials Express,* Vol. 1, # 3, pp. 252 – 256 (2011)
39. Shao, Y., Zhang, S., Engelhard, M.H., Li, G., Shao, G., Wang, Y., Liu, J., Aksay, I.A., and Lin, J., "Nitrogen-doped graphene and its electrochemical applications", *J. Mater. Chem.,* **20**, pp. 7491 – 7496 (2010)
40. Han, W., Kawakawi, R.K., Gmitra, M., and Fabian, J., "Graphene spintronics", *Nat. Nanotech.,* **9**, pp. 794 – 807 (2014)
41. Tombros, N., Jozsa, C., Popinciuc, M., Jonkman, H.T., and van Wees, J., "Electronic spin transport and spin precession in single graphene layers at room temperature", *Nature,* **448**, pp. 571- 574 (2007)
42. Han, Z., Wang, D., Qian, B., Feng, J., Jiang, X., and Du, Y., "Phase transistors, magnetocaloric effect and magnetoresistance in Ni-Co-Mn-Sn ferromagnetic memory alloy", *Japanese Journal of Applied Physics,* **49**, #1R (2010)
43. Bolotin, K.I., Sikes, K.J., Hone, J., Stormer, H.L., and Kim, P., "Temperature-dependent transport in suspended graphene", *Phys. Rev. Lett.,* **101** (2008)
44. Ang, P.K., Chen, W., Wee, A.T., Loh, K.P., "Solution-gated epitaxial graphene as pH sensor", *J. Am. Chem. Soc.,* 130(44) (2008)
45. Wang, T., Huang, D., Yang, Z., Xu, S., He, G., Hu, X.N., Yin, G., He, D., Zhang, L., "A review on graphene-based gas-vapor sensors with unique properties and potential applications", *Nano-Micro Lett.,* **8**, Issue 2, pp. 95 – 119 (2015)
46. Tzalenchuk, A., Lara-Avila, S., Kalabakhov, A., Paolillo, S., Syvajarvi, M., Janssen, T.J.B.M., Faliko, V., and Kubatkin, S., "Towards a quantum resistance standard based on epitaxial graphene", *Nat. Nanotech.,* **5**, pp. 186 – 189 (2010)
47. Ferralis, N., "Probing mechanical properties of graphene with Raman spectroscopy", *J. Mater. Sci.* (2010)
48. Liu, W.J., Chen, L., Zhou, P., Sun, Q.Q., Lu, H.L., Ding, S.J., and Zhang, D.W., "Chemical-vapor-deposited graphene as charge storage layer in flash memory device", J. of Nanomat., **2016** (2016)

49. Bertolazzi, S., Krasnoshon, D., and Kis, A., "Nonvolatile memory cells based on MoS_2/ graphene heterostructures", *ACS NANO,* **7**, #4, pp. 3246-3252 (2013)
50. Schwierz, F., Graphene transistors", *Nat. Nanotech.,* **5**, pp. 487 – 496 (2010)
51. Galashev, A.E., and Rakhmanov, "Mechanical and thermal stability of graphene and graphene-based materials", *Physics-Uspekh,* **57**, #10 (2014)
52. Lemme, M.C., "Current status of graphene transistors", *Solid State Phenomena*, Cornell Univ. Library (2009)
53. Wakabayashi, K., Takane, Y., Yamamoto, M., and Sigrist, M., "Electronic transport property of graphene nanoribbons", *New Journal of Physics*, **11** (2009)
54. Wang, X., Ouyang, Y., Li, X., Wang, H., Guo, J., and Dai, H., "Room-temperature all-semiconducting sub-10 nm graphene nanoribbon field-effect-transistors", *Phys. Rev. Lett.* (2008)
55. Fiori, G., and Iannaccone, G., "Ultralow-voltage bilayer graphene tunnel FET", *Electron Device Lett. IEEE*, **30**, Issue 10, pp. 1096 – 1098 (2009)
56. Ionescu, A.M., and Riel, H., "Tunnel field-effect transistors as energy-efficient electronic switches", *Nature,* **479**, pp. 329 – 337 (2011)
57. Wang, X., Tang, M., Chen, Y., Wu, G., Huang, H., Zhao, X., Tian, B., Wang, J., Sun, S., Shen, H., Lin, T., Sun, J., Meng, X., Show, J., "Flexible graphene field effect transistor with ferroelectric polymer gate", *Optical and Quantum Electronics* **48** (2016)
58. Fiori, G., "Negative differential resistance in mono and bilayer graphene $p - n$ junctions", *IEEE Electron Device Letters,* **32**, Issue 10 (2011)
59. Liu, Z., Ma, L., Shi, G., Zhou, W., Gong, Y., Lei, S., Yang, X., Zhang, J., Yu, J., Hackenberg, K.P., Babakhan, A., Idrobo, J.-C., Vajtai, R., Cou, J. and Ajayan, P.M., "In-plane heterostructures of graphene and hexagonal boron nitride with controlled domain sizes", *Nat. Nanotechnology*, **8**, pp. 119 – 124 (2013)
60. Chaudhari, S., Graves, A.R., Cain, M.V., Stinespring, C.D., "Graphene-based composite sensors for energy applications", *Proc. Of SPIE*, **9836** (2016)
61. Jessop, D.S., Sol, C.W.O., Xiao, L., Kindness, S.J., Braeuninger-Weimer, P., Lin, H., Griffiths, J.P., Ren, Y., Kamboj, V.S., Hofmann, S., Zeitler, J.A., Beere, H., Ritchie, D.A., and Degl'Innocenti, R., "Fast terahertz optoelectronic amplitude modulator based on plasmonic metamaterial antenna arrays and graphene", *Proc. SPIE*, **9747** (2016)
62. Koenig, S., Lopez-Diaz, D., Antes, J., Boes, F., Henneberger, R., Leuther, A., Tessmann, A., Schmogrow, R., Hillerkuss, D., Palmer, R., Zwick, T., Freude, W., Ambacker, O., Leuthold, J., and Kallfass, I., "Wireless sub-THz communication system with high data rate", *Nat. Photonics*, **7**, pp. 977 – 981 (2013)
63. Wallis, R., Degl'Innocenti, Jessop, D.S., Ren, Y., Klimont, A., Shah, Y.D., Mitrofanov, O., Bledt, M., Melzer, J., Herrington, J.A., Beere, H.E., and Ritchie, D.A., "Efficient coupling of double-metal terahertz quantum cascade lasers to flexible dielectric-lined hollow metallic wave guides", *Optics Express*, pp. 26276 – 26287 (2015)
64. Zhang, B., Song, Q., Wang, G., Gao, Y., Zhang, Q., Wang, M., and Wang, W., "Passively Q-switched Nd:$GdTaO_4$ laser by graphene oxide saturable absorber", *"Optical Engineering"*, 55(8) (2016)
65. Wang, Y.G., Chen, H.R., Wen, X.M., Hsieh, W.F., Tang, J., "A highly efficient graphene oxide absorber for Q-switched Nd:$GdVO_4$ lasers", *Nanotechnology*, **22** (2011)
66. Luo, L., Huang, Y.Z., Zhong, M., and Wan, X., "Graphene mode-locked and Q-switched 2 -µm Tm/Ho codoped fiber lasers using 1212 nm high efficient pumping", *Opt. Eng.* (SPIE), **5518** (2016)
67. Tang, Y., Yu, X.C., Li, X.H., Yan, Z.Y., and Wang, Q.J., "High-power thulium fiber Q-switched with single-layer graphene", *Opt. Lett.*, **39**(3), pp. 614 – 617 (2014)

68. Lin, C., Ye, C., Luo, Z., Cheng, H., Wu, D., Zheng, Y., Liu, Z., and Qu, B.,"High-energy passively Q-switched 2 μm Tm^3 + doped double – and fiber laser using graphene-oxide deposited fiber taper", *OSA Publishing*, **21**, Issue 1 (2013)
69. Wang, Z., Mu, H., Zhao, C., Bao, Q., and Zhang, H., "Harmonic mode-locking and wavelength-tunable Q-switching operation in the graphene – Bi_2Te_3 heterostructure saturable absorber-based fiber laser", *Optical Engineering* 55(8) (2016)

10 Summary

A series of experiments conducted by Geim and Novoselov has demonstrated the practical existence and importance of graphene as a two-dimensional phase of matter. There are some other examples of two-dimensional systems realized in 3D space, so called "2D in 3". They can propagate into the perpendicular to the 2D plane direction, and their resistance to this flexural dimensional distortion is small, not surprisingly for the small thickness. The 2D structure of graphene has new features due, for the most part, to the two distinct triangular sublattices, the structure similar to that of liquid helium with the highest ever mobility of 2090 m^2/Vs at T = 0.4 K. The system becomes an insulating Wagner crystal at T = 0.45 K with the electron density of 4.6 x 10^8 /cm^2. However, in practice, graphene does not show usual limitations of a 2D crystal. Another 2D-3 graphene feature is crumpling, i.e. out-of-plane shifts with the lengths comparable to the length (size) of the crystal. The available sample size free from crumpling is practically bigger than the sample itself. One important graphene quality is its softness in the transverse direction: any graphene plate of lateral dimensions of 1 μm or longer must have a support. On the other hand, graphene can give the thinnest possible conductor, at least, by a factor of 10. The graphene covalent bonding prevents the material from breaking into discontinuous islands that other materials cannot avoid. The minimal thickness is important for transistor logic (short-channel effect) flash memory (cross-talk adjusts cell reduction) and for on-chip interconnects (higher current density).

The most outstanding graphene characteristics: remarkably high conductivity, carrier mobility and high current density are impeded by the low availability of the material, only exfoliated graphene can be shipped in quantity[1]. At the present moment, for electronic applications, an epitaxial (e.g. grown by CVD on foil) process is required for manufacturing since micro-mechanical cleavage is suitable only for experiments. The necessary deposition temperatures are irreducibly high (e.g. for growth on SiC) but may be reduced for the CVD deposition if a plasma is implemented.

The high temperature processing facilities are more costly and are not used in the existing manufacturing methods (about 1000^0 C for graphene on Cu foil deposition). The typical growth time of a graphene layer in a CVD reactor may be minutes by the typical annealing process which is much longer than that for Si. Unlike in case of a Si wafer where surface layers can be removed to perform an annealing restoration of the Si crystal, a single graphene layer cannot be similarly restored.

A substantial development would be low-temperature CVD deposition methods that could be implemented *in situ* with the kinetic energy still sufficient for graphene growth. There have been reported several techniques that serve the purpose. One is graphene CVD growth on Cu at 300^0 C with benzene as the source gas[2].

The graphene high conductivity and current density can be utilized by various chemical or liquid exfoliation methods. The flakes' small size is critical to the material's low resistance, although the graphene refractory nature makes it impossible to sinter separate flakes together. Good electrical connection among flakes may be achieved by metal deposition, e.g. of potassium. The metal also improves the mechanical properties of the deposit. A principle of improving of the conducting (with respect to aluminum and copper) characteristics of long interconnects such as cables could be found in graphene deposition. Graphene is deposited on a catalytic layer of Cu. A large-scale application is the back-electrode of a solar cell, the graphene on copper (C/Cu).

Several major graphene applications may be incorporated within the existing silicon electronics manufacturing. They are optical modulators and flash memory, interconnects and RF transistors. It means that there should be an alternative to CMOS switching transistors.

The semi-metallic properties of graphene still do not preclude incorporating of graphene T-FETs (T = tunnel). Specifically, vertical graphene T-FETs are possible to create avoiding the "short-channel" effect. The T-FETs can also reduce the power dissipation that is one of the major problems for miniaturization. However, it is a small chance that the graphene T-FETs will replace the multi-gate silicon FET. However, there is a possibility that in the long run simple graphene structures will replace complicated silicon devices. As it was mentioned before, a replacement of the lightly developed silicon technology will be necessary for graphene switching devices manufacturing, a change which is costly and time consuming. The silicon technology replacement will likely occur first where the high power dissipation of silicon is a major problem. The first graphene incorporation is going to take place in optical modulators, photodetectors, flash memory and high-frequency devices. The graphene introduction may also make it easier to exploit other two-dimensional compounds, e.g. BN and MoS_2[3].

REFERENCES

Segal, M., "Selling graphene by the ton", *Nat. Nanotech.* **4**, 612-614 (2009)

Chen, Y.B., Liu, J.S., Lin, P., "Recent trend in graphene for optoelectronics", *J. of Nanoparticle research*, **15** (2013)

Li, Y., Rao, Y., Mak, K.F., You, Y., Wang, S., Dean, C.R., and Heinz, T., "Probing symmetry properties of few-layer MoS_2 and h-BN by optical second-harmonic generation", *Nano Lett.* **13** (7) (2013)

Index

Note: Page numbers followed by f and t refer to figures and tables, respectively.

D

E

H

T

U

V

W

X

Y

Z

Printed and bound by CPI Group (UK) Ltd, Croydon, CR0 4YY

01/11/2024

01782617-0006